PROCEEDINGS OF THE PLASMA SPACE SCIENCE SYMPOSIUM

ASTROPHYSICS AND
SPACE SCIENCE LIBRARY

A SERIES OF BOOKS ON THE RECENT DEVELOPMENTS

OF SPACE SCIENCE AND OF GENERAL GEOPHYSICS AND ASTROPHYSICS

PUBLISHED IN CONNECTION WITH THE JOURNAL

SPACE SCIENCE REVIEWS

VOLUME 3

PROCEEDINGS OF THE PLASMA SPACE SCIENCE SYMPOSIUM

HELD AT THE CATHOLIC UNIVERSITY OF AMERICA
WASHINGTON, D.C., JUNE 11–14, 1963

Edited by

C. C. CHANG AND S. S. HUANG

Department of Space Science and Applied Physics
Catholic University of America
Washington, D.C.

D. REIDEL PUBLISHING COMPANY / DORDRECHT-HOLLAND

GORDON AND BREACH / SCIENCE PUBLISHERS

NEW YORK

SOLE DISTRIBUTORS FOR NORTH AND SOUTH AMERICA
GORDON AND BREACH
SCIENCE PUBLISHERS, INC.
150 Fifth Avenue, New York, 10011 N.Y.

ISBN 978-94-011-7544-9   ISBN 978-94-011-7542-5 (eBook)
DOI 10.1007/978-94-011-7542-5

1965

PREFACE

Space, whether interplanetary or interstellar, is filled with plasma or ionized gas. The success of space exploration must count heavily on the study and understanding of the plasma. With this view, the Symposium of Plasma Space Science was held at The Catholic University of America in Washington, D.C., in cooperation with the National Aeronautics and Space Administration and the Goddard Space Flight Center, June 11–14, 1963. The symposium was organized by Dr. C. C. Chang as director and Dr. Y. C. Whang as assistant director.

The proceedings consist of four parts. The first part describes the solar phenomena: solar magnetic field, flare, atmosphere, cosmic rays, radiation, etc. The second part deals with the solar wind or interplanetary plasma. This includes both theoretical treatment and experimental measurements of Mariner II. The third part is concentrated on magnetosphere, with the measurements on trapped radiation from natural sources and high altitude thermonuclear explosions, and particularly with the magnetopause. The fourth and last part includes the ring current, geomagnetic storms, the aurora, the ionosphere and lunar surface phenomena.

There was round table discussion in each session. Many interesting comments and stimulating arguments emerged from the floor and are included in the Proceedings.

In the beginning, the symposium lectures were not intended to be published. However, during the meetings, the participants were urged to publish their lectures. It was certainly generous of the speakers who agreed to write up their lectures. The painful task of many speakers in finishing their manuscripts from incomplete notes and tape recordings were remarkably well done and deserve deep appreciation. However, a few of the lectures were withdrawn. In particular, the lecture on 'A Review of the Formation of Geomagnetic Cavity and Radiation Belts' by one of the editors (C. C. Chang) is omitted, because a large number of review papers have recently been published in the literature.

In the Appendix, two popular lectures, "NASA Space Science Program – Progress and Potential" by Dr. J. E. Naugle of NASA and "The Keynote Speech" by Dr. Leland J. Haworth, Director of the National Science Foundation, are included.

The editors are particularly grateful to Dr. C. Y. Fan, Dr. John M. Wilcox, Dr. Paul Nakada for their editorial help. Mr. Michael Schultz has contributed a great deal in collecting the manuscripts, and in assembling the needed materials.

The Rector, Right Reverend William J. McDonald, and Dean Donald E. Marlowe, of The Catholic University of America, Dr. Homer E. Newell, Associate Administrator for Space Science and Applications, NASA, and Dr. H. J. Goett, Director, Goddard Space Flight Center, NASA, all deserve the sincere thanks of the editors for their enthusiastic support.

There were many who contributed to the success of the symposium. In particular, the contributions of Dr. Y. C. Whang, Dr. H. W. Babcock and Dr. Wilmot N. Hess, Mr. J. R. Craig and Mr. J. C. Reese should be mentioned. The partial support of the U.S. Air Force Office of Scientific Research to the Editor (C. C. Chang) should also be mentioned.

CHIEH CHIEN CHANG
SU-SHU HUANG
Department of Space Science and Applied Physics
The Catholic University of America
Washington, D.C., USA

November 26, 1964 – Thanksgiving Day

CONTENTS

APPENDIX

WELCOMING ADDRESS

Ladies and gentlemen.

As Rector of the University, I am singularly privileged to extend a cordial welcome to you the members of this distinguished group of scientists, who in the interest of your profession have assembled here from near and far. Your participation is a tribute to the noble spirit of cooperation reflecting the humility with which you come to learn and the generosity with which you come to share. The acceptance of this attitude by all leaders would indeed enrich and be a source of great hope to our groping world.

It is most fitting that the Catholic University of America should be host to a gathering of this nature. In the early days of the University, Pope Leo XIII, its founder, wrote that "An education must be regarded as incomplete that does not embrace modern science. It is obvious that in the present keen competition of minds and the widespread, and in itself novel and praiseworthy passion for knowledge, Catholics ought to take the lead", he said, "not simply follow after. It is necessary, therefore, that we should cultivate every refinement of learning and rigorously train their minds to research and an investigation as far as it is possible of the entire domain of nature." In writing in this manner, Pope Leo XIII was but recalling the age old tradition of the Church. It is not surprising then to find that in the early history of the University we have many efforts which were indeed foremost in the mastery of science on this campus, that is, pioneering accomplishments, even in the area of your interest today.

At the turn of the century, two of our earliest professors were closely associated with Samuel Langley, Secretary of the Smithsonian Institution and encouraged him in his efforts to produce a heavier-than-air flying machine. One of those professors, Rev. George Surles, was appointed by Langley as Official Astronomer of the Astrophysical Observatory. The other, Rev. Clarence Woodman, was named Assistant Astronomer. Dr. Surles performed many experiments at the University on the resistance and the trajectory of projectiles, experiments which led to his invention of a range finder for the U.S. Navy. Aerodynamic research was actually commenced at the Catholic University of America in 1894 by Albert F. Zahm, who published in the scientific journals the results of his experiments on the resistance of models and skin friction. In fact in 1898, for purposes of aeronautical research, Dr. Zahm built the first scientific wind tunnel in the world equipped with measuring and recording apparatus. His laboratory was the first aeronautical laboratory associated with the school. To those, then, intimate with the early history of the University and its interest in aeronautics over the years, it came as no surprise that our School of Engineering and Architecture should be the first to announce a new Department of Space Science and Applied Physics. And we are very happy, indeed, that such a distinguished scientist as Professor C. C. Chang should be the Head of that Department. We are grateful also to Dean Marlowe, the Dean of the School for his constant interest in the promotion of this project, and, of course, to all those who helped us by their encouragement. So for these reasons we consider it most fitting that the Catholic

Chang & Huang (eds.), Proc. Plasma Space Sci. Symp. All rights reserved.

University of America should be acting as host to this distinguished gathering. Our tradition encourages our pride in welcoming you.

May I conclude, ladies and gentlemen, with a reference to your work voiced by one whose death we are now mourning, the late beloved Holy Father, Pope John XXIII. At a moment when one of the cosmonauts was orbiting the world he said, "Just as these historic events will take their place in the annals of the scientific knowledge of the cosmos, may they also become an expression of true and peaceful progress, contributing toward the sound foundation of human brotherhood." Were he talking to you today, I am sure he would have added as I do, "may your discussions during these days be exceedingly rich in their results". Thank you very much.

WILLIAM J. MCDONALD
Rector of The Catholic University of America

WELCOMING ADDRESS ON BEHALF OF
THE NATIONAL AERONAUTICS AND SPACE ADMINISTRATION

Monseigneur McDonald, Prof. Chang, ladies and gentlemen. It is a pleasure for me on behalf of the National Aeronautics and Space Administration to join in welcoming you to the Washington area to participate in the Symposium on Plasma Science. We appreciate the opportunity of cooperating with our neighbor university, the Catholic University of America, through our Goddard Space Flight Center which we hope all of you will have the opportunity of visiting.

The mere list of titles of papers to be presented at this Symposium demonstrates the major role of plasma physics in the interpretation of phenomena in the upper atmosphere and in interplanetary space. The papers also include important new observational data, obtained by means of sounding rockets or satellites and space probes on the planetary scale which contribute to extension and improvement of the theoretical aspects of plasma physics as applied to planetary and eventually cosmic distances and to the range of other parameters such as density and electrical conductivity, and magnetic field strength experienced in space.

Your field of knowledge is relatively new. I believe that the use of the word "plasma" in its presents sense is due to Irving Langmuir. In 1928 during his investigation of ionized gases he proposed to distinguish two states in the following terms: "except near the electrodes where there are sheaths containing very few electrons the ionized gas contains ions and electrons in about equal numbers so the resulting space charge is very small. We shall use the term 'plasma' to describe this region containing balanced charges of ions and electrons". While the subject is new, the domain of its application is very great. Dr. Biermann began a paper on the relations between plasma physics and astrophysics a few years ago with the observation that most of the matter in the universe exists in the form of a plasma. Celestial bodies like the earth and the other planets contribute as far as we know only a minor fraction by mass. This aspect receives additional emphasis by the recent observations according to which large scale magnetic fields pervade interstellar space and at least parts of interplanetary space, while the magnetic fields on stars especially on the sun have been known somewhat longer. The interaction of the interstellar, the interplanetary, and the stellar material of these cosmic magnetic fields belongs to the most interesting application of plasma physics and magnetofluid dynamics to astrophysics. Such is the broad scope of the field of knowledge to be opened up by our discussions in the next few days with the aid of such distinguished colleagues from abroad as Prof. Alfvén, Prof. Biermann, Prof. Chapman, Dr. Dungey, and Prof. Kopal.

In conclusion I wish merely to note that other aspects and applications of plasma physics are of interest in the activities connected with the exploration of space. These include the problems associated with the re-entry of bodies into the earth's atmosphere, including the interaction of the plasma with electromagnetic waves, the so-called communications blackout problem, dragged force of a body moving through a plasma,

Chang & Huang (eds.), Proc. Plasma Space Sci. Symp. All rights reserved.

plasma accelerators for laboratory similation of astrophysical phenomena, of hypersonic flight, and for use as controlled thrusters for distant space missions. It is hoped eventually that plasma physics research may lead to the realization of thermonuclear fusion, and its use for power and space vehicles. My own technical field in years past was aerodynamics and special reference to boundary layer flow and turbulence. As an aerodynamicist I have been interested in plasma physics which has its own problems of instability and its special varieties of turbulent diffusion and turbulent electrical conduction. As in the development of aerodynamics ingenious experiments and correlated ingenious theoretical concepts seem to be the way to progress.

Best wishes for stimulating discussions in this exciting field and again welcome to Washington.

HUGH L. DRYDEN
Deputy Administrator of the
National Aeronautics and Space Administration

PART I

SOLAR PHENOMENA

THE SOLAR MAGNETIC CYCLE

H. W. BABCOCK

Mount Wilson and Palomar Observatories; Carnegie Institution of Washington, Washington D.C. and California Institute of Technology, Pasadena, Calif.

Abstract. A theory of the sunspot cycle, based on a hydromagnetic model, is outlined. The theory rests largely on recent findings with the solar magnetograph, which has provided new data on the sun's poloidal field and on the development of local bipolar magnetic regions. The model involves distortion of a shallow poloidal field by solar differential rotation; the resulting toroidal component is amplified to the point of instability, with consequent production of sunspots. The theory plausibly relates various types of solar activity and provides a derivation of Spörer's law of sunspot latitudes.

Everyone is aware that solar activity, as shown by the number of sun-spots, ebbs and flows in a cycle of ten or eleven years. Not only the number of sun spots but also the associated flares, prominences, plages and coronal activity vary from a level near zero at minimum, to a maximum in the same eleven year cycle. What is the nature of this cycle? What goes on within the sun to cause these changes which, incidentally, through solar terrestial relationships, have some influence on our environment here on the earth.

We know that sun-spots have strong magnetic fields and that the sun has a weak poloidal field, a so-called general field. Sun-spots tend to occur in bi-polar groups with leading and following members of the group showing opposite magnetic polarity. The polarity relationships reverse themselves every eleven years, showing that the complete cycle is a magnetic one of 22 years duration. Likewise the main poloidal field of the sun reverses its polarity. This reversal, so far observed only once, but presumably repeating, occurs near the phase of maximum sun-spot activity. We know also that sun-spot activities are associated with plages, prominences, flares and other forms of optical activity on the surface of the sun. All the evidence points to the fact that the solar cycle is a magnetic or a magnetohydrodynamic one. That the changes in the magnetic pattern occur as rapidly as they do suggests that they come about through the circulation or motion of plasma carrying the magnetic field with it.

In exploring some of the current ideas regarding the solar cycle, one may recall that a magnetic field in the plasma is amplified by relative motions in a material that draw out or extend the lines of force. Such an increase of magnetic energy occurs at the expense of energy in some other form, frequently kinetic energy of large motions of large sections of the plasma.

In the case of the sun we are dealing with a plasma of very high electrical conductivity. Therefore, in most of the considerations to follow, the magnetic lines of force can be considered as elastic bands that are firmly embedded in a viscous fluid, such that transverse motion through the fluid is extremely slow. At some stages and under some conditions, however, the conductivity and the time scale are such that

Chang & Huang (eds.), Proc. Plasma Space Sci. Symp. All rights reserved.

the transverse slipping of the lines of force does become significant and important. These conditions seem to prevail in the turbulent convective layer at the sun's surface and in the corona.

One of the well-known facts about the sun is that it does not rotate like a solid body. It rotates faster at the equator than at higher latitudes. Thus, the period of its rotation at the equator is about 25 days. With increasing latitudes the period increases smoothly to about 34 days near the poles. These facts about the differential rotation have been known for years from observations of thousands of sun-spots as well as by spectroscopic studies of the Doppler effect.

We now turn to our particular model. The magnetic lines of force here are assumed to lie initially in meridian planes, and at a shallow depth. We think of the lines of force of the main poloidal field looping far out from the sun and we concentrate for the moment on the submerged portions of these lines of force. We think of them as shallow and initially in meridian planes. The sun's differential rotation draws out the submerged lines of force from east to west. Each line forms an extended loop with the part at the equator preceding. As time goes on the drawing-out process continues until each loop wraps around the sun several times. The submerged lines of force take on the configuration of tightly wrapped spirals in each hemisphere on opposite sides of the hemisphere. In this process the field strength becomes greatly amplified. The amplification is brought to a halt when the strength of the magnetic field, first in intermediate latitudes, reaches a critical value sufficient to produce instability. This means that the field strength grows until the magnetic pressure $H^2/8\pi$ becomes comparable to the gas pressure. This gives the plasma locally a buoyant effect causing it to rise to the surface with the production of a center of activity; that is, a magnetic region at the surface of the sun.

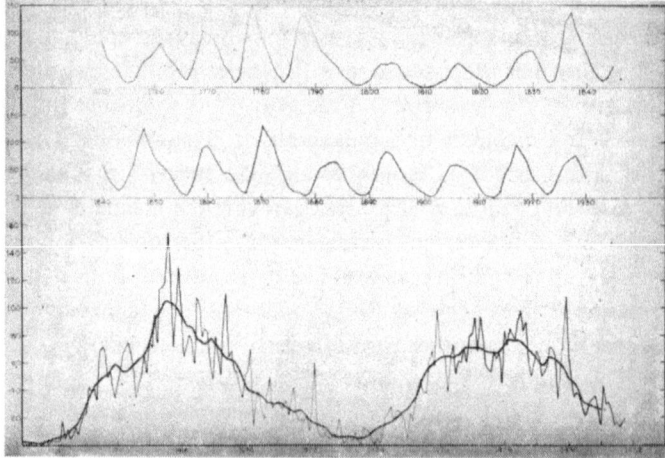

Fig. 1.

From these magnetic regions, the lines of force break out into the sun's atmosphere. The areas in which the field's lines intersect the sun's surface are in general strong field regions and produce sun-spots. The numerous instabilities of this sort bring to a halt the amplification process at the particular latitude but the onset of instability is a gradual one showing a decrease in latitude as time goes on. The occurrence of thousands of spots results in the transfer of the initially submerged field into the sun's atmosphere with liberation of the lines of force from the sun and also with the resultant reversal of the main poloidal field. The whole process on this model is then repeated but with opposite magnetic polarity throughout.

Figure 1 is the sun spot activity curve running from 1730 to about 1930. The sun spot number is plotted against time. The curve is not strictly periodic and, of course, there are pronounced non-uniformities in amplitude. Evidently the cycle is not well stabilized but we shall be concerned here only with the gross characteristics of the cycle and shall ignore the non-uniform features of it.

Carrington in 1858 first emphasized the latitude dependence of sun spots. In each eleven year cycle the first spots tend to occur in the vicinity of 30° heliographic latitude.

Figure 2 shows that as the cycle progresses the mean latitude decreases until after

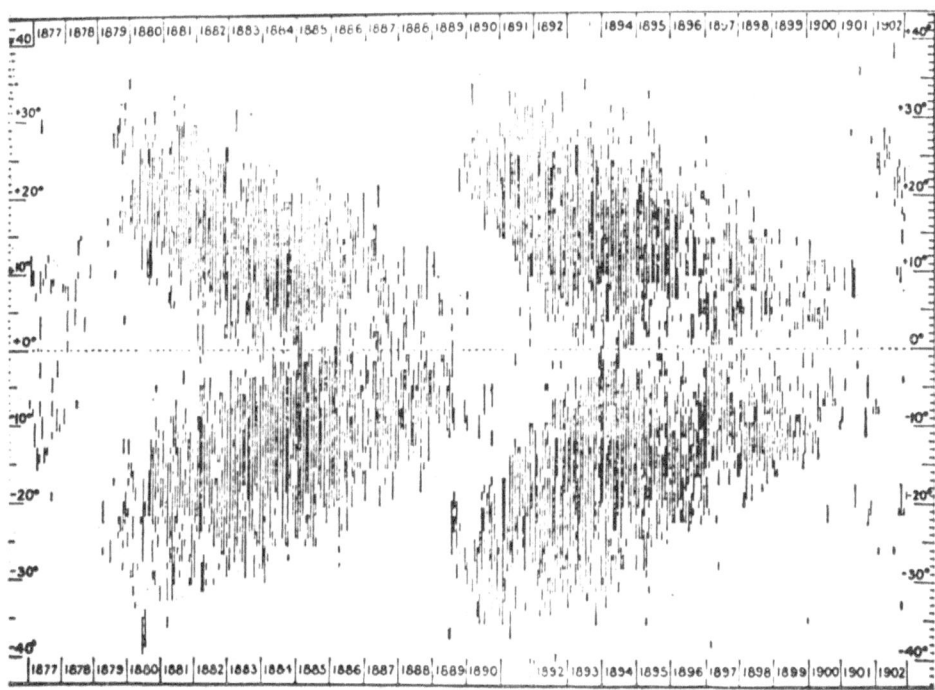

Fig. 2.

about eleven years when the number of spots has diminished. They are appearing at latitudes of 6 or 8 degrees. The dotted line across the middle represents the equator.

The heliographic latitude runs up and down from that, the first spot appearing in the vicinity of 25 or 30 degrees.

Much of the work on this effect was collated by Spörer and the facts represented here are sometimes referred to as Spörer's Law. The diagram referred to is known for obvious reasons as the "butterfly diagram". One of the virtues of the model that we are discussing is, as we shall see later, that it gives a representation of Spörer's Law or the butterfly diagram. And I think that one might bear in mind the general features of this diagram for comparison later with plotted curves.

Figure 3 illustrates data by d'Azambuja on the differential rotation of the sun.

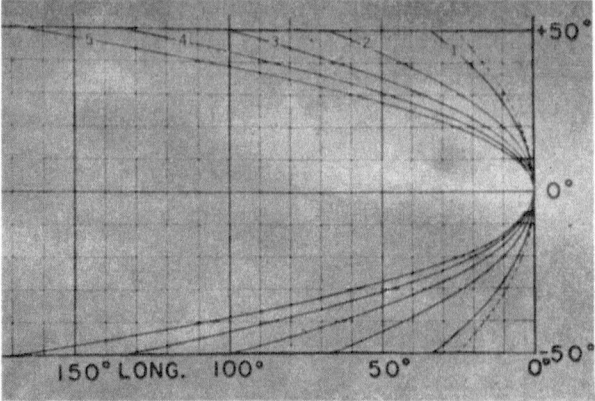

Fig. 3.

Here again the ordinate is heliographic latitude. A meridian on the sun is represented on the right hand side of the diagram. Because of the slower rotation in the higher latitudes, after varying numbers of whole turns of the sun, the meridian becomes distorted in a curve somewhat resembling a parabola. And after five whole rotations the displacement, the lag in solar longitude, at a latitude of 50° amounts to approximately 180° or one-half a turn around the sun.

Figure 4 is a vertical cross-section of the sun showing schematically three lines of force in a meridian plane. These lines are taken to be typical of the general field. They loop out to a great distance. It is assumed that similar field lines are uniformly distributed around the axis of the sun so that the whole pattern is really a symmetric one. The total magnetic flux involved, as indicated in the upper right hand corner, is of the order of 10^{22} maxwells. This is an observational result (although a crude one) obtained with the solar magnetograph. It is also an observational result that the lines of force of the general field, emanating from the sun, are limited to high latitudes generally above 50 or 55 degrees. This observational fact also comes from a study with the magnetograph. In other words, the sun does not exhibit a pattern characteristic of a uniformly magnetized sphere, but these lines of force of a general field are limited almost to the polar caps. On our model now we are going to assume that the lines of force below the surface are limited to a thin and rather uniform layer, a magnetic sheath lying at a depth of perhaps one-tenth of the solar radius and having a thickness of the order of 5 per cent of the solar radius. On this assumption the field strength of this uniform sheath in the vicinity of the equator is about 5 gauss. This

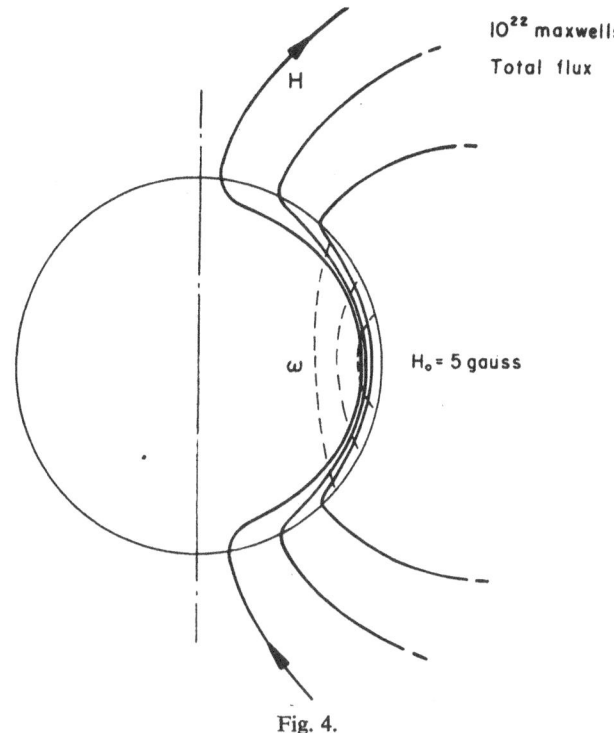

10^{22} maxwells

Total flux

$H_o = 5$ gauss

Fig. 4.

comes simply from the dimensions of the sheath and from the total magnetic flux involved. Also, in the diagram are indicated some dash lines labelled by the symbol ω. These are surfaces of constant angular velocity. The belt at the equator is rotating most rapidly. The other surfaces represent cross-sections of slower angular velocity. These cut more deeply into the sun than does the magnetic field patterns. As the sun rotates, then, the belt at the equator moves most rapidly and the shear effect draw out the lines of force from east to west.

Figure 5 shows schematically the result of the drawing out. By the "onion-skin" pattern one sees the configuration of the submerged field and we are assuming here for simplicity again that the magnetic layer is a completely uniform one. Because the lines of force merge up to the surface in the polar cap, the field intensity diminishes at latitudes above 45 or 50 degrees. This means, in fact, that the field strength in the sheath will be greatest in the moderate latitudes and somewhat less in the polar caps than at the equator. In this part of the process the field is still too weak to effect the motion of the plasma. In fact, the lines of force are simply following the motion of the plasma itself. The energy going into the magnetic field which is on the increase comes from the kinetic energy of differential rotation.

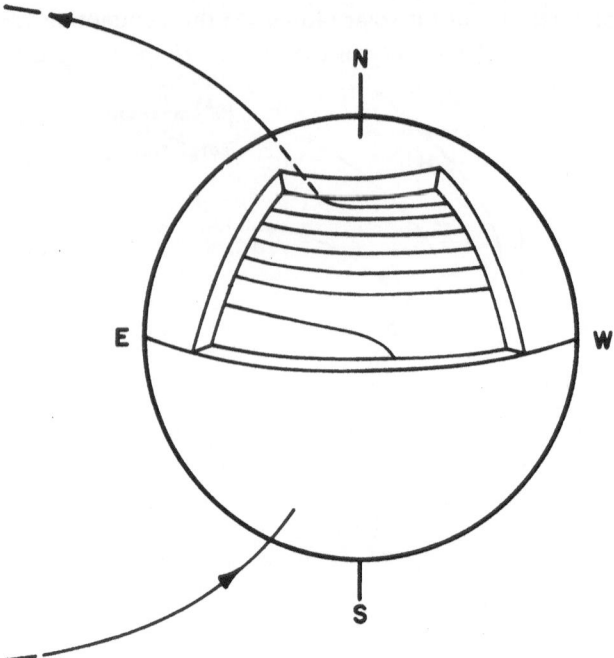

Fig. 5.

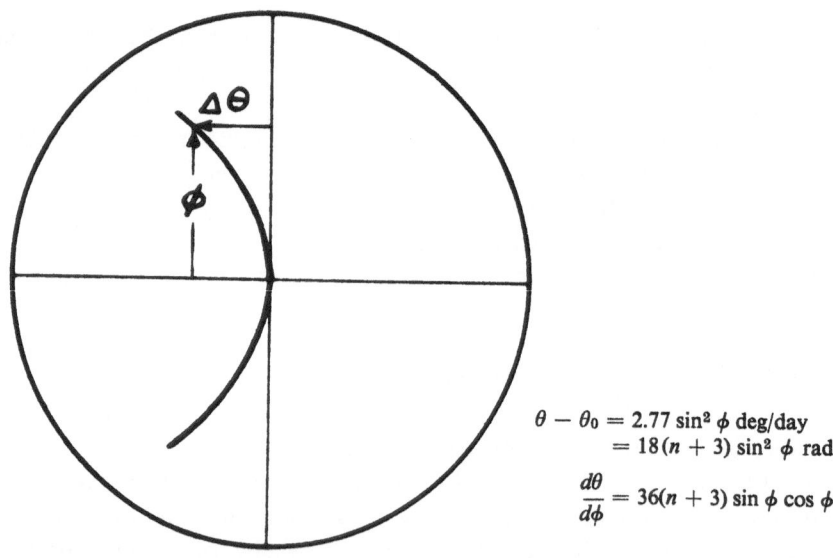

$$\theta - \theta_0 = 2.77 \sin^2 \phi \text{ deg/day}$$
$$= 18(n + 3) \sin^2 \phi \text{ rad}$$

$$\frac{d\theta}{d\phi} = 36(n + 3) \sin \phi \cos \phi$$

Fig. 6.

In Figure 6 we now wish to derive the slope of the spiral winding as a function of the latitude (the heliographic latitude ϕ) and of time. The first equation there is simply an observational result. It comes from the work of W. H. Newton who studied thousands of sun spots and derived the mean result $2.77 \sin^2 \phi$ in degrees per day for the lag in latitude. In the second line this result has been converted to a lag in radians. The quantity $(n+3)$ represents the time in years since the initial stage when the lines of force lay in meridian planes. I have taken the quantity $(n+3)$ because we are assuming that process goes on for three years before any instabilities set in. And the quantity "n" will then be used to represent the number of years after the onset of the sun spot cycle. In the last line we simply differentiate the expression. This will be useful in what follows.

Figure 7 shows how we can write the expression for the slope of the curve. The angle is the angle simply between the lines of force and the meridian plane and it is given by the expression at the bottom.

In Figure 8, at the top, we represent a bundle of lines of force showing variation in latitude. First of all before the rotation effect sets in the lines of force lie essentially in meridian planes and the field strength at the equator H_0 is about 5 gauss. There is

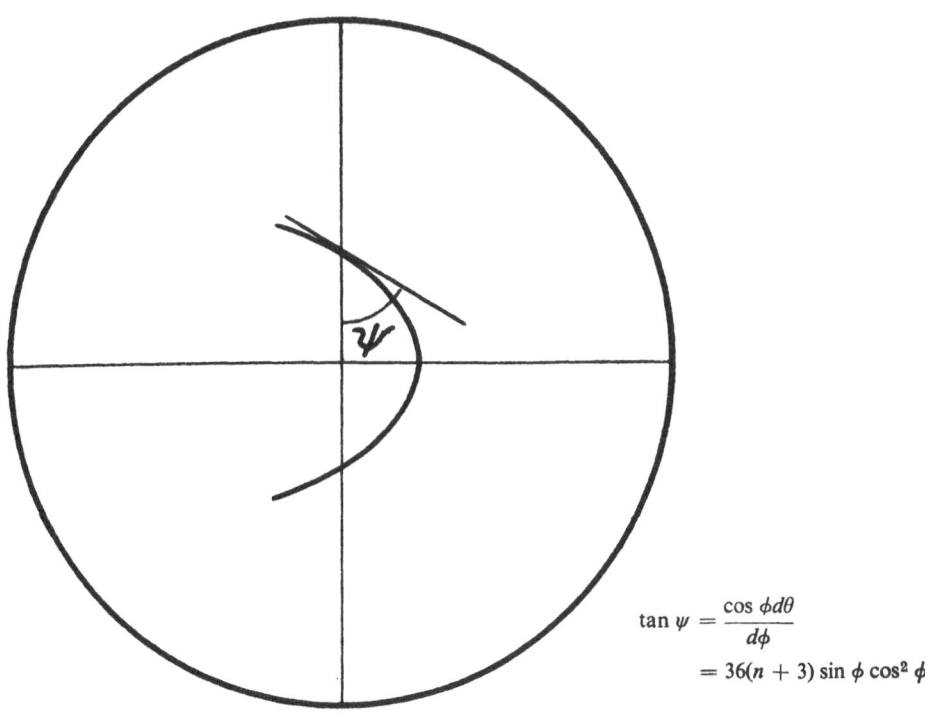

$$\tan \psi = \frac{\cos \phi \, d\theta}{d\phi}$$

$$= 36(n + 3) \sin \phi \cos^2 \phi$$

Fig. 7.

a factor secant ϕ because the radius decreases as one goes up in latitude. A much more important factor is the secant of the angle which represents the amplification due to the drawing out in solar longitude. And the expression in the third line to the right is then the general expression for the field strength effected by these factors. In

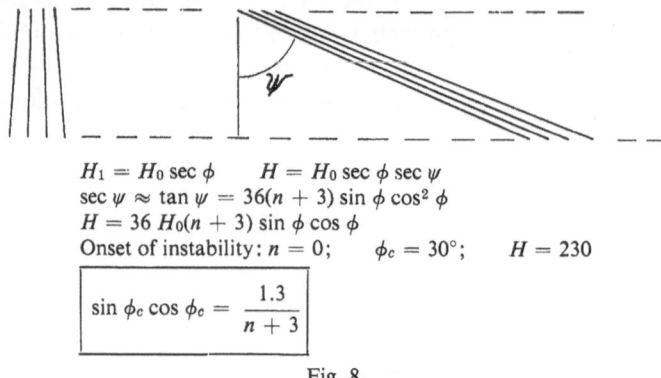

$$H_1 = H_0 \sec \phi \qquad H = H_0 \sec \phi \sec \psi$$
$$\sec \psi \approx \tan \psi = 36(n + 3) \sin \phi \cos^2 \phi$$
$$H = 36 H_0(n + 3) \sin \phi \cos \phi$$
Onset of instability: $n = 0$; $\quad \phi_c = 30°$; $\quad H = 230$

$$\sin \phi_c \cos \phi_c = \frac{1.3}{n + 3}$$

Fig. 8.

the second line we merely substitute from the result of the preceding figure and derive the third line, an expression for the field strength, in terms of the time in years $(n+3)$, and of the heliographic latitude. Now the time of reference here is the onset of instability and we know from observation that this occurs at a latitude of 30°. We have assumed on rather plausible grounds that this process has been going on for about 3 years before instability sets in. Substituting then $n=0$ and $\phi = 30°$, in the third line we come out with $H=230$ gauss. Now this is a figure representing the field strength in our fictitious uniform sheath when instability sets in. In the last line, then, we have a parametric equation relating the heliographic latitude where the field is becoming critical to the time in years (n) after the onset of the instability. This applies to those portions of the field where amplification is still going on, where the field has not yet reached the critical value.

In Figure 9 is plotted essentially the last equation from the preceding figure. We have here the same coordinates as were used for the butterfly diagram. I would like to emphasize that the curved lines here represent the *locus of the latitude at which the field strength is reaching the critical value*. They are, of course, not representations of the magnetic field lines. The vertical dashed line at the left at time $(n= -3)$ is merely the time when the cycle had its initial conditions with the lines of force in the meridian plane. Three years amplification went on during which time the lines wound around the sun some 5 or 6 times. Finally, at $n=0$ instability sets in at $\phi = 30°$. With the onset of instability, the field pattern is interrupted and amplification can be thought of as terminating at that latitude while it continues at lower latitudes. So as time goes on the latitude at which the critical value is being reached is diminishing. One has then, I think, really a representation of Spörer's Law. Because of the many irregularities due to turbulent effects the adherence to mathematical relationships will be rather poor and one can plausibly allow a good deal of leeway in the actual appearance of spot groups and unstable magnetic areas at the sun's surface so that the rather large scatter which is in fact observed is readily understood. These lines may be taken to represent the skeleton of the pattern so to speak. One sees that at the end of each

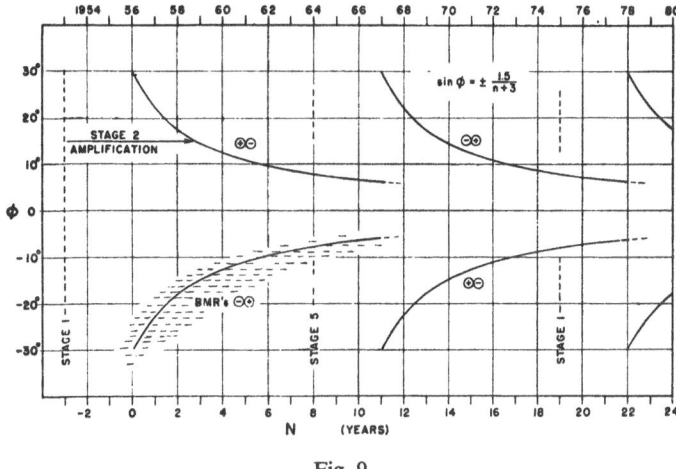

Fig. 9.

cycle the activity has gone down to something in the order of 6 or 7 degrees in latitude. Figure 10 shows schematically a magnetic flux rope. And now we are giving up

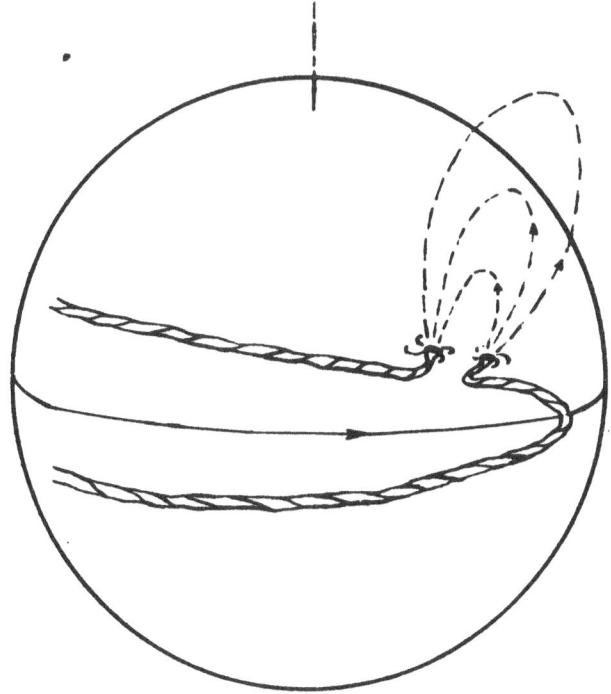

Fig. 10.

our fiction of a uniform magnetic sheath. In fact, it is more reasonable to think of the lines of force as being bundled up or twisted by the differential rotation. So that instead of a uniform sheath one thinks of a number of magnetic ropes. And because of the differential rotation there is an overriding effect tending to induce a twist in the rope and the twist will be of opposite sense for those portions on opposite sides of the equator. Where a loop of this rope has become buoyant and has reached the surface there is then formed a bipolar magnetic region from which the lines of force loop up to high elevations in the solar atmosphere, increasing as time goes on. This is the sort of magnetic arch that occurs above a center of activity. There is also observed in connection with sun-spots an effect known as chromospheric whirls. They are not consistent as to configuration and sense but still there is a statistical predominance of one sign in each solar hemisphere. They were studied years ago by G. E. Hale. One sees perhaps a plausible means for accounting for these whirls due to the twist of the magnetic flux ropes. One could point out, however, that the sign of the twisting at the surface where the rope is broken is not unambiguous. There may be a relaxation effect resulting in a spiral in one sense, or the main twist being induced at the equator may be transmitted up to the surface in which case the sense of the whirls would be of the other sign. This is not too much in discord with the observations.

Figure 11 shows the general tendency for the flux lines to reach higher and higher into the solar atmosphere as time goes on. There is here a time sequence from the

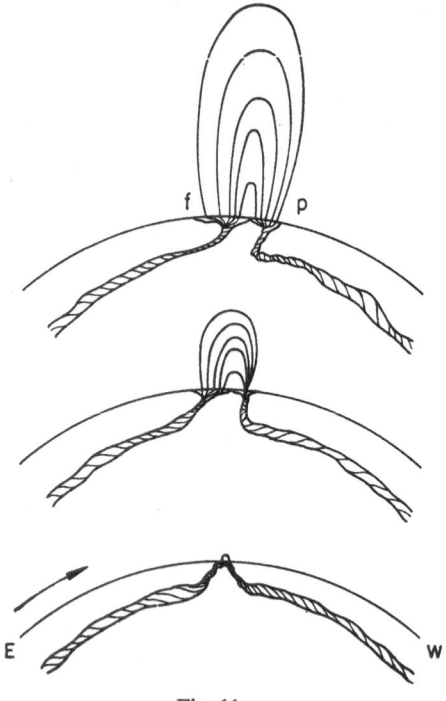

Fig. 11.

bottom to the top. First of all, the loop barely reaches the surface, then it emerges, the magnetic flux lines grow higher and at the same time the magnetic areas at the surface increase their size. This is an observational result and, I think, a very important one. That the magnetograph shows invariably that magnetic areas continue to expand. They do not disappear by contracting and submerging again. And I believe this has far reaching consequences. At the stage in the development of a group when the sunspots are largest, a maximum is reached in respect to the total amount of magnetic flux. After this, with further expansion, the mean field strength in the magnetic areas goes down. And they tend usually to merge with other areas east or west on the surface of the sun.

The plasma can, of course, flow along the magnetic ropes and because of the increasing area of the magnetic region one is led to suspect that there is an upward flow of plasma along the flux rope into the surface region with a consequent spreading. This, if it really occurs, would lead to a diminution of the diameter of the flux rope in the vicinity of this decaying region. Then a new center of activity may form because the flux rope has a smaller than average diameter and a consequently higher field strength. In this way one could try to make an argument for the recurrence of activity in solar longitudes where it has already commenced, – where there have already been magnetic regions. Because of the differential rotation again the overriding effect of the shallower layers tends to straighten up the submerged flux rope below the leading side of the group, and to give a greater inclination to the flux rope on the following side, or in the following part of the magnetic region. This might be thought of as tending toward a breakup of the spots in the following part of the region whereas the vertical nature of the flux rope below the leading part would tend to extend the lifetime of the leading spot. In this way, one can attempt to explain the observational

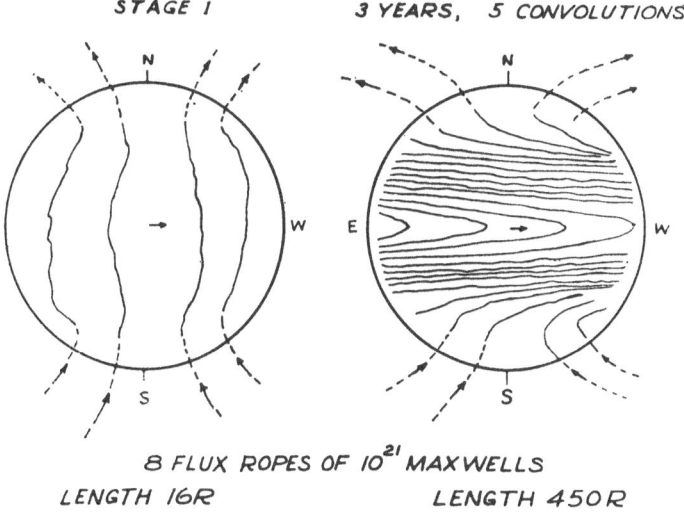

Fig. 12.

result that leading spots do, in fact, predominate; they have a longer lifetime, and are generally more prominent.

What about the total length of the magnetic flux rope and its capability for producing the observed number of active regions during a solar cycle? (See Figure 12). Let's imagine that initially we have eight flux ropes, each on the order of 10^{21} maxwells of magnetic flux. This would give the observed amount of flux for the general field. The initial length when these ropes lie in meridian planes is of the order of 16 solar radii. After three years of differential rotation, however, and drawing out of flux ropes with some five convolutions, the total length has reached a value of something like 450 solar radii. Now the amplification will to some degree be terminated; that is, at latitudes near 30°; nevertheless, it will continue at lower latitudes for nearly another decade. Therefore, it is reasonable that a total length of magnetic flux rope equal to about 1000 solar radii would be formed in the whole 11-year sunspot cycle. Each cycle results in the appearance of some 2000 to 3000 spot groups, each having on the average a magnetic flux of 10^{21} maxwells. So that if each unit length of flux rope (R) produces 2 or 3 bipolar regions this is ample to account for the observed amount of activity.

I'll run through very briefly just two or three sample magnetograms without stopping to explain much about them. These are obtained by measuring the Zeeman effect. In general, the brighter parts of the pattern correspond to stronger field strengths, and the field strengths run from about 2 gauss in the weaker parts of the pattern to 60 gauss in the stronger part. The really strong field regions where it is over 60 gauss are, however, represented by completely black areas (See Figure 13). I show these merely to indicate that a study of these magnetic diagrams leads to the conclusion that magnetic areas on the sun do expand as they grow older. And I think that this forces us to the conclusion that the flux loops originally formed by drawing out below the surface of the sun eventually are allowed to slip through the photospheric layers and are liberated in the solar atmosphere.

One has to compare a considerable number of these to gain observational conclusions.

In Figure 14 I have attempted to show the developing configuration of the field lines as magnetic regions on the sun have become older. This curve at the bottom represents a cross-section of the sun perpendicular to its axis. The field lines represent here the stage "a" at an early stage. As time goes on the field loops are thought of as rising to higher altitudes in the solar atmosphere. This is in line with observations of coronal features; it is also in line with the expansion of the magnetic regions over the disk of the sun. So finally a stage "b" would be reached. I believe that we are forced to the conclusion that as these growing loops approach each other there must at some stage be a severing and a reconnection of the lines of force with in many cases a neutralization of anti-parallel field lines which would result in the liberation of some magnetic energy in the sun's atmosphere. Through a long development of this sort, then, one visualizes the initially submerged lines as working their way through the photospheric layers and being liberated in the sun's atmosphere. The

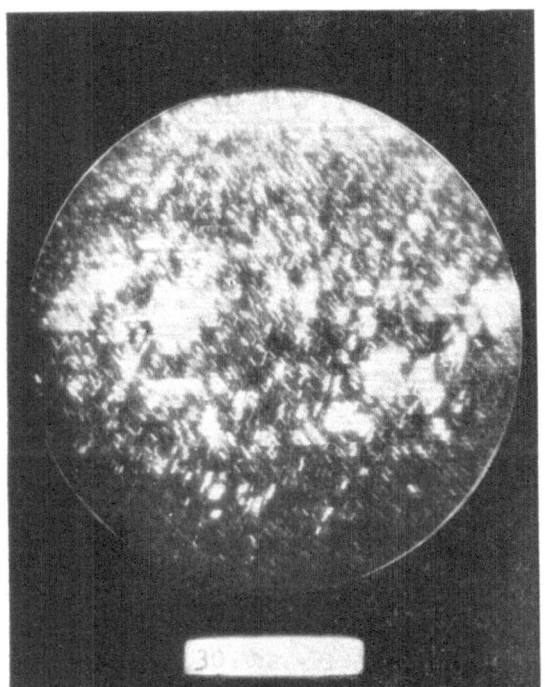

Fig. 13a.

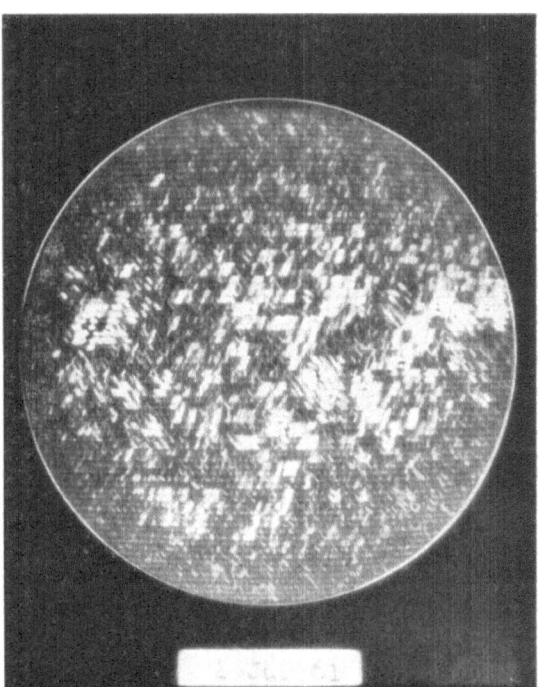

Fig. 13b.

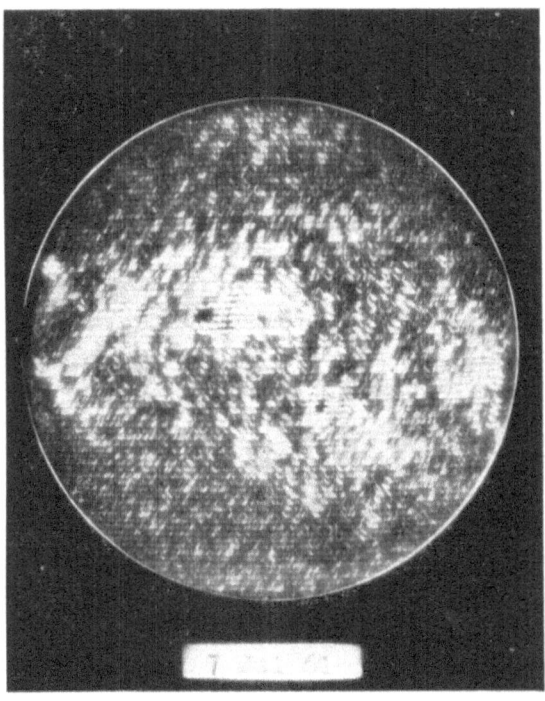

Fig. 13c.

energy involved here comes (as far as our model is concerned) from the differential rotation, and is converted into magnetic energy in the submerged field. Finally, after liberation of the field lines in the corona, much of the magnetic energy reverts to thermal energy in the plasma of the corona.

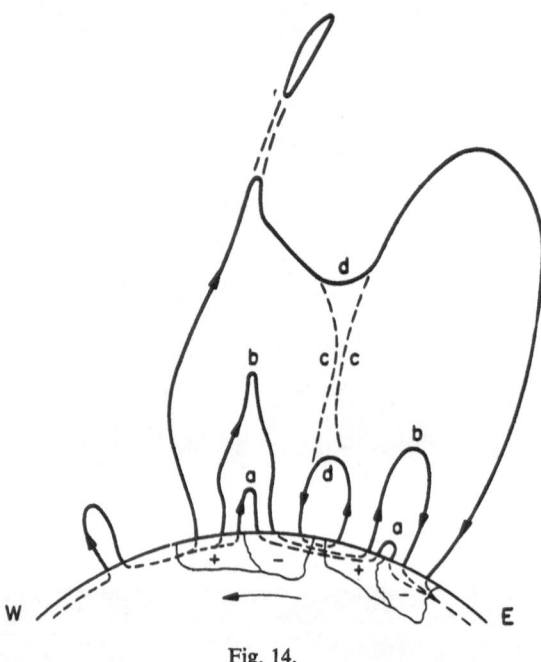

Fig. 14.

Figure 15 shows again a slightly different effect involved in the expansion of the field lines. I have not said much about the general field here and its reversal. The fact seems to be on this model that the initial general field is cancelled as the sun-spot cycle progresses. We shall see how this comes about in the following way. Bipolar regions in intermediate latitudes have a leading and following part. There seems a slight tendency for the preceding part to move toward the equator and for the following part to move toward the poles. The following parts in the sense of the solar rotation move toward the poles, and their magnetic polarity is such as to cancel the magnetic field in the polar cap. About 2 per cent of the emergent magnetic flux of all bi-polar regions is sufficient to cancel the initial poloidal field, and to supplant it with a reversed poloidal field; that is, one of opposite polarity which will then be available for the next magnetic cycle after this present one has decayed. The preceding parts of these bi-polar regions on opposite sides of the equator tend to merge together and to cancel each other. Now again this process involves severing of the lines of force. It involves annihilation of anti-parallel lines of force in the regions indicated by the small letter "a" and this results again in the liberation of a flux loop in the high solar atmosphere.

I have left many features in an unsatisfactory state here. In particular, one must at some point account for the maintenance of the differential rotation. I think that it can be shown that if this is not maintained by some other process that it would have enough energy to continue the solar magnetic cycling for only a few thousand years. There are many other features that have been treated here in a very hasty and very

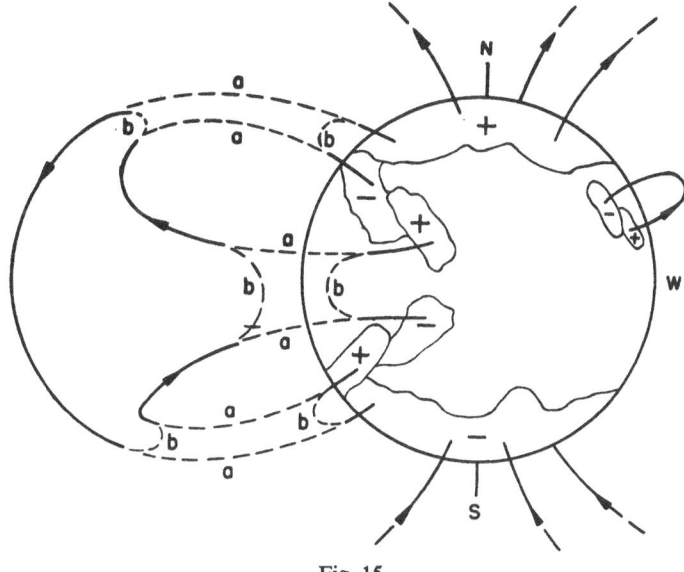

Fig. 15.

qualitative way, and which need to be put on a more quantitative basis. We do have observational evidence for the main poloidal field and for its reversal at the last sunspot maximum. We have observational evidence for the spreading of the magnetic areas over the surface of the sun as they age. But there is a very great need for firming up and increasing the quantitative nature of the observational results.

Discussion and Questions

L. Tonks: I would like to ask about the bearing of the conservation of momentum on this phenomenon. Just off the top of my head I would think that the loss of magnetic field to space might result in a loss of angular momentum of the sun. Insofar as the lines collapse back the angular momentum is preserved so that maybe the differential rotation is preserved to that extent. Would Dr. Babcock comment on this?

H. Babcock: I can't put it on any quantitative basis. I think that perhaps there is no necessity in having the lines of force collapse back to the sun. I would also suspect that in terms of angular momentum the outflow of material in the solar wind would have much more important effects than the angular momentum of the magnetic field itself.

C. Y. Fan: Can you explain how the lines of force are so straight at phase one?

H. Babcock: Well, I don't believe that they are in fact straight by any means. I think that on the average in phase one we have something resembling this condition where the lines of force lie more or less in the meridian plane. I believe that if we really analyzed the situation that this would be just an approximation. So I am describing a model and I am perfectly willing to admit the fit of the model to reality in many cases is not at all exact.

E. G. Fontheim (University of Michigan): Has there been any attempt to derive or guess at the current necessity to cause the magnetic distortions which give rise to the magnetic activity?

H. Babcock: The currents can be computed using Maxwell's equations. One can find references to this aspect of the problem in various papers such as those of Cowling.

Unidentified (Honeywell): I have two questions. One, I'd like to have you expound a little bit more on the statement given at the end of your talk related to the differential rotation. It is obviously a key requirement and did I understand you to say that there is no explanation for this differential rotation? The second question. You mentioned severing and reconnecting of lines of force but this isn't allowed in electromagnetic theory. I wonder if you would elaborate on that point.

H. Babcock: As regards to the equatorial acceleration. First of all it is simply an observed fact and as far as our model here goes, we are starting from that fact. It is a very obvious thing physically. Now to account for it there are theories. I am not at all prepared to go into them now, but Mustel', for example, has considered the differential rotation of the sun, starting with the outward flow of heat along a solar radius in a rotating frame of reference. Now as for severing and reconnection, this is a concept that is difficult to accept in this context. But here again there is evidence that in a highly turbulent regime such as exists near the solar photosphere, the conductivity is very much reduced. Sweet and Elsässer have shown that one can expect greatly reduced conductivity under certain conditions where the turbulence is high. Now beyond that point I am not going to say anything on the theoretical side of it but I think that the observations force us to the conclusion that the magnetic lines of force are carried through the photospheric layer and are liberated in the atmosphere of the sun. I see no other conclusion that can be drawn from the fact that magnetic regions on the surface disappear by expanding and merging with one another. They never disappear by contracting and submerging again.

L. Biermann: I would like to make a comment on this question of differential rotation. It can be shown that whenever a rotating body such as the sun rotates non-uniformly then this state is kept stationary by the emission of radiation. For example, stars in radiative equilibrium (that is for radiative energy transport) it has been shown a number of years ago by von Zeipel. And for convective layers it has been shown in a contribution to the Stockholm symposium on electromagnetic phenomena in 1956 and there have been various theories since that time. So I think that on theoretical grounds there is some basis for the assumptions made in the theory by Dr. Babcock.

J. Wilcox: Could you make any comment about the return of the lines of flux, both of the poloidal field and the unipolar field?

H. Babcock: The so-called unipolar regions are observed in a limited number of cases. Now of course they are not really unipolar but they appear in that sense on the observational record. By this I mean that we observed on the magnetograms, areas of rather limited size showing one magnetic polarity and there was no associated area of opposite polarity. This simply means that the lines of force emerging in a bundle from this unipolar region must diffuse presumably high in the solar atmosphere then return to the sun in very weak widespread areas where they may not be detected. Nevertheless the best of these unipolar regions which was observed in 1953, maintained its existence for about six or seven solar rotations. It could easily be identified. And it is an interesting fact that shortly after the meridian passage of this unipolar region there were distinct disturbances in the geomagnetic field. That is, there was a very distinct terrestial result. Apparently the bundle of lines of force from this unipolar region extended very far out from the sun, far enough to interfere with the earth. There was a 27 day sequence of geomagnetic storms. Also there was a disturbance of cosmic radiation as shown by Dr. Simpson's neutron monitor. We have been looking for further unipolar regions but haven't found any good persistent ones during the present solar cycle. I am not sure that I understood your question about the poloidal field but to observe this requires high sensitivity and slow scanning. It requires a minimum of disturbing effects, instrumental or otherwise. Nevertheless long sequences of observations have shown it and the existence of it has been confirmed by von Klüber in England. There are certain observations from radio astronomy which confirm the order of magnitude of the poloidal field (Högbom).

J. Wilcox: With regard to the poloidal field, in one pole you would have lines of one polarity leaving say and at the southern pole returning. Would you have any comment on how these points link together?

H. Babcock: Well, only that I suspect that the field current is very irregular and it is distorted by magnetic disturbances at the lower latitude and by turbulence in the high solar atmosphere. Of course all our observations at Mt. Wilson are optical ones and pertain to the surface of the sun.

KINEMATICS OF SOLAR FLARES*

G. E. MORETON**

Solar Observatory, Lockheed-California Company, Burbank, Calif.

Abstract. The occurrence of high velocity motions in flares and the chromosphere constitutes one of the most important problems of solar physics. One method of acquiring data on transverse or horizontal motions in the solar atmosphere is by means of high time-resolution ($\sim$ 10 sec) monochromatic cinematography. Results of a study employing this technique indicate that flares fall into two distinct catagories. Most flares are characterized by gradual brightening to maximum intensity, and slow ($\sim$ 10 km/sec) expansion of the flare borders. Flares in this group commonly initiate the flare-surge event, i.e., ejection of matter in a brush-like shape, with a typical velocity of around 150 km/sec. Flares of the second category are less frequent than the former, but they are characterized by more violent development and usually exhibit a distinct acceleration phase in their rise to maximum intensity and area. The flare expansion is often in a preferred direction with a velocity $\sim$ 100 km/sec. In a significant number of cases, extensive cloud or wave-like disturbances are ejected at this "explosive" phase. These disturbances are observed to propagate over distances as great as a solar radius, with a characteristic velocity $\sim$ 1000 km/sec. The fact that Type-III and centimetric radio bursts are closely associated with the explosive phase may imply that this phase is the relevant time for the major energy release and particle acceleration processes in the flare region. Furthermore, the fact that *all* eight high energy ($>$ 20 keV) X-ray events measured by rocket and ballon experiments are explosive-phase flares lends support to this suggestion, and emphasizes their geophysical consequences.

1. Observational Techniques

The analysis of high-velocity motions in solar flares constitutes one of the most important problems of solar physics. Direct observation of such motions are important in formulating theoretical ideas concerning the acceleration of electrons to relativistic energies and the acceleration of protons and heavier particles to sub-relativistic energies. The principal data on motions in flares are: (1) the gross Doppler shift of spectral lines, from which motion in the line of sight can be deduced; (2) the shapes and profiles of spectral lines, which provide information on density, temperatures, electric and magnetic fields, and turbulence; (3) time-lapse photographs of flares, which permit direct measurement of transverse or horizontal motions, as well as geometric and morphological aspects.

The latter is the simplest of the three data to obtain. The most common techniques employ a simple refracting telescope arrangement, a monochromator of the Lyot birefringent filter type, and a camera. The filter usually has a bandpass around 0.5 Å centered at the wavelength of the strong Balmer hydrogen line, Hα (6563 Å). The Hα image of the solar disk is usually about 17 mm in diameter, and is photographed on 35 mm film. Many observatories throughout the world use this method and photograph the sun on a patrol basis, registering flares and other chromospheric phenomena at a rate of four frames per minute or slower.

* This research is supported by the National Science Foundation and the Lockheed Basic Science Laboratory.
** Present address: Sterrewacht 'Sonnenborgh', Utrecht.

Chang & Huang (eds.), Proc. Plasma Space Sci. Symp. All rights reserved.

Some of the observations to be discussed here have resulted from a special study directed at the problem of obtaining more precise data on fast motions in flares, and high-speed propagation phenomena associated with flares. New observational techniques have been employed, such as (1) photography at rates $\geqslant 6$ frames per minute; (2) displacement of the filter bandpass to positions plus and minus 0.5 Å into the wings of Hα to detect Doppler shifted and broadened features, (3) use of optical systems to obtain resolution $\approx 5''$ of arc and other techniques to maximize image quality, and (4) special techniques for viewing and enhancing the resultant film images.

2. Flare Motions and Associated Ejecta

There is no proof that the visible Hα solar flare is the seat of the flare phenomenon, but it is certainly the most evident and important visual manifestation. In the total flare process some 10^{32} ergs of energy are released over a period of hundreds of minutes, and some 10^{32} electrons and a like number of protons are accelerated and ejected in periods of seconds to minutes. From this one might easily conceive of the flare as a violent eruption. In striking contrast is the fact that, in general, what is defined as the Hα flare is a relatively stationary phenomena.

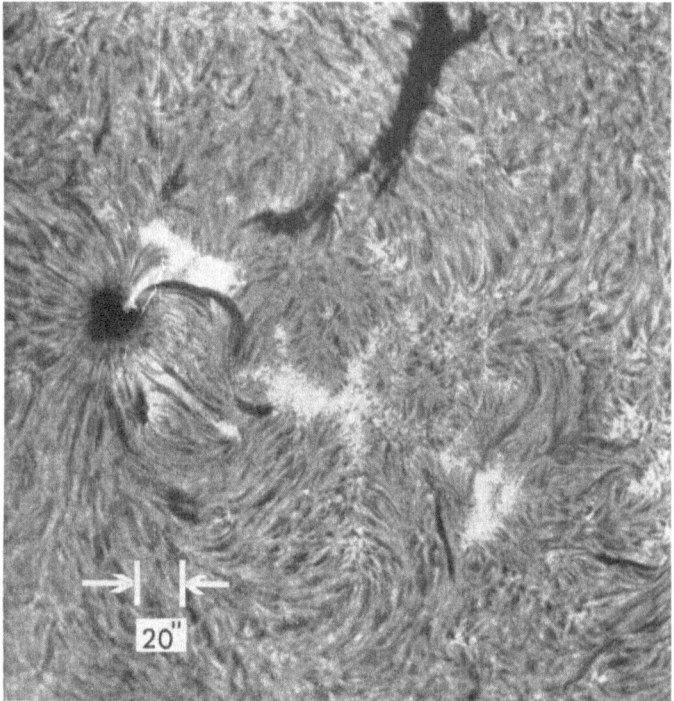

Fig. 1.   High-resolution Hα filtergram made 6 Sept. 1962 by J. W. Harvey, using 7-inch objective lens and 0.5 Å passband Halle filter. Part of the small 1⁻ flare developed along chromospheric fibril structure near umbra of sunspot.

Flares rise gradually to maximum brightness and intensity in a period of 5 to 10 minutes (SMITH and SMITH, 1963). High-resolution filtergrams, such as shown in Figure 1, indicate that flares appear as a nearly simultaneous brightening of already existing chromospheric structure. Often, different regions will brighten at the same time, grow in area, and then merge. The spread in emitting area, particularly if the spread is directional, or involves a rapid successive brightening of nearby regions, may give the deceptive appearance of mass motion. In the majority of cases such expansion is probably due to the propagation of excitation. On the other hand, actual mass motion may be involved in events where slow, lateral displacements of whole regions of a flare take place. Such displacements generally occur in large flares which consist of double, parallel filaments, and the motion is such as to increase the separation of the filaments (MALVILLE and MORETON, 1963). In general, the apparent expansion of bright flare borders, and displacements of whole regions, usually proceed at rates <10 km/sec.

About 30% to 40% of the flares observed against the disk of the sun are *associated* with the ejection of material in the form of a feature called a surge. The surge (more properly flare-surge, since all surges are initiated by flares) is a particle stream ejected during the premaximum stages of a flare. The expelled material has a density of 10^{11} protons/cc, and usually moves out from the flare with a velocity around 150 km/sec. It is emphasized that the main body of the flare does not partake in the surge motion.

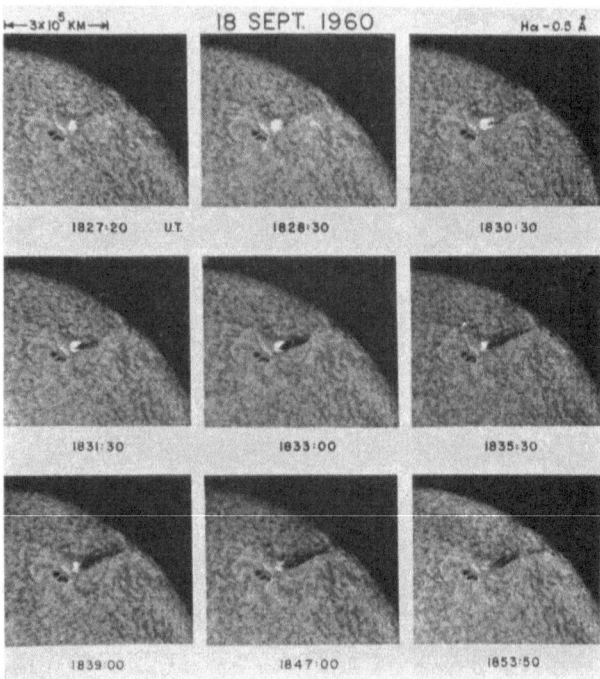

Fig. 2.   Flare-surge, photographed in the violet wing of Hα, Sept. 18, 1960. Surge starts from small point within flare region and travels outward with a tranverse velocity of 300 km/sec.

Instead, with adequate time and spatial resolution it is apparent that the surge is ejected from a localized region in or near the flare, as is shown in Figure 2.[1] In the initial moments of the ejection the surge often appears to be short, finger-like extension of flare brightness, and is sometimes mistaken as fast, directional expansion of the flare. In later moments the surge may become transparent, subsequently becoming visible as a long absorption feature, often reaching 10 times the linear dimensions of the flare. These successive changes from a bright to a dark feature are caused by varying conditions of excitation as the particle stream moves into the corona. Such conditions are evident in Figure 3, which shows a flare-surge against

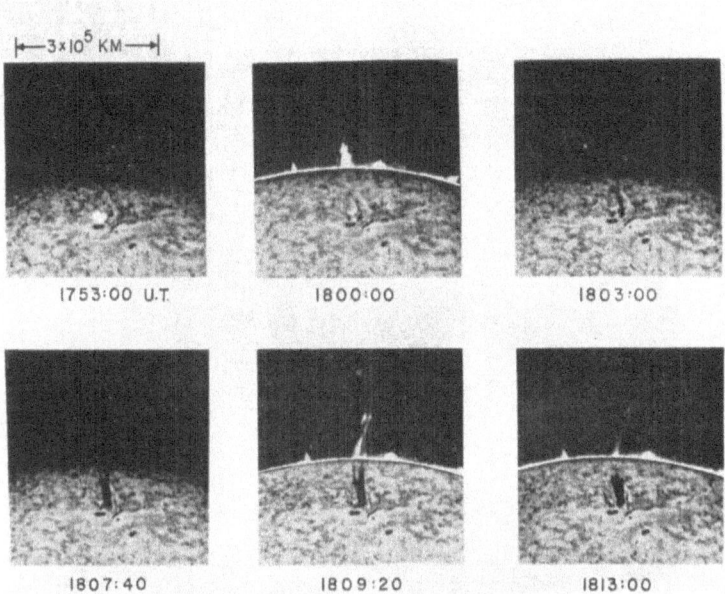

Fig. 3.   Flare-surge, May 1, 1960, photographed at Hα + 0.5 Å. The first frame shows the flare three minutes after maximum, at which time the surge is invisible against the disk. Overexposed filtergrams, made to record faint limb features, show that the surge first crosses the limb between the first and second frames. The second, fifth, and sixth frames are a montage of the disk and over exposed limb frames.

the disk and against the sky as it moves into the corona. Surges are more frequently observed by displacing the Hα filter bandpass to positions 0.5 Å into the wings of Hα. This technique enhances the contrast between the Doppler shifted and broadened components of the surge, against the chromospheric background. Off-band surges often appear as irregular dark blobs which flow along irregular lanes between the background mottling. The complicated motions may be due to deflections caused by radial magnetic fields extending up from background plage, which tend to oppose the kinetic energy of the surge material ($H^2/8\pi \sim NkT$) (MALVILLE and MORETON, 1961).

Observations of flares at the limb provide data on their cross-section geometry and radial outward motion. Most limb flares appear as small mounds 2000 to 5000 km in height (SMITH and SMITH, 1963). Statistical analysis of limb flares show that only 25% exceed 10000 km in height. The infrequent occurrence of large flares at the limb indicates that most of the flare region is at low chromosphere levels; the mound and loop structures of limb flares probably correspond to the bright and dense nuclear regions of flares observed on the disk. Limb flares appear to expand and rise, but seldom with velocities greater than 10 km/sec. As with disk flares, fast ejecta are often *associated* with flares at the limb, but the flare remains relatively stationary in the chromosphere. A good example of how the question of flare motion can be clouded by the misinterpretation of associated ejecta is shown in Figure 4. Careful

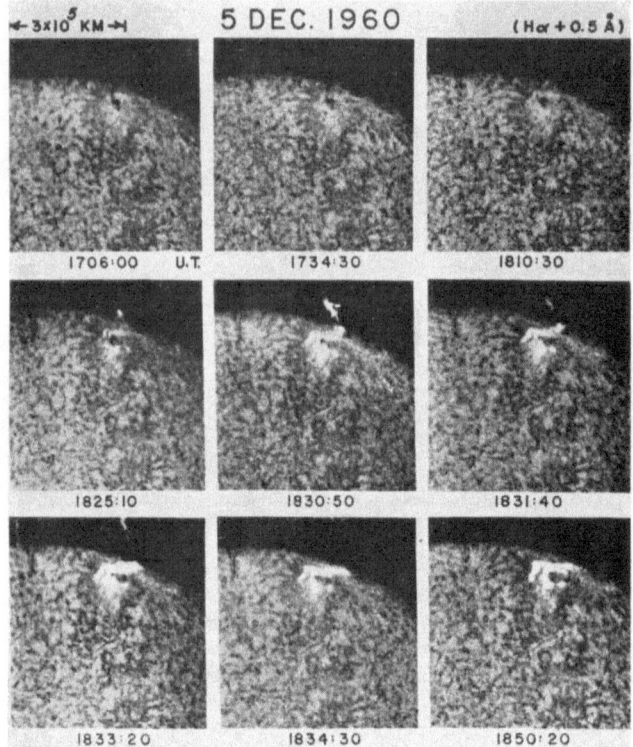

Fig. 4.   Rising Flare-Filament Activation, December 5, 1960. An absorption filament near the flare site started ascent into the corona before the flare was first visible. The rate of ascent accelerated as the flare increased to maximum. The rising material is visible as a flare-brightness prominence against the sky on normal disk exposures.

inspection of 4 frames/min on-band Hα filtergrams presented what, at first, appeared to be convincing evidence that a large part of the flare was ejected beyond the limb. However, inspection of red-wing filtergrams obtained at 6 frames/min revealed that what appeared as part of the flare was actually a rising absorption filament, originally located adjacent to the flare region, and appearing in emission as it rose beyond the limb. The filament began its ascent *before* the flare appeared, and its ascent was accelerated as the flare increased to maximum brightness.

SMITH and RAMSEY (1963) show that about 50% of major flares are associated with such filament activations.

3. Explosive Flares

In general, the majority of flares show lateral and vertical motions < 10 km/sec, and are commonly associated with ejecta having velocities ~ 100 km/sec. Recent studies employing fast-rate cinematography, good spatial resolution, and systematic observations in the Hα wings, have revealed a new class of flares which are characterized by motions and moving disturbances which have velocities an order of magnitude greater than the majority of flares. We have applied the term "explosive" to these flares, because of their exceptional appearance when viewed in Hα films (ATHAY and MORETON, 1961 and MORETON, 1961).

In contrast to most flares that show continuous area and intensity development to maximum brightness, the explosive flare exhibits a distinct pre-maximum phase, marked by an abrupt increase in the Hα flux (area × intensity). The principal factor in the abrupt increase in Hα flux, and the appearance of explosive development, is the rapid expansion of the flare borders. In explosive flares all or part of the peripheral bright regions expand or extend, often in a preferred direction, with velocities of several 100 km/sec, and in some cases as great as 500 km/sec. The period of accelerated development – the explosive phase – can be as short as 10 sec, and is commonly less than 30 sec, after which the flare may show little or no further areal development. The increase in area and the increase in intensity appear to develop independently: maximum intensity may be reached before, or during the explosive phase; however, sufficient photometric studies are not yet available to prove this point.

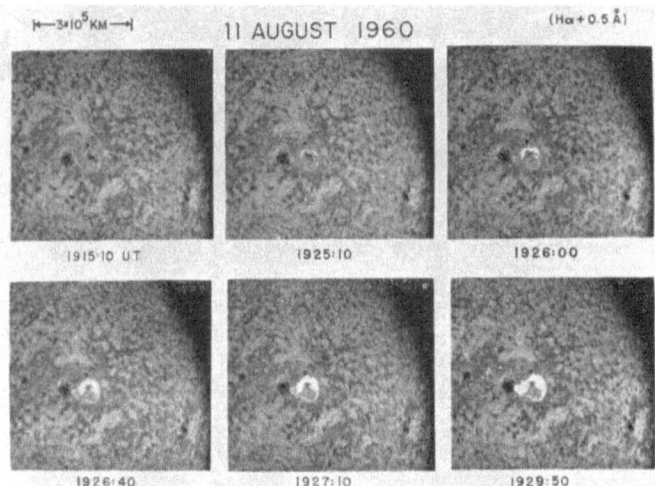

Fig. 5.   Explosive flare of August 11, 1960.

Prior to the explosive phase the bright flare regions spread and brighten in a normal fashion, and the flare may nearly reach its total maximum area; the region undergoing explosive development may be limited to less than 30% to 40% of the total flare area.

An example of an explosive flare is shown in Figure 5. The sequence was selected from 10-second interval filtergrams obtained at Hα+0.5 Å. This flare developed slowly and reached maximum intensity at 19.26.00 UT. The explosive phase occurred during the subsequent 40 second interval, during which time the flare emission spread, preferentially, toward the small sunspot at the top and the small sunspot group at the left. During the short interval of the explosive phase the emission spread at a rate of 220 km/sec. After the explosive phase the expansion decelerated abruptly, and the flare spread slowly to maximum areal development at about 19.30 UT. A small explosive flare is shown in Figure 6. Although not immediately apparent

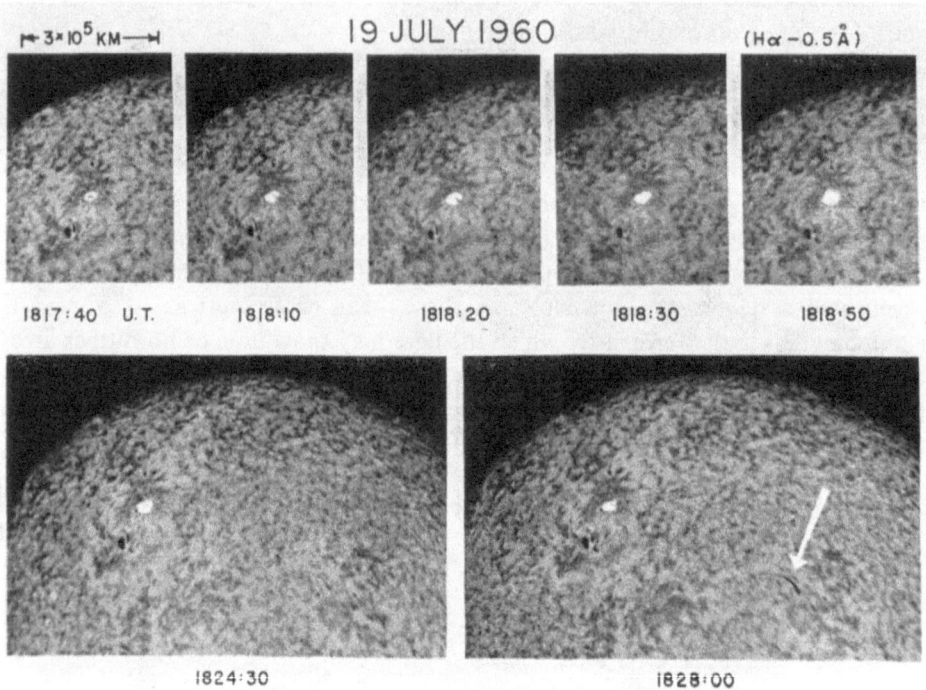

Fig. 6.   Explosive flare of July 19, 1960 and activation of distant filament.

in the small scale photographs, the flare undergoes sudden expansion at the rate of >250 km/sec during the short 10-second interval between 18.18.20 and 18.18.30 UT.

4. High Speed Propagation Phenomena

Comparing the explosive flares in Figure 4 and 5 with the previous figures showing non-explosive events, we notice that the explosive flares show no indication of ejecta near the flare site. However, explosive flares may initiate moving disturbances with velocities and ranges greater than heretofore detected for any optical solar feature. These fast disturbances may be detected in two ways; the first is demonstrated by the lower two filtergrams in Figure 6. 427 seconds following the explosive phase a filament, 430×10^3 km distant from the flare, suddenly becomes visible. From this type of event we infer that a disturbance travelled across the sun from the flare with a speed, in this case, of 1000 km/sec. Photographs made at the center of Hα (not shown) show the filament to be the quiescent type, which, in its inactive state, has insufficient Doppler shifted or broadened components to absorb in the Hα wings. When the here invisible disturbance encounters the filament it undergoes a sudden "winking" appearance: At Hα the filament may momentarily disappear; in either wing the filament momentarily appears. The duration of the winking is ~100 sec. Recent observations show that the winking is due to a Doppler shift, and that the activated filaments first move down, and then up.[2]

The second method of detecting 1000 km/sec disturbances is by direct photographs of faint, cloud- or wave-like ejections from the explosive expansion of flares. Well documented cases are rare, for two main reasons: these disturbances are visible only in the wings of Hα, and because of the faintness and low contrast of the disturbance. These features are usually of such low contrast as to be impossible to locate on single

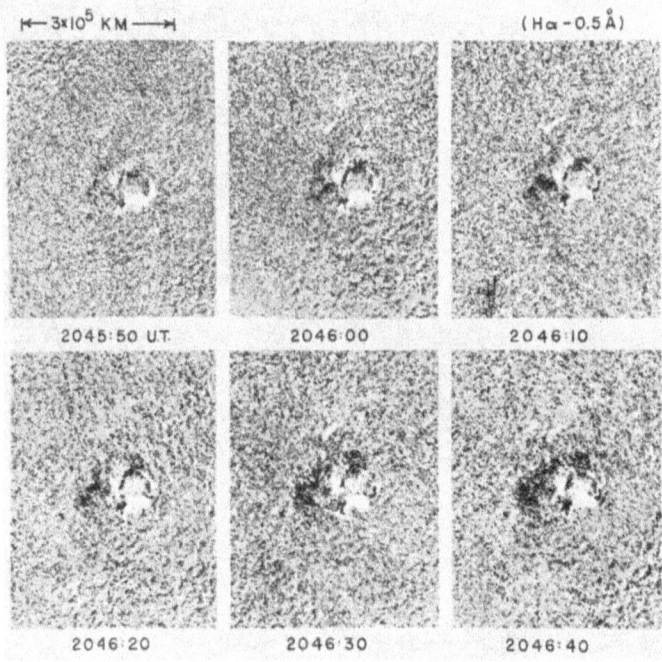

Fig. 7. Explosive flare of June, 25, 1960. Photographic subtraction is used to enhance the faint, fast changing features of this event. The 10 sec interval sequence reveals an absorption feature which moves out from the flare site at 1000 km/sec.

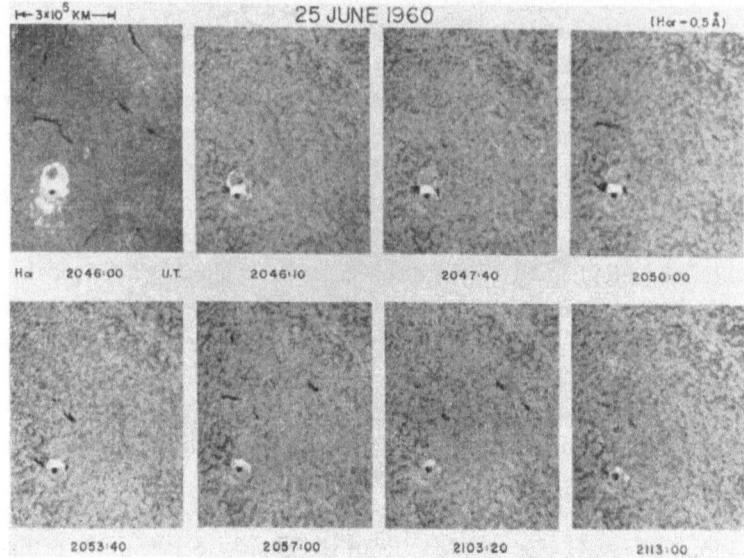

Fig. 8.   Filament activations caused by the explosive flare of June, 25 1960. Filaments visible in the first Hα frame are not visible in the subsequent off-band frames until encountered by the shock fronts initiated at the explosive phase.

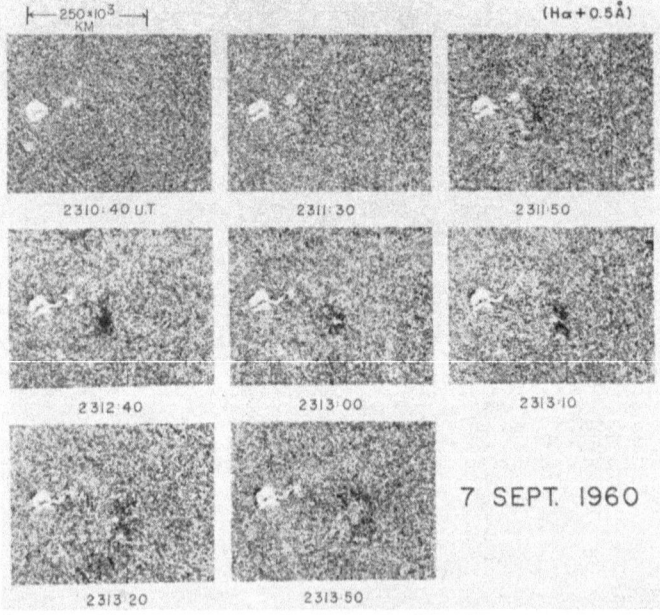

Fig. 9.   Explosive flare and moving disturbance, Sept. 7, 1960. Photographic subtraction reveals a faint cloud-like feature which moves with a speed of 1100 km/sec.

frames of the original photographic negatives, but they are easily seen by cine-projection techniques. In Figure 7 the technique of photographic subtraction is employed to enhance the moving feature by cancellation of the relatively static background. The faint absorption cloud moves from the flare with a transverse velocity component of 1000 km/sec. Figure 8 shows the subsequent winking appearance of filaments effected by the disturbance. (A faster front, moving at 2000 km/sec preceded the one shown, but is too faint to reproduce on individual frames.) Selected cancellation prints in Figure 9 show another fast disturbance which moves with fairly constant velocity of 1100 ± 50 km/sec. The disturbance is first visible at a position 97×10^3 km from the flare site. Extrapolation of the disturbance to the flare position results in a time zero near the explosive phase.

The paucity of data, combined with the complexities of detection and measurement, prevent any definite conclusions to be drawn regarding the nature of these new propagation phenomena; however, some facts are fairly well established:

(1) In all cases the flare from which the disturbance is ejected is an explosive flare. The occurrence of a disturbance is independent of the magnitude of the flare, as is shown by the importance classification of the responsible flares:

Flare importance	3	2	1	1⁻
Number of events	3	6	6	2

(2) The speed of the disturbance is determined by either accurately measuring the time interval and distance between the explosive phase and subsequent first activation of a distant filament, or by directly measuring the displacement of the disturbance as it moves from the explosive region. In three cases both methods were possible, and the direct measurement confirmed the validity of the speed inferred from the measurement of filament activation. The speed is a measure of the transverse velocity of the disturbance, (since the radial component cannot be detected), and in the majority of cases is around 1000 km/sec:

Number of events	Transverse velocity (km/sec)
1	350
5	500–750
9	1000–1300
1	2000

(3) These disturbances are able to propagate over projected (against the solar disk) distances of one solar radius (7×10^5 km) and, in a few cases, greater. The disturbances are highly directional, and move away from the flare site through an

arc with included angle usually around 60°. The data are really insufficient to properly assess other propagation properties; however, from some events we may tentatively suggest that (a) the disturbances propagate with fairly constant velocity and, (b) the visible Hα disturbance is an *effect*, caused by the source as it depresses the chromospheric background, and similarly, causes the initial downward motion of distant filaments. Observation of these disturbances provides the impression that they may be guided by magnetic lines of force. If this is the case then the transverse component certainly does not follow the general poloidal field, since 80% of the disturbances move from east to west, in the direction of the sun's rotation. Perhaps the disturbances follow strong local fields, which are drawn out in an west-east direction by the equatorial acceleration of the sun; however, there is no observational evidence for such fields.[3]

4. Explosive flares accompanied by 1000 km/sec disturbances are outstanding events. Further insight into their nature may be provided by the fact that the majority is associated with an equally outstanding event in the radio spectrum: the Type-II (slow drift) radio burst. Type-II burst are characterized by slow (1 Mc/sec systematic drift of the main emission features from high to low frequencies (WILD *et al.*, 1963). The drifting radio emission is generally accepted as resulting from plasma oscillations excited as a disturbance moves through the chromosphere-corona. Velocities derived for the disturbance causing the Type-II bursts lie in the range 1000–1500 km/sec, which is about the same as measured for the optical disturbances. Only about 3% of common flares are associated with Type-II bursts (ROBERTS, 1959) in contrast, 60% of the explosive flare disturbances had such bursts:

Flare importance	1	2	3
Percentage of 702 flares with Type-II's (ROBERTS, 1959)	2	7	31
Percentage of 15 explosive-disturbance events with type-II's	30	80	100

The reverse correlation, the number of Type-II bursts with flares shows that out of 21 Type-II events, occurring during 10-sec interval filtergrams, 62% were caused by explosive type events. (Only six were accompanied by visible disturbances, however.)

Although the limited evidence suggests that the Hα disturbances are the visible result of a moving source which also excites the Type-II bursts, there is little detailed correspondence between specific features of particular events. This is probably due, in part, to the relatively crude optical data, and also the fact that some Type-II bursts are multiple, with sources that begin at slightly different times and move in different directions with different speed (WEISS, 1963). In general, the supersonic velocity of the disturbance, some ten times the velocity of sound in the corona, suggests that the exciting agency is some kind of a shock wave. A magnetohydrodynamic wave is usually invoked to explain the characteristics of the Type-II bursts, and no evidence to the contrary is provided by the optical data. Indeed, the disturbances, when viewed in movie projection, give the strong impression of a moving shock front.

5. The Explosive Phase as a Non-Thermal Phenomenon

Flares exhibiting the property of explosive development are more common than those accompanied by Hα shock fronts, probably because of the observational difficulties in detecting such fronts. A partial survey of high time-resolution filtergrams has revealed some 270 explosive flares, and indicates that these flares show more frequent association with high energy, non-thermal, processes than do common flares.

The most common type of flare-associated radio emission occurring in the centimeter wavelength region is the impulsive burst at 10 cm. Under special conditions some impulsive bursts can be attributed to thermal processes in a 10^6 °K condensation. However, for the great majority of bursts temperatures in the range 10^7–10^9 °K are required, and seem to preclude thermal origin (WILD et al., 1963). The 10-cm burst is usually attributed to a combination of synchrotron and bremsstrahlung processes. In a survey of 183 explosive flares, all (24) flares of importance greater than 1^+, and 36% of the remaining flares of lesser importance, produced 10-cm bursts (ANGLE, 1963). In comparison, only a 75% and 16% association, respectively, would be expected if these were common flares (HARVEY, 1963). Type-III bursts (fast drift), usually attributed to plasma waves excited by a stream of electrons moving through the corona with speeds of 0.4c, are also highly associated with explosive flares, being about twice as frequently associated as common flares. Furthermore, studies of a single importance class (1^-) shows that the probability of association of an explosive flare with a Type-III burst is doubled if the flare also is accompanied by a 10-cm burst.

The relevance of the explosive phase to the production of centimetric and Type-III bursts is further enhanced by comparison of the time of the start of the explosive phase and the start of the bursts. In common flares the radio bursts show no preference for any particular phase in the pre-maximum development of the flare. In contrast, about 75% of the Type-III and 10-cm bursts start in a 90-second interval around the explosive phase (MALVILLE and MORETON, 1961b).

Another explosive phase feature is the production of high energy (> 20 keV) X-rays. Using balloon and rocket methods, eight events have been detected during times when 10-sec interval filtergrams are available. ANDERSON and WINCKLER (1962) attribute X-ray bursts of this nature to bremsstrahlung electrons, and KUNDU (1961) suggests that the hard X-ray bursts have their origin in common with the impulsive 10-cm burst. Recent investigation of the optical characteristics of these flares show them *all* to be of the explosive type (MORETON, 1964). In five of these events where the time history of the hard X-ray flux is available, the burst tends, on the average, to lag the explosive phase by about 2 minutes.

The suggestion that *some* flares, the explosive type, may produce bremsstrahlung

X-rays, while common flares do not, also follows from the work of POUNDS, *et al.* (1963). They have studied three flare events in the 4–14 Å region detected by Satellite Ariel. Slow X-ray enhancement caused by two of the flares can be accounted for by a thermalized plasma of chromospheric density at a temperature around 10^7 °K. The third flare, however, produces a "burst" with a harder spectral distribution that the authors suggest is caused by bremsstrahlung radiation. Similarly, NESHPOR (1963), from analysis of ionospheric effects caused by flares, suggests that only *some* flares are characterized by a considerable increase of intensity in the hard X-ray region.

In conclusion, there is evidence for two types of flares and two types of processes for production of flare radiation. Most flares are relatively stationary phenomena, which produce low velocity ejecta and a variety of gradual emissions, such as enhancement of 10-cm emission and low energy X-rays. These enhancements accompany a similar slow rise in the Hα flare. SMERD (1963) suggests these "glow-like" emissions are electromagnetic only, and probably arise from thermal or quasi-thermal processes, such as compression by excess magnetic pressure $(H^2/8\pi > NkT)$.

Explosive flares are characterized by the brevity and the high velocity of associated phenomena. During the explosive phase part of what otherwise might appear as a normal Hα flare undergoes accelerated development. Streams of fast electrons $(\sim 10^5$ km/sec) produce strong Type-III bursts, and some 10^{35} electrons are accelerated in the explosive region. The burst of electrons, with energies up to 100 keV, interact with the magnetic fields and particles in the chromosphere, and produce bursts of centimetric radiation and hard X-rays by synchrotron and bremsstrahlung processes. Magnetohydrodynamic waves are excited, and in turn excite plasma oscillations as they move into less dense coronal regions, giving rise to Type-II radio emissions. On occasion, perhaps because of strong local fields, the MHD wave moves tangentially, and can be detected in Hα as a disturbance moving from the explosive region with a speed of around 1000 km/sec.

PARKER (1958) has shown that about 2 minutes is required to accelerate particles to relativistic energies by the Fermi mechanism. This time is in agreement with the approximate time lag between the initiation of the explosive phase and subsequent high energy X-ray bursts, cited above. On the other hand SMERD (1963) has pointed out that the brevity of the discharges producing successive Type-III bursts (≤ 1 sec) requires a faster acceleration process. Perhaps more extensive comparison of optical and radio data with satellite measurements will provide further important information on this problem, and the problem of flare dynamics in general.

References

ANGLE, Karen: 1963, *Astron. J.* **68**, 533.
ANDERSON, K. A. and WINCKLER, J. R.: 1962, *J. Geophys. Res.* **67**, 4103.
ATHAY, R. Grant and MORETON, G. E.: 1961, *Astrophys. J.* **133**, 935.
HARVEY, Gladys: 1963, *Astrophys. J.* **139**, 16.
KUNDU, M. J.: 1961, *J. Geophys. Res.* **66**, 4308.
MALVILLE, J. M. and MORETON, G. E.: 1961a, *Nature* **190**, 935.

MALVILLE, J. M. and MORETON, G. E.: 1961b, *J. Geophys. Res.* **66**, 2546.
MALVILLE, J. M. and MORETON, G. E.: 1963, *Publ. Astron. Soc. Pacific* **75**, 176.
MORETON, G. E.: 1961, *Sky and Telescope* **21**, 145.
MORETON, G. E.: 1964, *AAS-NASA Symposium on Solar Flares* (W. HESS, ed.), NASA, Washington, D.C., p. 209.
NEŠPOR, Y. I.: 1963, *Izv. Krimsk. astrofiz. obs.* **29**, 152.
PARKER, E. N.: 1958, *Phys. Rev.* **109**, 1328.
POUNDS, K. A., WILLMORE, A. E., BOWEN, P. J., NORMAN, K., and SANFORD, P. W.: 1963, *Proc. Roy. Soc.*
ROBERTS, J. A.: 1959, *Australian J. Phys.* **12**, 327.
SMERD, S. F.: 1963, 'Solar Radio Emissions' to be published in the Proceedings of the Symposium on Results of the IGY-IGC.
SMITH, H. J. and SMITH, Elske v. P.: 1963, *Solar Flares,* Macmillan.
SMITH, S. F. and RAMSEY, H. E.: 1964, *Z. Astrophys.* **60**, 1.
WEISS, A. A.: 1963, *Australian J. Phys.* **16**, 240.
WILD, J. P., SMERD, S. F., and WEISS, A. A.: 1963, *Ann. Rev. Astron. Astrophys.* **1**, 291.

Notes Added in Proof

1. S. I. GOPASYUK and M. B. OGIR (*Izv. Krimsk. astrofiz. obs.* **24** (1963) 185) have studied the flare-surge by means of 45 mm image diameter H-alpha filtergrams combined with sunspot photo-heliograms. They provide convincing evidence that many surges have their point of origin in the umbra or penunbra of sunspots.

2. New results by H. W. DODSON and E. R. HEDEMAN (in *AAS–NASA Symposium on Solar Flares* (W. HESS, ed.), NASA, Washington, D.C., 1964) using swept wavelength spectroheliograms confirm that the "winking" filaments darken first in the red wing of H-alpha, then in the violet wing; this effect may be attributed to a Doppler shift caused by the filaments moving first down (toward the photosphere), then upward. Further confirmation is provided by sequential filtergrams obtained in H-alpha and in both wings of the line (G. E. MORETON, Astron. J. **69** (1964) 145); the latter observations show two filaments that appear to undergo three distinct down-up oscillations, with a period of oscillation of about 7 minutes.

3. The same sequential filtergrams mentioned in Note 2 reveal another example of a shock-like disturbance that appears to move across the chromospheric background with an apparent transverse velocity component of 750 km/sec. When the disturbance reaches two distant filaments they undergo the "winking" activation. The off-band filtergrams show that the leading edge of the disturbance is darker than the chromospheric background in the red-wing, and brighter in the violet-wing. Hence, the effect may be the same as observed in the filaments, i.e., what is recorded on the filtergrams as a "disturbance" may be the result of an invisible agency that momentarily depressess the chromospheric structure (mottles) as it passes tangentially through the solar atmosphere.

SOLAR FLARES AND CONCURRENT PHENOMENA IN THE SOLAR ATMOSPHERE

H. ZIRIN*

High Altitude Observatory, Boulder, Colo.

Abstract. The observation of solar flares in visual light is mostly made in the light of hydrogen-α, which shows only a small part of the interesting phenomena which accompany the flare. We discuss observations of coronal and prominence phenomena accompanying the flares, and their connection with non-optical phenomena.

At the moment of onset of a large flare a number of phenomena occur. These include (1) a shock wave, (2) a radio burst both in decimeter and later in decameter frequencies, (3) ionospheric fade-outs and absorptions, (4) a great increase in the density and temperature of the corona and (5) a brightening of an area of the surface as observed in hydrogen-alpha (Balmer-α). These phenomena, particularly (2), (3), and (4), are shown to be connected with the production above the flare of a very hot coronal region, with a temperature about 4 million degrees, and with a high energy non-thermal component with fast electrons with mean energies around 2 keV. The density of the coronal condensation is between 3×10^{10} and 3×10^{11} hydrogen atoms per cubic centimeter. A little after the flare cosmic rays and low energy protons are noted at the earth in some cases. A day or two later, the particles which produce a geomagnetic storm arrive. These various effects are discussed, and several flares are examined in detail.

We see that before attempting to understand the reasons for the occurrence of flares, or to predict them, it is important to find out just exactly what a solar flare is.

I have enjoyed very much hearing Mr. Moreton's presentation because I think we are both working on various ends of the same elephant, or the same complex phenomenon which we call the solar flare. The body of our knowledge of solar flares, until 10 or 15 years ago, had been accumulated principally from one type of observation, namely, observations with the Lyot filter in the Balmer line of Hydrogen-α. In a way this has perhaps prejudiced our picture of flares and, therefore, the various groups that have been studying them in the last ten years. Other observations are beginning to show us there is much more to the flare than the Hα phenomenon itself. I might first talk about the kinds of phenomenon that accompany the flare, and I think when I present a few flares in detail you will see that at a certain point of the development of this phenomenon a number of things occur within a few minutes of one another (with a statistical spread of a couple of minutes partly due to observational errors). At the onset of a flare several events are seen to occur in the solar atmosphere, and other things happen which presumably are due to goings on in the solar atmosphere. One obvious phenomenon is the expulsion of material, which you have seen in Mr. Moreton's pictures. First, material is blown out from the regions near the flare, typically with velocities anywhere from 200 to 1000 km/sec. Second, at the same time a radio burst takes place, characteristically in the decimeter or 3000 Mc/s region. Many different radio bursts are observed, but these may also be seen when no flare whatever occurs. So we shall limit ourselves at the moment to the phenomenon which essentially only occurs at the onset of a large

* Present address: Mt. Wilson and Palomar Observatories, Pasadena, Calif.

Chang & Huang (eds.), Proc. Plasma Space Sci. Symp. All rights reserved.

flare, namely, the decimeter burst observed at 2800 Mc/s. The largest body of material in this regard has been recorded by Covington at Ottawa. The third phenomenon is the ionospheric event which is known by many different terms depending on the way in which you observe it. The sudden ionospheric disturbances or shortwave fade-outs are detected by absorption of signals from distant stations. The SCNA (sudden cosmic noise absorption) indicates the ionization in the D-layer by measuring the attenuation of cosmic radio noise, i.e., radio noise from astronomical sources which normally is constant but at the time of the flare is absorbed by D-layer ionization. A third method is the phase anomaly (SPA) which measures the lowering of the D-layer. Finally, there is the sudden enhancement of atmospherics (SEA), that is, an enhancement of the atmospheric noise at 18 or 30 kc received from distant thunder storms and other atmospheric phenomena.

All of these ionospheric phenomena are closely connected with X-rays from the sun, presumably produced by energetic electrons produced in a solar flare. These solar phenomena which may be connected with this are often seen when we make coronagraph observations above solar flares at the limb. At such times we find a great increase in the density and temperature of the solar corona in the region just above the flare. We have concentrated on this point because we operate corona-graphs and are interested in studying what is happening just above the flare. I don't want to go into the solar corona in great detail and we will look at some of the spectra in just a moment. Finally, near solar flares we see a brightening in Hα. Sometimes, but not always, these are the phenomena which occur at the beginning of the flare and you can see the fact that very many of our flare observations which have been limited to Hα can sometimes distort the picture of the flare a little bit.

There are other phenomena which occur a little later after the flare – the arrival of cosmic rays in the form of low-energy protons, and much later the occurrence of the magnetic storm. Sometimes one does not always observe these effects. I would like to talk about a few events in detail and the interpretation of them that go a little further into what Mr. Moreton discussed concerning the various phases of the flare. Perhaps we do not always look at them the same way but we may learn things by looking at them in various ways. I might point out from the start that one reason we are interested in the solar atmosphere above the flare is that if X-rays are pro-duced in the flare they must be produced in the corona since it would be very hard to get them out to the surface. We are, therefore, very much interested in what is going on in the atmosphere, and as you will see there are many ways in which we can look at this. Although I am throwing rocks at the Hα study of solar flares, most of the pictures I show are in Hα, and this is for a very good reason. One can only produce a picture of a flare in this line (or in the line of ionized calcium). The other phenomena are much harder to see, but we will try to look at some of the other subtle phenome-non. Finally, I might say that although a solar flare is an extremely large and important phenomenon which disturbs all interplanetary space as far as the earth, it is essentially invisible except for a few of the very largest flares, which you might, with luck, see by conducting very careful observation in integrated light for a hundred years. In order to make the Hα pictures you have seen, one needs a special filter which is very difficult to keep in adjustment. Additional requirements include good photographic and guiding techniques. Thus, the visual aspect of the solar flare in the wavelength range that our eyes see is very small. On the other hand, if we could look at the sun with X-ray eyes or with extreme ultraviolet eyes, the flares would appear truly important.

Figure 1 shows an Hα filter picture from Sacramento Peak. It is a large eruptive

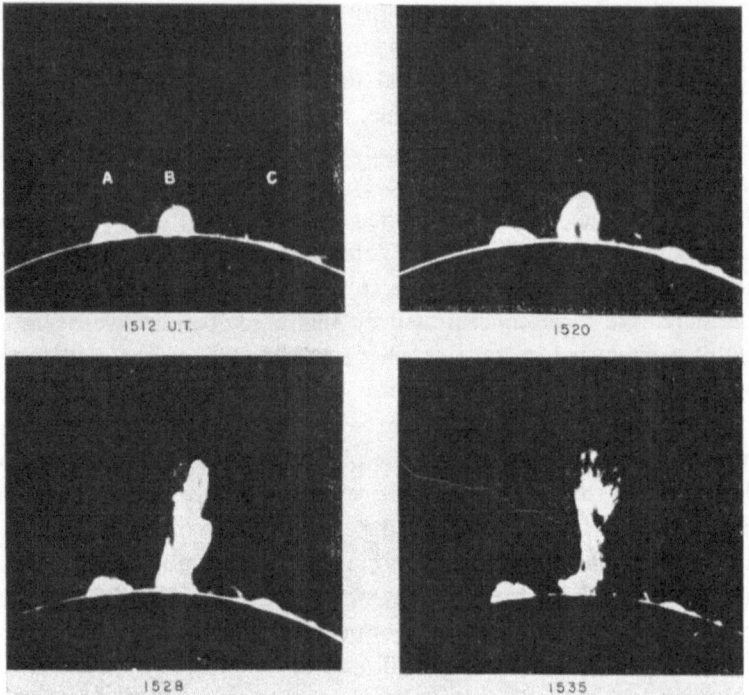

Fig. 1.   Eruptive prominence photographed in Hα, June 14, 1956. (Sacramento Peak Observatory photo.) *A*, *B*, and *C* denote quiescent prominence, eruptive, and flare, respectively.

arch which occurred June 14, 1956. The arch started out like a small loop but gradually reached a height of 300000 km. Spectra were obtained of this event with the Climax coronagraph (Figure 2). With the coronagraph we simply put the slit of the spectrograph across the phenomenon that we were observing. We move the sun away from the slit and make a succession of spectra. Although the direct picture in Hα, which is limited to a narrow band around the center of the Hα line, shows only the eruptive arch and two small prominences on either side, the spectrum in Figure 2 shows that the prominence to the left actually had an extremely wide Hα emission line characteristic of solar flares. This phenomenon was actually much brighter than the eruptive prominence here pictured, but because of the large velocity along the line of sight, most of its radiation was shifted out of the pass-band of the narrow band Lyot-filter. Incidentally, the spectrum of the top of the eruptive arch shows that it was expanding with a velocity of approximately 800 km/sec. There is a very turbulent regime at the top of such expanding loops or in the typical material expelled by

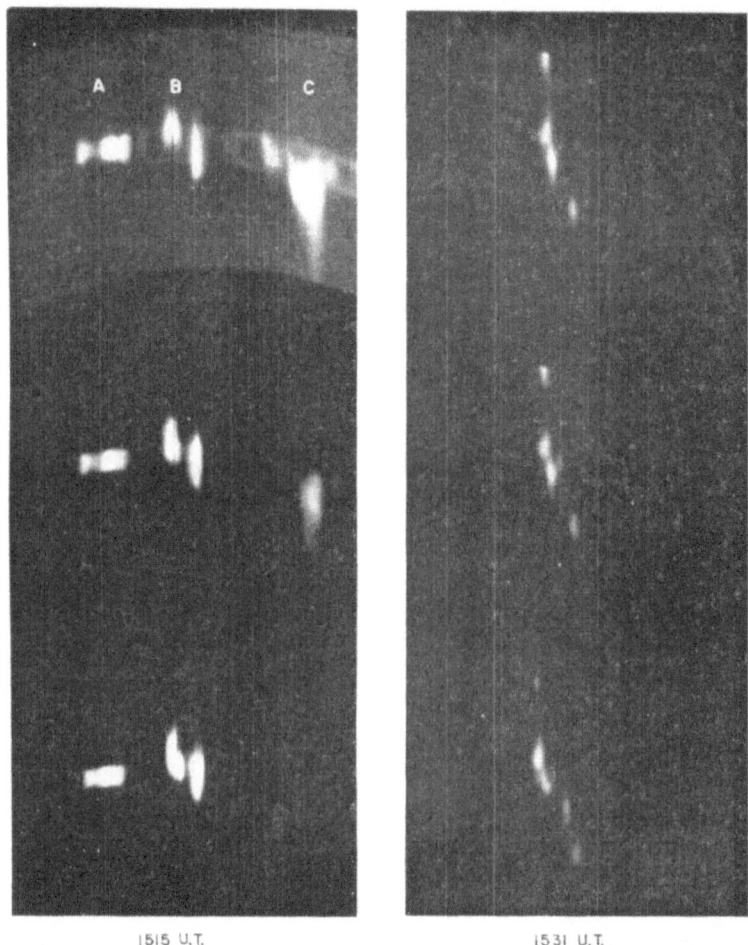

Fig. 2. Hα spectrum of same obtained at Climax. *A*, *B*, and *C* denote quiescent prominence, eruptive, and flare, respectively. Second spectrum made at top of eruptive.

a large flare, the so-called flare spray. This event shows how difficult it is to observe flares with narrow-band Hα filters. The effect seems to be most strongly marked near the limb of the sun, because the material is mostly horizontal. The problem would not be so difficult if one could always take a spectrum and ascertain if there is material radiating outside the pass-band of the filter.

As many of the particle physicists here know, in November 1960 there was a very large sunspot group which came over the limb on November 5, and produced two extremely large cosmic ray flares on November 12 and November 15, 1960. This

H. ZIRIN

region went over the limb on November 19 in quite a blaze of glory, with loops, surges, etc. On November 20 there was a third cosmic ray flare, considerably smaller than the first two, but still a distinct event, as reported by CARMICHAEL *et al.* The Canadian observers (CARMICHAEL *et al.*, 1961; COVINGTON and HARVEY, 1961) pointed out that this flare occurred after the spot group was over the limb, and that this was the first observation of this type. HANSEN (1961) examined the Climax flare patrol observations for that day and found that, in fact, the region was not completely out of sight, and that a large event had occurred on the limb at just the time indicated. Figure 3 shows a series of pictures of this event made in Hα at Climax. A picture of

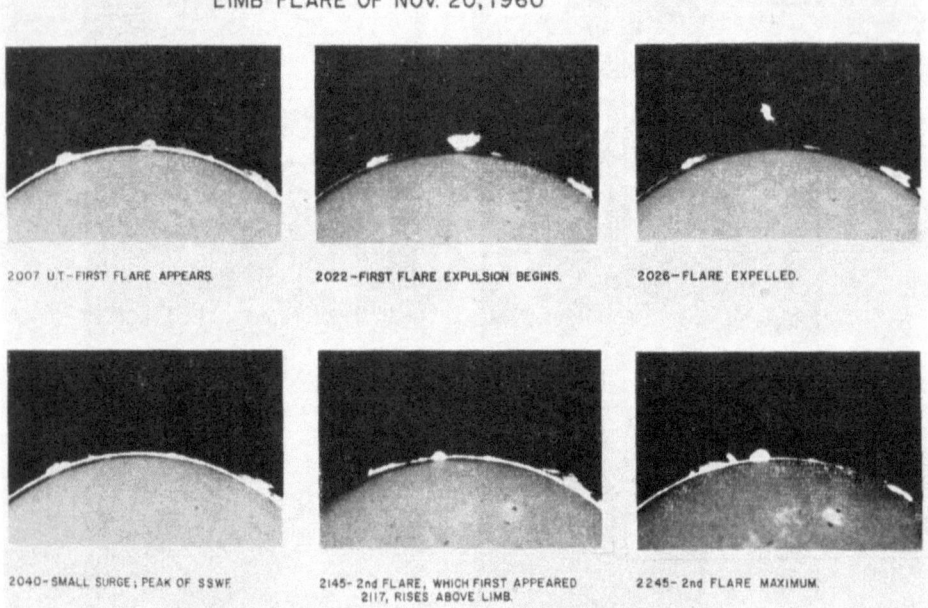

LIMB FLARE OF NOV. 20, 1960

2007 U.T–FIRST FLARE APPEARS.

2022 –FIRST FLARE EXPULSION BEGINS.

2026–FLARE EXPELLED.

2040– SMALL SURGE; PEAK OF SSWF.

2145– 2nd FLARE, WHICH FIRST APPEARED 2117, RISES ABOVE LIMB.

2245– 2nd FLARE MAXIMUM.

Fig. 3.   Limb flare, November 20, 1963 in Hα (Climax).

the disk is taken in the center of Hα with the band-pass of 0.75 Å; then the end polaroids of the Lyot filter are removed to increase the pass-band to 3 Å, and an occulting disk replaces them. The limb picture thus obtained is superposed directly on the narrow-band picture of the disk. This method gives us a chance to look both at the disk phenomena which are normally classified as flares, and at the goings on in the atmosphere above.

The following sequence of events happened on November 20, 1960. At 1945 there was a small shortwave fadeout, accompanied by a small increase in the 2800 Mc/s radiation from the sun as measured by Covington, and a decrease in the observed radio radiation at 18 Mc/s (called a sudden cosmic noise absorption – SCNA) because of the absorption in the D-layer. This ionospheric effect tells us that there was a small flux of energetic X-rays coming from the sun at that time.

At about 1950 UT a bright cloud of material rose above the limb. The cloud of material grew brighter and brighter until at 2022 a very dramatic change occurred. In a period of four minutes this large ball of gas was rapidly expelled from the sun with a speed of about 500 km per second. At this time a number of violent phenomena which we normally associate with large flares occurred. In particular, there was a great outburst of 2800 Mc/s radiation observed by Covington. There was the beginning of a large and deep cosmic noise absorption on 18 Mc/s which reached a maximum about 2045 UT, at which time about 90% of the flux incident on the ionosphere was being absorbed. At the same time that the first flare was expelled, there also was a great brightening of the corona (although we cannot fix the time of this exactly), and apparently the acceleration of the cosmic rays (later observed by Carmichael *et al.*) occurred. All of the ionospheric and radio effects occurred within about one minute of the time at which the first cloud of material began to move upwards. There are other cases in which such explosive events have been observed. Another good one is July 20, 1961. In that case an active pillar prominence appeared before the flare, however, it was not of flare brightness, and was not accompanied by a small SCNA. When the flare occurred, only a small brightening was seen in the Hα observations, but the pillar prominence which had appeared shortly before was blown off the sun, and the entire series of ionospheric and radio events occurred, all within a minute of the beginning of the expulsion. The cosmic rays in either case came 20 or 25 minutes later, simply due to time of flight delay.

After the first flare on November 20 was expelled at 2023 by a second flare which we think took place in the large sunspot region just over the limb of the sun, there was nothing visible in Hα, on the limb of the sun or on the disk, yet the maximum absorption of radio waves occurred at 2045 UT. At that time there was also a peak in the 2800 Mc/s emission, and the maximum ionospheric effect. I would guess that had we taken coronagraph spectra at this time, we would have seen an intense coronal condensation. However, the Climax observers did not take spectra at this time because in the Hα patrol nothing at all was visible on the limb of the sun. At 2117, with the ionospheric fade-out still in progress, another bright flare appeared. It seems to have been the top of the large flare itself which had occurred at 2023 UT, and was growing in Hα size during this period. This conclusion is based on the fact that it occurs just over the point at which the sunspot group had gone over the limb. The new cloud of material grew larger and larger in size, and a surge appeared which seems to have left the sunspot region at about the time of the main flare. We obtained quite a few spectra of this second bright ball, and it showed the highest excitation conditions.

Our conclusion is that a very large flare occurred in the sunspot region behind the limb at 2023 UT, which blew off the first flare. This flare was not visible to us in Hα until 2117, but the ionosphere, which looks, so to speak, with X-ray eyes, could see the flare. This is because the energetic phenomena occur in the corona above the flare. As this hot coronal material cooled, there was an expanding Hα cloud which eventually grew high enough to be seen over the limb. The sunspot at this time was between 15 and 20000 km below the limb, so we are talking about events that are really not very high above the sunspot. Figure 4 shows some of the spectroscopic observations made during the flare. The first spectra were made at 2007 UT, during

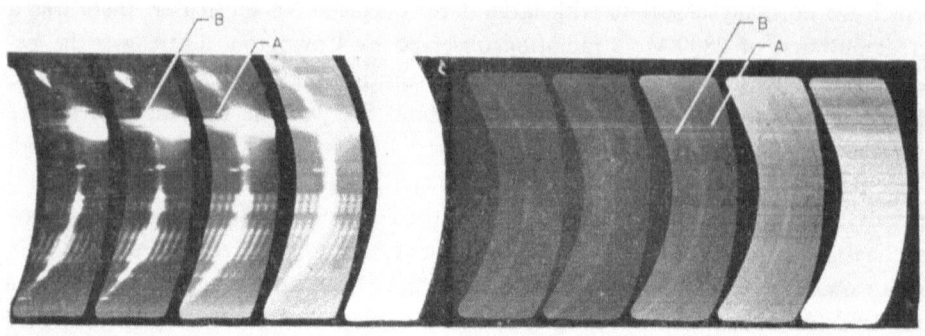

the first small flare. They are taken at successive heights above the limb in Hα and in the red coronal line 6374. This spectrum was made at 2007, the time at which the first cloud of gas of the first flare appeared above the limb. The first flare shows as a large bright emission in Hα, with a small amount of continuum. Nearby, at point A, we see an even stronger continuum emission which is exactly above the place where the sunspot group was and at the point where the flare later occurred. The first event was hard to see without a broad-band filter, because of the great displacement of the Hα emission. The effect of the enhancement of continuum in the region before the flare occurred is something we have previously observed in a flare on December 18, 1956 (ZIRIN, 1959). The continuum that we see is produced by Thompson scattering of the photospheric radiation by electrons in the corona. Where we see this continuum, therefore, we know that there is a substantial density of material at coronal temperatures. The brightening of the continuum in this case, as in the other case, has no bright counterpart in Hα. That is, we have here a clear cut case of a coronal condensation independent of any prominence or pre-flare. Figure 5 shows a spectrum made at one height at 2145 UT in the second event. In it we can see some of the many elements that are present in a solar flare. Remember that hot tenuous material radiates very poorly and cool dense material radiates very well. The gas at several million degrees does not radiate well in the visible region but mostly produces emission lines in the extreme ultraviolet. The background of curved absorption lines here is merely the spectrum of the sky, and should be disregarded. The streak across the middle of the spectrum is the photospheric continuum scattered by the free electrons in the flare condensation. Measurement of the intensity of this continuum gives us a very good way of determining the total number of electrons along the line of sight.

We know the Thompson scattering cross-section and the over-all geometry involved, and from the measured brightness may deduce the total number of electrons along the line of sight. If we then assume that the flare is not made up of long fingers

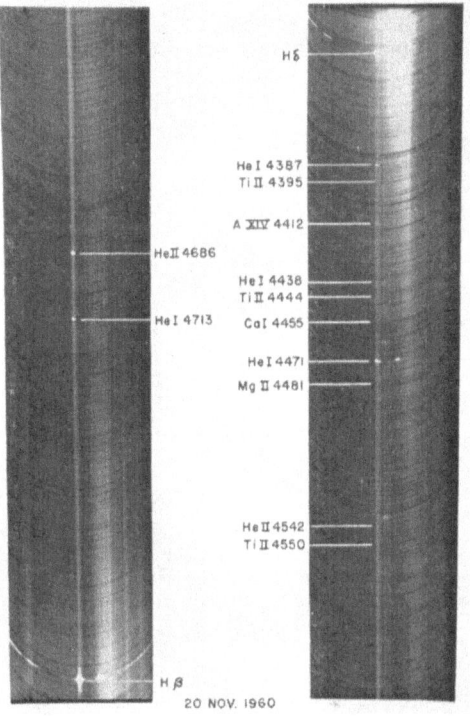

Fig. 5.   Same, 2145 UT, Hα to Hγ, ht. 20000 km (Climax).

pointing at us or away from us, but is more or less symmetric, we may determine the average electron density. Of course, if there is any filamentation at all the local density is increased proportionally. This approach is quite independent of the temperature of the gas so long as it is all ionized, because all electrons scatter with the Thomson scattering cross-section. The density of the coronal cloud deduced in this case is about 2×10^{11} electrons per cubic centimeter. Since the number of protons is assumed to equal the number of electrons, this would also be the density of hydrogen atoms per cubic centimeter. This spectrum is also dominated by lines of neutral and ionized helium, as well as the very bright hydrogen lines. There are some very weak lines of ionized metals. In a typical quiescent prominence these lines are nearly as bright as the neutral helium lines. In a flare, however, the temperature is much higher and the metals are in higher stages of ionization which we do not see in this region of the spectrum. The only metallic line whose intensity appears to increase in the flare spectrum is the 4481 line of ionized magnesium, which is a high excitation line quite strong in the spectra of B and A stars. We also may see one coronal line in this part of the spectrum, the 4412 line of argon XIV. 650 volts are needed to ionize argon XIII and produce this ion. This gives us another guess at the temperature of the coronal material. The theory of the ionization equilibrium in the corona was first studied by Professor Biermann in 1947, (BIERMANN, 1947; ELWERT, 1954), and we can tell that temperatures of at least 2 or 3 million degrees are required to produce these high stages of ionization. Figure 6 shows another spectrum obtained at 2227 UT,

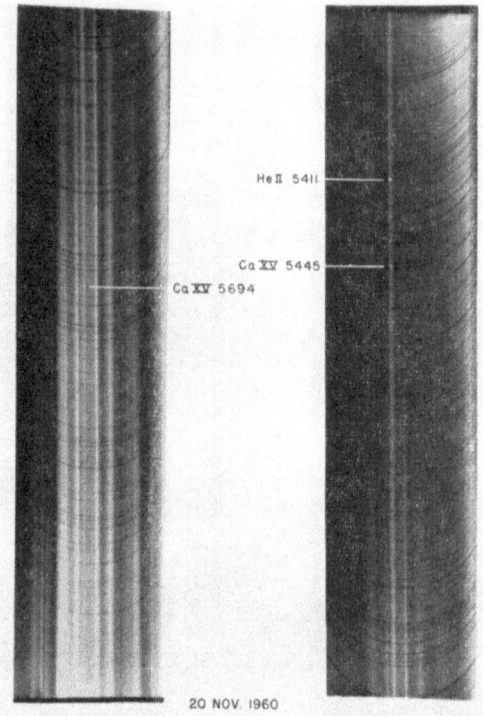

20 NOV. 1960

between 5400 and 5800 Å. Here we see two other important coronal lines characteristic of flares, the two lines of Ca XV at 5694 and 5445. These ions require 814 volts for their production, and are the highest stages of ionization anyone has yet observed anywhere. We also see the 5411 line of ionized helium. The spectrum of the flare in this region is dominated by one line of ionized helium and two lines of Ca XV. The broadening of the Ca XV lines by thermal motions gives us another measure of the temperature in this flare. In this case we find about 4 million degrees. Note how closely the positional distribution of the coronal lines and the continuum emission are correlated. It is generally true that one does not observe these high ionization lines without strong continuum. This is mostly true because calcium is not an abundant element and we will only see the radiation when there is a high electron density along the line of sight. We have many spectra taken at many heights in this flare which show that the coronal line emission and the continuum emission (or scattering) go hand in hand. In the case of hydrogen and helium the connection is much weaker. The ionized helium and neutral hydrogen emission is weak high in the flare but it becomes very strong lower down. This is because hydrogen and helium may be emitted very efficiently by cool gas. Study of the gradient of the electron density shows that the bright helium and hydrogen emission is produced by only a small fraction of the gas involved. In these lower areas where the helium emission is very strong, the ionized helium lines are quite narrow, showing that they come from a cool component. Higher up, where the weak helium emission apparently comes purely from recombination from the hot coronal gas, these lines are much broader.

The one question which we have not completely resolved is whether the spectra we show here pertain to a flare, or post-flare loops. Very often solar limb flares of this type are followed by bright loops of material condensing out of the corona. The Hα structure seen here has the form of loops, and is probably not different from the post-flare normally seen, except that they are far brighter than usual. At the time these spectra were taken, the flare was still in progress, as is shown by the

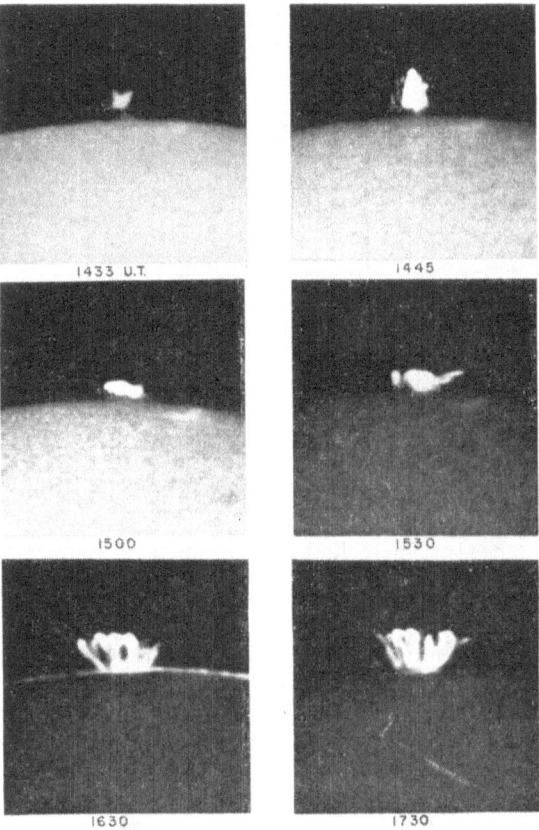

Fig. 7.   Limb flare, November 10, 1961, in Hα (Climax).

fact that decimeter emission of an impulsive nature was still being produced, and that the shortwave fadeout was still in progress. So although we can not be sure that the Hα phenomena that we are looking at is part of the flare in the strictest sense, we can be fairly sure that we are looking at the region in the solar atmosphere where a solar flare is going on, albeit in its later stages.

Figure 7 shows another solar flare on November 10, 1961, which didn't show an explosive phase, although it did produce a substantial proton event and polar blackout. The flare appeared as a bright protuberance above the limb at 1433 UT. At this time there was a large decimeter outburst recorded by Covington, Figure 8,

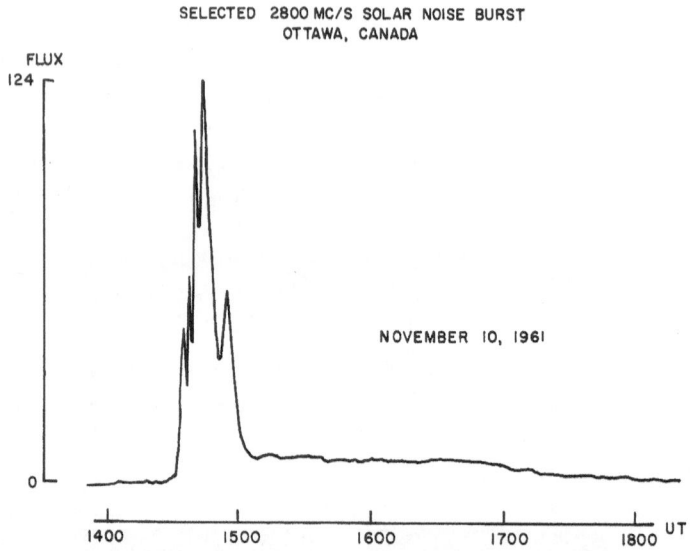

Fig. 8.   2800 Mc/sec emission from above as recorded by Covington at Ottawa.

as well as a small sudden cosmic noise absorption recorded at Boulder. The SCNA was presumably small because the sun was very low in the sky, and the X-rays do not penetrate deeply into the D-layer under those circumstances. The Hα flare appeared as a bright blob above the limb at 1433 UT, exactly the moment at which the radio outburst began. There had been a small increase in the radio radiation beforehand, an effect which has been noted in many other events of this type. This increase is probably connected with the increase in coronal continuum just before the flare to which I have referred. This bright blob of material appeared to be "boiling" and changing its shape rapidly, without any particular transverse motion. It faded out about 1500 UT and was replaced by some very bright and elegant loop prominences, which lasted until 1800 UT. These loops are characteristic following limb flares, as I remarked before. Although the loops were the brightest phase of the limb flare in Hα, the radio and ionospheric events were over by the time they were formed. There is some correlation between the individual spikes in the decimeter emission and the pulsations of the blob of Hα emitting material. We only were able to obtain spectra of this event during the loop phases, which showed the usual bright emission of the Ca XV yellow line. The increase in coronal density was also measured by the K-coronameter, an instrument which measures the electron density in the corona by the polarization of the corona. The K-coronameter measured a substantial increase in density at a height around 200000 km above the limb. A large outburst was also observed in decameter wavelengths between 15 and 40 Mc/s by Dr. Warwick at the HAO radio observatory.

We see that in this case, as in many others, the energetic phase of the flare ended before the brightest Hα phase was seen. Presumably the radio emission and the ionospheric event were produced by energetic electrons accelerated in the small blob of material we see above the limb, or at least in the near vicinity. Presumably, the solar protons were also produced in this short intense event. The bright Hα loops seem to be just an aftermath, during which the hot coronal material cools, condenses, and drains down to the surface along the magnetic lines of force. A similar development is seen in the Hα movies, for example those of Mr. Moreton that we saw. The flash phase is seen on these films to occur very early in the flare event. We then see a spreading out of the Hα emitting area to a maximum brightness phase which presumably corresponds to the development of these very bright groups. This last statement is somewhat conjectural, as we still do not have a conclusive proof of what an event like that of November 10 would look like when seen directly on the disk.

In connection with our interest in the coronal phenomenon during and after the flare, it is useful to call attention to the so-called post-burst increase in the decimeter emission. This appears very clearly in Covington's record of the November 10 event (Figure 7) as a long tail-off in the emission. If we measure the electron density in the post-flare loops by using the Thompson scattering, we obtain an electron density of about 3×10^{10}. With the apparent geometry that we see, this condensation would be opaque in 10 cm, and its radiation could be approximated by the Rayleigh-Jean law. When we calculate the radio emission in this way, we find that it agrees exactly with the observed decimeter emission if we use a temperature of about 3 million degrees. So the long tail-off of emission can be explained as a purely thermal phenomenon. Many times one observes flares accompanied by only a gradual increase in the decimeter emission, and these may be explained by flares in which high energy electrons are not produced, but only a hot coronal condensation is formed, which gives rise to the observed radiation. The bursty phenomenon appears to be connected with the production of high energy electrons, which radiate by bremsstrahlung. These energy electrons also produce the X-rays which cause the ionospheric effect. The acceleration mechanism must be a fast and violent one, because of the characteristic difference in time-scale between the burst, which is always spiked and rapidly varying, and the normal gradual rise and fall, which is a slow phenomenon.

We have a few other ways of measuring the electron density in the flare condensation. We may measure the intensity of the coronal lines and by measuring the ratio of this intensity to the scattered continuum, derive an excitation function which depends on the microscopic electron density. By use of a few physical parameters one may thus obtain the electron density. Another way of determining the electron density is by the intensity of the ionized helium lines which are seen to arise from coronal recombination. This is perhaps not so good a measure, because the helium recombines preferentially in the cooler parts of the condensation. In any event, these various measurements give us about the same order of magnitude for the coronal condensation either in or following the flare. Measurements near the peak of flares give densities of the order of 2×10^{11}, whereas measurements late in the flare give densities around 3×10^{10}. The temperature may be measured as we have indicated before from the ionization equilibrium in the corona, which gives us the right amount of Ca XV at about 2.5 million degrees, or from the Doppler broadening of the lines, which gives about 4 million degrees. This is not the temperature of all the material in the flare. A small amount of the material is cooler, and produces the very intense radiation in Hα and the other low ionization lines. An even smaller amount, perhaps one electron in 10000 is at a very high energy, at least 20 kilovolts, and gives rise by brehmsstrahlung to the decimeter and X-ray emission. It is interesting to note that protons moving with the same velocity would have energies in the mev range, and could contribute to the proton shower produced by the flare.

We can now summarize the events which occur in the corona around a solar flare. By direct coronal observations of condensation, and by the small increase in the decimeter radiation, we see that there is often an increase in density in the corona before the flare outburst. On occasion this is observed in the form of a pre-flare, namely an active prominence or small flare which may appear above the sunspot group. At the time of the flare itself a blast wave is produced which will blow this pre-prominence off the sun. Sometimes, however, no blast wave is produced. The flare itself appears to be a cloud of very hot material at a height of 10 to 20 thousand km above the surface. In this hot dense cloud there are many electrons at energies above 20 kilovolts which produce the radio and X-ray emission which we see. The high energy protons which come to the earth are presumably also produced in this region. It is not clear what the percentage of the cloud is at these high energies, and what percentage is at coronal energies. The evidence indicates that most of the material is only at 4 million degrees. As the flare energy is dissipated, the coronal condensation spreads out, the material cools, and drains down to the surface in the form of bright coronal loops radiating in Hα.

We, therefore, are beginning to have a fairly consistent picture of what goes on above solar flares. We now have at our disposal so many ways of measuring the emissions at different wavelengths – in the X-ray and extreme ultraviolet regions, in the radio spectrum, and in the visual region – that we are now to have an opportunity to put together a good picture of the energy distributions in a flare. Probably the next solar cycle will see us making great strides forward in this field.

References

BIERMANN, L.: 1947, *Naturwiss.* **34**, 87.
CARMICHAEL, H., STELJES, J. F., ROSE, D. C., and WILSON, B. G.: 1961, *Phys. Rev. Letters* **6**, 51.
COVINGTON, A. E. and HARVEY, G. A.: 1961, *Phys. Rev. Letters* **6**, 51.
ELWERT, G.: 1952, *Z. Naturforsch.* **74**, 432.
HANSEN, R. T.: 1961, *Phys. Rev. Letters* **6**, 260.
ZIRIN, H.: 1959, *Astrophys. J.* **129**, 414.

Discussion and Questions

L. Biermann: In this new picture of what a flare is like, is the expenditure of energy larger than previously thought?

H. Zirin: I am not sure of the answer to that because much of the previous evaluation of energy expenditure has tried to estimate what energy is radiated in other regions of the spectrum. I think that Dr. Warwick can answer this better than I. The particle emission, if it goes in all 4π solid angles and if you extrapolate it down to lower energies, might be the biggest thing anyway. I am not sure. The Hα radiation is about the same as the X-ray and particle emissions.

OBSERVATIONAL STUDY OF THE DYNAMICS OF THE SOLAR ATMOSPHERE

JOHN W. EVANS

Sacramento Peak Observatory, Air Force Cambridge Research Laboratories, Sunspot, New Mexico

It has always been evident that the classical picture of the solar atmosphere as a smooth spherically symmetrical layer was only an approximation, since horizontal inhomogeneities, like the granular appearance of the photosphere, are plainly visible. Theoretical requirements of mechanisms for the mechanical transfer of energy from the granulation layer into the high chromosphere and corona have spurred the study of these inhomogeneities, a most important aspect of which is the motion of material in the solar atmosphere.

Since the pioneer work of Richardson and Schwarzschild in 1950, the study of atmospheric motions in the sun, as shown by local Doppler shifts in spectral lines, has been very rewarding. The directly observed characteristics are as follows:

(a) At any one time the vertical motions in the solar atmosphere are approximately randomly distributed, with an rms velocity which increases with height, from about 0.22 km/sec at the lowest levels to 2 km/sec in the middle chromosphere 3000 to 5000 km higher.

(b) At any one point on the sun, the vertical motions in the photospheric layer are quasi-oscillatory. Oscillations at a point characteristically endure for two or three cycles and die out, only to begin again with a phase shift. The strongest oscillations have a preferred period of 240 sec, but the power spectrum, which includes strong and weak motions, has a peak at 300 sec.

(c) The motions at different heights are very nearly coherent and simultaneous, although the characteristic periods decrease slightly with height (from 300 sec in the lower photosphere to 280 sec in the lower chromosphere, perhaps 500 km higher). Thus an oscillation begins with an upward velocity of propagation considerably in excess of the velocity of sound, and quickly degenerates into a standing wave, with the same phase at all heights, but amplitudes which increase with height.

(d) The oscillating elements vary in size from about 2000 km up to more than 10000 km in horizontal extent.

(e) The observed motions are only slightly correlated with the brightness of the underlying granulation layer. However, there is a very strong correlation between the brightness of a line and its Doppler shift. The Sacramento Peak observations show maximum brightness in a line occurring at the time of greatest compression, in contrast to the Mt. Wilson finding that maximum brightness comes at the time of maximum velocity (or neutral compression). Since this is a matter of considerable theoretical importance, the discrepancy must be removed by more critical observations.

Chang & Huang (eds.), Proc. Plasma Space Sci. Symp. All rights reserved.

(f) As we go up through the chromosphere, the vertical motions change in character, with a broader power spectrum which shows a high frequency tail, peaked at a period of about 160 sec, and very poorly correlated with the motions at lower levels. Although not nearly as clear-cut as the photospheric oscillations, some short lived periodic motions are definitely present at the highest observable levels.

(g) The horizontal motions are quite different, showing no periodicities of less than 100 minutes. They are non random, however, in that much of the motion occurs in large cells of "super granulation" of the order of 15000 km in radius. These are areas within which material moves outward from a center with velocities of about 0.5 km/sec. They appear to cover some 30% of the solar surface uniformly in all latitudes.

(h) Because of its monotonic increase with height in the solar atmosphere (except at the lowest levels), the instantaneous rms velocity measured in any given line is a useful height parameter, which has the convenient property of independence from atomic characteristics or theoretical assumptions about the structure of the solar atmosphere and the process of line absorption. A study at the Sacramento Peak Observatory shows conclusively, for instance, that the level of origin of lines in the λ 3850 region is very much higher than that for lines of equal strength in the $\lambda > 4200$ region, presumably because the overlapping wings of the closely packed ultra violet lines prevent our seeing down to the level of true continuous radiation.

The foregoing picture may appear to be a formidable problem for the theoretical hydrodynamicist. It turns out, however, that the calculated properties of the solar atmosphere agree with the observations surprisingly well, and differ mostly in details (some of which are very significant). Perhaps the most troublesome outstanding problem is the almost instantaneous establishment of standing waves by mechanical excitation from below. The observations suggest instead some form of excitation which acts nearly simultaneously at all heights, similar in nature to a blast of absorbable radiation.

Discussion and Questions

Unidentified: Is there any relation between period of oscillation and lifetime of solar granules?

J. Evans: Well, we don't really know. When we correlate the motion with the intensities of the granules, the relation seems to be non-existent. There is practically no correlation. There is a strong suggestion in our visual examination of these films that when we see the sudden appearance of a granule on the surface of the sun, it is followed about 40 seconds later by an oscillation at that point. And beginning the oscillation with an upward movement. But this is a very difficult observation and I for one am not really convinced it is right. That is the only relation between granule brightness and the velocity that we observe.

ROUND TABLE DISCUSSION

Participants. J. Dungey (chairman), H. W. Babcock, J. Evans, C. Y. Fan, H. Laster, G. Moreton, J. W. Warwick, J. Wilcox, and H. Zirin.

J. Dungey: Since much of this discussion will be related to solar flares, I propose that we begin with other topics. I hope that the speakers will inform us of their future plans, possible new techniques, etc. We have already heard about one from Dr. Zirin regarding how to approach the magnetic intensity. I mentioned to Dr. Babcock that the Russians are beginning to measure the component of the magnetic field perpendicular to the line of sight. Perhaps Dr. Babcock would like to comment on this.

H. Babcock: I am sure that, technically, it is much more difficult to measure the transverse component of the field using the Zeeman effect. However, it is possible. Dr. Howard is now handling most of our work with the magnetograms at Mt. Wilson and I think he has plans along these lines. However, I do not think that they are imminent and it may be some time before we are operating along that particular line.

He is also working toward the improvement of the speed and resolution of recording magnetic patterns for using the line of sight component. This is a matter in which the technique can be improved a great deal. Dr. Howard is particularly interested in achieving sequences of scans with good resolutions on active regions in order that the history of magnetic field patterns, for example, during various stages of development can be followed in detail. I think that the results, to date, on this are rather crude and there is much room for improvement.

J. Dungey: Are there any other speakers who would like to discuss any topics other than flares?

H. Zirin: I believe that the material that Dr. Evans presented this morning is of great importance. The work which has been completed by his group and Dr. Leighton on motion of the chromosphere and the photosphere is very important. One of the most important facets of this work has recently been completed by several of Dr. Leighton's students on the chromospheric network. They have found, as you saw in the excellent film that Dr. Evans showed, that the chromosphere breaks up into cells of a linear dimension of the order of 30000 km where the material is flowing upward toward the edges of each cell. Presumably it completes the circulation by going in a downward direction. The edges of these cells appear as bright patches which are faculae. Therefore, you find that the chromosphere shows structures, the edges of which are emission patterns and which, as Dr. Howard has shown, are the loci of magnetic fields.

Some of the best pictures made at the center of Hα at Mt. Wilson show what appears to be solar spicules. This, as well as clusters, jets, and material coming up only from these regions, is something which we have not discussed much. Some of the pictures which were recently taken by Dr. Howard and me in the helium line reveal that the helium absorption is limited to the edges of the 30000 km cells. A great deal of attention is now being paid to the cellular structure which you clearly saw in Dr. Leighton's photographs.

J. Dungey: There is one point about which I am concerned. People do not seem to observe calcium K lines very much. I wonder if you would expect to see any difference in the motion of an ion than that of a neutral hydrogen.

H. Zirin: Actually, Dr. Leighton's group spends most of its time with the K line. It works with Hα also, but there is a great deal of work going on in the K line. Virtually, you see the same structure in the K_2 line. Dr. Howard and I took a K_2 plate and an Hα plate and looked at them under a blink microscope, which is used for examining variable star plates, and tried to match elements. We saw the same elements. The contrast is different, but the pattern of K_2 and that of Hα are practically identical.

J. Dungey: Are there any other questions or comments?

R. Hales (Brigham Young University): In the latent picture, the cells were noticeably larger farther from the center. The cells toward the center were approximately of the same order of magnitude as were the granulation cells of the photosphere. There have been no comments regarding this. Since Dr. Evans is not present, would Dr. Zirin care to comment on this?

H. Zirin: I am able to do this because Dr. Simon just lectured to us at Boulder about that picture. Apparently, the horizontal velocities of the hydrogen atoms at this high level in the chromosphere are much greater than the vertical velocities. In the center of the picture, there is not much shown in the way of velocities. Thus, they are left with a very fine structure. That was a cancelled picture done by cancelling positive and negative sides of the line. Therefore, I think that what is seen is the fact that there is a very small vertical velocity. I asked Dr. Evans whether he could observe any velocities there. His answer was to the effect that it is not clear.

R. Hales: I think that the question is, is there any attempt to tie in to the fact that the apparent size of the cells of the vertical center is approximately equal to the size of the granulation in the photosphere?

H. Zirin: I do not know.

J. Warwick: Dr. Babcock has mentioned some of Dr. Howard's plans for the future. One might briefly mention another subject which falls into this category. In connection with deep space probes such as Mariner and the future ones to Mars, one would like to compare the measurements of a magnetometer and a plasma detector with various geomagnetic indices. One would also like to compare them with the solar magnetic field as much as possible. To this end, there are plans to make special solar magnetograms that would be suitable for this type of comparison. The standard solar magnetogram has been done for many years by techniques worked out by Dr. Babcock; one examines the entire disc of the sun. It takes about an hour to complete this. Normally, one is done per day.

In connection with the Mariner-II flight, we made special observations in which we wanted to get a faster repetition rate than one hour, e.g., 5 minutes. We limited the disc to the equatorial region, near the western part of the sun. We brought the data out on the chart recorder and on tape for machine reduction, simply taking one scan after another as long as the seeing permitted. This gave fairly detailed information about the time variation of the magnetic field on certain sections of the sun. We hoped to see the correlation between plasma and magnetic field conditions in the interplanetary space.

J. Dungey: Are there any other questions besides any regarding flares?

Unidentified (directed to Dr. Zirin): In your discussion this morning, you showed in some of your spectrograms, lines due to helium I and II, hydrogen lines, and also some lines of calcium XV, etc. You had low temperature or low energy lines together with extremely high energy lines (approximately 600 electron volts for the calcium lines). Would you explain how these low-temperature lines are mixing with the high-temperature lines?

H. Zirin: That is a question which has been the concern of many for a long time. I think that we should leave the problem of setting up conditions where the conductivity across the lines of force is low enough to enable these temperature differences to occur to the magnetohydrodynamicists. In reality you see much of the material cooling. In the various stages of the cooling process, you get a time cross-section of the process, emission lines corresponding to the most popular temperatures you might say. Of course, this does not explain how to get the high and low temperature lines side by side. It must be due to the fact that we have strong magnetic fields which inhibit the mixing of each individual tube of material with the neighboring one which may be cooler or hotter, as the case may be. To the best of our resolution, we can only state that these things are normally occupying very closely in the same region of space.

E. Hones (IDA): I have a question for Dr. Babcock. In the model which you discussed this morning, you indicated that plasma is flowing along the magnetic ropes you showed. I imagine that sunspots are the exit of these ropes into the region outside the surface of the sun. Therefore, in a sense, you are looking down into the sun along the rope into a sunspot. I should like to know what is the reason for sunspots being dark compared to the surface. I always thought that the region of the spot was colder, but on the other hand, looking down into the sun along the magnetic rope, the region might appear to be hotter.

H. Babcock: You should ask Professor Biermann. I do not know if he is present at the moment. However, I believe that the fact that the sunspot is colder is true and the reason for this, according to Dr. Biermann's theory, is: the strength of the magnetic field is sufficient inside the spots to inhibit the convection. Therefore, the heat that would normally be transported outward by convection does not come out as fast. There is still radiation. Since convection is inhibited, there is less outward transfer of heat and it appears cold.

Falls (General Electric Company): I should like to direct this question to Dr. Moreton. We have seen many pictures of the flares, but we are not satisfied with what is actually occurring there. I believe that Dr. Warwick suggested that there was a pouring in of particles previously injected into the corona. I was wondering if the theory was confirmed by the Doppler measurements of Dr. Moreton, or if this was an interpretation that had not been agreed upon.

J. Dungey: Are you asking the definition of a flare?

Falls: My question is, is this "pouring in" interpretation of Dr. Warwick the accepted view of what is occurring there?

J. Warwick: I am not certain of your question. I did not present a model of a flare at all. What I suggested was that an active center is a continuous source of fast particles which we observe in the form of radio bursts at large distances from the surface of the sun. I suggested that particles from the active center are emitted and flow away from the sun rather than toward it. I am not sure that they would be visible in any event in optical measurements.

G. Moreton: I would also like to bring up a point regarding that. You specifically mentioned the explosive phase of the flare. I am wondering what bearing the results which I showed you this morning may have on that. I also pointed out this morning, the manner in which the various phenomena are very closely associated with the explosive phase including X-ray bursts.

J. Warwick: The point is well taken. The work to which Mr. Moreton is referring is the association of the rapid brightening and expansion phase, the flash phase of the flare and Type-3 bursts. They do associate with this phase. Therefore, one might ask if the active centers are continuous sources of particles. As I see it, these active centers are abundant sources of Type-3 bursts, also at times of flares. The same active center will be a source of bursts before and after the flare. In fact, before the work of Mr. Moreton and his associates was carried out, the very existence of an association between Type-3 bursts and flares was in doubt. I think some people, even today, would question this, although the particular result was convincing to me.

G. Moreton: I think a very important point to be mentioned is that there are considerably more Type-3 bursts than visible flares. On the other hand, I do not think that we can definitely say that we see all the flares which occur. There is, perhaps, a strong possibility of something which might be called an invisible flare, particularly in terms of the bomb phenomena which we are just beginning to survey. The Ellerman bomb is a microflare that is, perhaps, one, two, or three seconds apart and has flare-like characteristics although it does differ from normal flares considerably.

H. Zirin: I disagree with the other speakers. It seems to me that although there is a considerable degree of association between Type-3 bursts and flares, there are many more Type-3 bursts, as you pointed out. The one phenomenon that seems to be more closely tied to large flares than anything is the decimeter bursts. If there is a large decimeter burst, 2800 mc/s, to my knowledge, there is always a large flare. The longer wavelength emission is closely related to the complexities of the outer solar atmosphere. In the short wavelength region, you are looking straight down into the region on the surface of the sun. We see in Hα, the region of the sunspots, which I would like to identify with the flare. In the long wavelength regions, you are looking at a very remarkable phenomenon occurring in the upper atmosphere, which depends upon magnetic fields, the ability to store electrons, the synchrotron resonance in one case, and the ability to excite plasma oscillations in the case of the fast bursts. Therefore, I think that one has to separate the radiations coming from the center of the flare and the secondary radiations that are tied up with the fast particles coming out to a certain degree.

J. Dungey: I think you said this morning that we tend to define a flare in terms of Hα. When you mention the kernel of the flare, do you mean the Hα?

H. Zirin: The Hα and coronal region are limited to a very small region, e.g., 4000 or 5000 km in diameter, rather than a solar radius as, for example, the burst shown by Dr. Warwick.

J. Dungey: I understand. You meant the smallness of the region. I was not sure if you meant that this was primary.

J. Warwick: The point which I wanted to emphasize is that a solar active region, in fact, the seat of the flare within this 4000 or 5000 km limit on the surface of the sun, is often a continuous source of relativistic or near relativistic electrons and protons. These may be emitted for days from the evidence of the metric and decimetric bursts. We observe the bursts as Dr. Zirin points out, at very high altitudes in the solar atmosphere, but the source of them is the very same active center that occasionally also produces, but rarer, an optical flare. I only ask that we take them into account when we discuss the phenomenon of an active center. We certainly must take them into account when we speak of the energetics of that region. One cannot simply pick out a flare and state that we are dealing with an active region, when that same region may have been the source of one thousand Type-3 bursts. I think that these bursts are continuously generated in the very regions of the sun that occasionally produce flares.

G. Moreton: I have a question. With the sunspot minimum approaching, there was a lack of active centers on a few days. Do you find a correlation between the sporadic Type-3 bursts which you record in your spectrum and the lack of active centers?

J. Warwick: If there were no active centers, we would get almost no Type-3 bursts at all. Even in May, we have had Type-2 bursts and Type-4 bursts.

G. Moreton: When did we have Type-2 burst?

J. Warwick: The date was May 23. There are good days at sunspot minimum. It is just that they occur less frequently. I would be interested to know how close an association could be brought between, for example, this solar activity and fast particles measured in outer space. I failed to mention it, but on September 6, which was the date of the last Type-4 burst I showed, was associated with a weak proton event observed outside the magnetosphere by Explorer XII. There was no flare at the time of this event. Therefore, we have here a remarkable negative incident which would suggest that the fast particles were associated with an optical flare.

G. Moreton: Is this the September 1961 event?

J. Warwick: Yes, September 6, 1961. There was no optical flare on that day. The same region produces flares and protons events again on September 10, and again, one rotation later, on September 26, on the eastern hemisphere. So it looks to me like a bonafide association.

Unidentified: I would like to take advantage of the contribution to scientific knowledge here, and ask if there is anyone either on the panel or in the audience who can provide me with more information on the possible photospheric disturbances during flares. The only reference I have been able to obtain on this is a French publication, and I have not been able to follow it to its source. Can you give me any information on photospheric phenomena associated with flares?

G. Moreton: Are you speaking of white flares in particular?

Unidentified: Yes.

G. Moreton: I can make a comment here. We do have a person in our group who did manage to observe the flare which, incidentally, I showed as a negative X-ray event. There have been observed a few of these. The reason why I want to say a few words is that there is a controversy. I recall that the first time I looked at the sun, the director of the observatory informed me that every time there was a flare, one of his observers was able to see a white light flare. Later, in another observatory, I heard that not many were seen. I also recall reading that every cosmic ray flare was a white flare. I think, therefore, that this entire matter of white light flares is very interesting. I myself am interested in it, and we hope to have a white light patrol running within a month or so for the purpose of looking for them. I should call attention to the fact that Becker, in 1961, ran a white flare patrol for about four months with negative results.

H. Zirin: Actually what I wanted to mention was something different. Since you brought this up, I should like to direct a question to you. It seems to me that you can see white light faculae around the limb, the plages around the sunspot. It seems to be very strange to me that you do not see the flares because they are presumably at the faculae. Do you want to answer that before I mention what I originally wanted to say?

G. Moreton: Yes. There was one point which I did not understand in your paper. Can you account for a white light flare going over the limb and yet not being visible against the background? This, granted, has not been 100 percent verified photographically, but Miss Engle saw it on September 3; i.e., she saw that it was moving out over the limb, and, as it went up or over, it just disappeared. I was curious because of your observations of a continuum and your explanation of this in terms of Raman scattering.

H. Zirin: The continuum that we see in our pictures is extremely faint. One could never detect it. It is 10^{-4} of that of the solar disc. When one talks about a white light flare, he is not talking of that continuum alone; he is talking about many other things. I do not know how to explain that observation except in contrast.

What I originally intended to say is that there is a white light phenomenon which was reported by one of Leighton's students. Perhaps Dr. Babcock might have some comment on this. They made pictures of sunspots before and after flares and they claimed that, invariably, there were displacements of the sunspots relative to one another before and after the flare. They believed that the sunspots ended up closer to one another, and in fact, went further and tried to evaluate the changes in magnetic energy of the total system before and after the flare. This observation is capable of being tested. It would be interesting to look further into it.

H. Babcock: I do not have any specific comments to add, but I believe that the fact that flares do tend to occur in spot groups of mixed polarity, and where the spots were changing most radically was established during Hale's work in 1930. I do not know about measureable differences in sunspots just before and after flares.

J. Dungey: Are there any comments on Hale's work, particularly where he has found these large velocities going in opposite directions, the rapid stretching of material?

H. Zirin: I am not very familiar with the stretching of the material. I do know the changes in field configuration that were observed before and after flares. Dr. Babcock and Dr. Howard looked into and made measurements of themselves and found somewhat different results. Actually, when I was there, I did not go into this in as much detail as Dr. Howard did, when he went there and found some convincing material. It is really very difficult to make these observations. In particular, the maps that they made were not made by an automatic scan. For that reason, it was really a little difficult to comment on whether the fields had changed. Again, if the flares occur in the most rapidly changing part of the sunspot region, it would be surprising if there were not any change. Perhaps we are letting our intuition push us. Dr. Howard seems to have been satisfied with the material he saw there.

J. Dungey: What I was referring to was the interpretation of the tremendous line broadening, in which he said that one wing came from higher up than the other one. That the upward going wing was above the downward going wing.

H. Zirin: Dr. Babcock and Dr. Howard have made many observations of the broad lines that are found in flares. I am not familiar with the differences in height between the blue and red wings. They have also made some interesting comparisons with spectra from a large linear pinch that they have been operating at the observatory and can produce the same type of shift in hydrogen lines. I do not remember very much about the up and down motion of the different wings. Therefore, I cannot comment on that.

F. Sylvestri (USAF): I have a general question which involves many of conditions. It is about the dispersion and propagation of various frequencies through the plasma. In any great event of the sun, which line will get to the earth first, considering all the dispersive media on the way?

J. Dungey: As far as actual propagation goes, many of these things go at the speed of light. The radio frequencies go at the speed of light once they are out of the plasma region. The condition is that they are emitted from a height where the refractive index is important for them. They have to go a good distance before they reach the speed of light. The only other things are, of course, the particles.

F. Sylvestri: Of the various frequencies in the radio range, which will get here first?

J. Warwick: I think that Dr. Dungey's answer was a good one in that respect.

To mention one point in detail which he mentioned momentarily, there is a delay of frequencies from a narrow band burst resulting from the fact that the burst is generated near the plasma level. This leads to relative delays of frequencies within the burst over a range of 10 megacycles at 100 Mc/sec. Frequencies within that range may be delayed several seconds relative to one another. In terms of the overall development, the Type-3 burst appears completely in its proper time sequence to the optical flare, in that the delays are only a few seconds. The optical radiation is observed faithfully as it was generated at the sun. The particles have a lower velocity. The times of propagation may be 20 or 30 minutes depending upon the convolutions of the trajectory in interplanetary space and near the sun.

J. Harris (Boulder Laboratory, National Bureau of Standards): Up until now, we have been discussing the high part of the solar cycle, i.e., flares of large magnitude and high speed particles. In a period during May and right now, we are experiencing low solar activity. We still have magnetic storms. I was wondering what the arguments are against discussing low velocity particles instead of high velocity particles.

J. Dungey: For a magnetic storm which is initiated by a flare, the velocity of the flare particles is about 1000 km/second. During the solar minimum, we get magnetic disturbances which are not associated with flares, but they may be associated with M-regions. Perhaps Dr. Babcock would like to discuss unipolar regions.

H. Babcock: I think that I revealed, in my talk this morning, all that I know about polar regions. Near the minimum of the solar cycle, there is still usually some magnetic activity observed on the solar disc with a magnetograph even when there are no sunspots observable. One can find weak magnetic regions, remnants of dipole regions, which at one time were, perhaps, stronger. During the last minimum, there were only a very few days when the magnetograms did not show some weak local activity generally of a dipolar nature in the lower latitudes. Yet, there were, perhaps, a half-dozen days when the disc was free from any detectable fields except weak fields at the poles. Therefore, on almost all the occasions, there might be some reason to expect the ejection of particles.

J. Dungey: On that point, we may look to Dr. Snyder's paper tomorrow morning.

J. Warwick: One has the impression that a unipolar region has a very great influence on interplanetary space conditions. Could you comment on that?

H. Babcock: Our evidence was for the best unipolar region in 1953. Although it only persisted for about six months, it definitely did seem to affect the terrestrial field. It did affect, likewise, the space between the sun and the earth. We had altogether about three well-identified unipolar regions, at that time. We would hope to keep track of conditions during the approaching solar activity minimum, as it is just before minimum that one has the best chance to identify these unipolar regions.

J. Warwick: There is, perhaps, some evidence from the Mariner's magnetometer. Will Dr. Snyder mention that tomorrow?

C. Snyder: I am not going to comment much about the magnetometer.

J. Warwick: Are you going to mention the relation of this unipolar region and the 2-week cycle as they seem to have?

C. Snyder: I do not think I know enough about that unipolar region to comment much about it.

J. Warwick: Let me ask a question about unipolar regions. I was a little puzzled by Dr. Babcock's comment this morning about the magnetic effect of the unipolar region in 1953. It seemed to extend along out into space, because it produced effect in the geomagnetic field. It seems to me that possibly a plasma stream was coming out from that unipolar region and it was the interaction of the plasma with the earth magnetic field which produced the disturbance. It was not necessarily the magnetic field itself. Did I misunderstand you this morning?

H. Babcock: Well, you may be quite right, although I did, at one time, have the picture in mind that a long bundle of lines of force might actually extend out from this unipolar region. Perhaps one can make an equally good case for some sort of ejection process.

C. Snyder: I think that when the magnetic field data and the plasma data on Mariner II are improved in order that we can correlate one with the other, we can make some remarks on this. At the moment, the magnetic field data are not quite acceptable.

D. Matthews (Defense Research Corps, Canada): The subject of M-regions was passed over and this is one important topic on which nothing has been said about during the course of the day. Are there any comments from any members of the panel on whether there has been any progress on observing M-regions? May I add a short comment? The auroral radio wave absorption at Ft. Churchill has been more frequent during 1962 than in 1961. This involves several events and the statistics are significant. As far as that is concerned, we hardly seem to be approaching solar minimum. In a recent case, it had a 26 or 27 day recurrence extending over 11 cycles which has just ended.

J. Harris: A moment ago, I was speaking of particles of lower velocity. Dr. Warwick has indicated that these active regions may be continuously emitting particles even though there is no indication of flare activity. I was wondering if there are any arguments against assuming these as lower velocity particles?

J. Warwick: Do I infer that you suggest that there is a continuous source of plasma from an active center? Yes, I think so. This conclusion was drawn from the study of the association of the radio noise sunspot regions and geomagnetic storms. These radio noise regions were in his data noise storm centers. Now, these noise storm centers are also much more abundant sources of Type-3 bursts than activity centers of other types on the sun. Therefore, I would say, that there is a positive association between the plasma and the effects on the geomagnetic field. Whether this has anything to do with the physics of the ejection of fast particles or the generation of plasma streams, I do not know. But I think a positive correlation is indicated.

Capt. Stevens (USAF): Dr. Moreton, I have a question concerning the moving disturbances which we saw. It appeared to me that all these disturbances moved off to the east and, I believe, more specifically toward the northeast. I know there were effects in the other directions, as occasionally I saw filaments off to the west which were near equal distances from some effects to the east. Was this just coincidental or a valid observation?

G. Moreton: No, it was coincidental. We have checked this out and they go all ways. The important point is, as you pointed out, that there is a highly directional type of disturbance. It goes in one direction and then gets a large filament immediately sitting behind a large flare.

C. Fan: It is true that the filaments you observed are emission from neutral hydrogen atoms? What is the reason that their motion is affected by the solar magnetic field?

G. Moreton: The point was that the effect itself might well be propagated along magnetic fields which might be assumed to interconnect active regions.

H. Zirin: I might add one remark to that. Although you observe the filament in the light of neutral hydrogen, the filament is actually highly ionized and you are only seeing a smaller percentage of the atoms that are neutral. The filaments wouldn't be there if there were no magnetic fields holding them up. A change in the magnetic field could produce great changes in the filament. In this connection, I wanted to get back to the subject that Dr. Moreton discussed earlier, the line broadening. I have always thought that the line was being shifted because the filament appears in the center of the line and disappears off the center of the line. I do not understand why you do not like that way of thinking.

G. Moreton: Do you mean a gross shift of the filament? We have tried to detect with negative results any displacement of the filament. The reason that the broadening mechanism was invoked was that we have nothing else.

H. Zirin: For the displacement, one would need a radial motion toward or away from us, and I do not see how broadening explains the disappearance.

G. Moreton: Do you mean the disappearance of the center line?

H. Zirin: Yes.

G. Moreton: Well, this is due to the charge exchange process which depopulates the ionized hydrogens atom, and makes it transparent in the center of the line. The actual broadening is a secondary effect, due to the change in the kinetic condition in the filament. For a shift of half an Ångström, it is very sensitive. I see, like most others, you are very skeptical. I don't know. We are still trying to get more data on this.

F. Sylvestri: Does anyone have a feeling for the plasma frequency on the surface of the sun?

J. Warwick: Yes, at three thousand megacycles you are observing the chromosphere. If you want to observe down into the photosphere you have to go down to millimeter wave lengths.

PART II

INTERPLANETARY PLASMA AND COSMIC RAYS

INTERPLANETARY SOLAR-WIND MEASUREMENTS
BY MARINER II*

CONWAY W. SNYDER and MARCIA NEUGEBAUER

Jet Propulsion Laboratory, California Institute of Technology, Pasadena, Calif.

Abstract. The interplanetary spacecraft Mariner II was launched on August 27.29, 1962, and it transmitted scientific and engineering data back to Earth almost continuously until January 3.29, 1963. In addition to making the first close-up measurements of electromagnetic radiation from the planet Venus, it collected data on the magnetic fields and the charged atomic particles in space for the equivalent of 104 days, extending over more than 4.5 solar rotations.

The solar plasma probe, consisting of a cylindrical electrostatic analyzer and a sensitive stabilized electrometer, measured the flux and energy distribution of positively charged particles moving radially outward from the Sun over a range from 230 to 8200 eV/unit charge. Approximately 40 000 such energy spectra were obtained. The plasma detected appears to consist of two components, assumed to be nuclei of hydrogen and helium.

The results show that expanding solar plasma was present in space during the entire mission, with average flux in agreement with the measurements of Luniks II and III and Explorer X. The velocity is seldom constant for more than a few hours, varying between approximately 3.5 and 8.0×10^2 km/sec and showing a rather striking correlation with the daily K_p index of geomagnetic activity. This fact, together with the clear tendency for periods of high velocity to recur after 27 days, indicates that most of the plasma is emitted by localized regions on the Sun.

The experimental data will be summarized, and their correlation with solar events, geomagnetic events, and other interplanetary data will be discussed.

1. Prior Solar Plasma Research

Until recently, the existence and the properties of the interplanetary plasma were inferred from observations made on the solar corona, the tails of comets, the scattering of sunlight, the variations in cosmic-ray intensity, and the fluctuations of the geomagnetic field (ROSSI, 1963). Three years ago, the first direct measurements of plasma fluxes in space were made by Soviet lunar and interplanetary space stations (MUSTEL', 1963), but these observations were not very prolonged and apparently gave little detail as to the nature of the plasma.

Somewhat later, in March 1961, the American space probe, *Explorer X*, made more detailed measurements of the plasma just outside of the magnetosphere for about one day (ROSSI, 1963). This experiment had severe limitations. Its directional resolution was poor, and it required 20 min to obtain one energy spectrum, and a very rough spectrum, at that. Despite these limitations, and the fact that one could not be sure that the properties of the interplanetary plasma were not seriously distorted by the proximity of the magnetosphere, it was an extremely significant experiment. It determined the approximate magnitude of the principal parameters of the plasma, which was all that it had been intended to do. In particular, it measured fluxes of

* This paper presents the results of one phase of research carried out at the Jet Propulsion Laboratory, California Institute of Technology under Contract No. NAS 7-100, sponsored by the National Aeronautics and Space Administration.

the order of 10^8 ion/cm^2/sec (confirming the earlier Russian values), it bracketed the velocity between approximately 200 and 400 km/sec, and it suggested (but by no means proved) that the direction of flow is radially outward from the Sun. The rather striking success of the plasma experiment on Mariner II is due in part to its having been designed with the benefit of the Explorer X results.

2. Mariner II

The general features of the *Mariner* spacecraft and mission are undoubtedly familiar (NEUGEBAUER and SNYDER, 1962). From the standpoint of the plasma measurements, the important facts are these. The orientation of the spacecraft was controlled by an attitude-stabilization system, which kept the entrance aperture of the plasma probe pointed to within 0.1 deg of the center of the Sun at all times except during the orbit-correction maneuver (when no scientific instrument was operating). Data transmission was continuous from liftoff to the final loss of signal 129.0 days later. Transmission of data from the scientific instruments was continuous from August 29.6 (when Mariner was 721 000 km from Earth and 1.01 AU from the Sun) until January 3.3 (at 86 650 000 km from Earth and 0.708 AU from the Sun), except between October 30.9 and November 8.9. Reception of data was also continuous at one or more of the three tracking stations from liftoff to December 17, with the exception of a few gaps ranging in length from a few minutes to as much as 6 hour, and was 10 hour per day thereafter. The total scientific data received amount to 104 days' worth, extending over more than 4.5 solar rotations.

Mariner carried seven scientific experiments, only two of which will be mentioned here (NEUGEBAUER and SNYDER, 1962). The magnetic field experiment was conducted by P. J. Coleman, L. Davis, E. J. Smith, and C. P. Sonett. We are grateful for permission to quote certain of their results. Three components of the magnetic field were measured every 37 sec by a triaxial fluxgate magnetometer which was capable of detecting changes of 0.7 gamma on its most sensitive scale.

3. The Plasma Spectrometer

The instrument chosen to make the solar plasma measurements was a positive-ion spectrometer. As illustrated in Figure 1 and shown schematically in Figure 2, its principal components were (1) a cylindrical electrostatic analyzer, which separated positively charged ions according to their energy per unit charge (E/Q); (2) a programmer and a high-voltage sweep amplifier, which applied the proper balanced potentials to the plates of the analyzer; and (3) an electrometer, which measured the current from the charge collector at the output of the analyzer. It was designed, built, and tested by C. Josias, J. L. Lawrence, Jr., and H. R. Mertz of the Jet Propulsion Laboratory (JOSIAS and LAWRENCE, 1963). We have found no reason to doubt that the entire instrument operated faultlessly throughout the four months of the mission.

The electrostatic analyzer was constructed of gold-plated magnesium, with its interior coated with gold black to minimize the amount of ultraviolet light which could get into the charge-collector region. The deflecting plates were 120 deg in length and separated by 1.3 cm. The entrance aperture was 5.0 cm² in area and rectangular

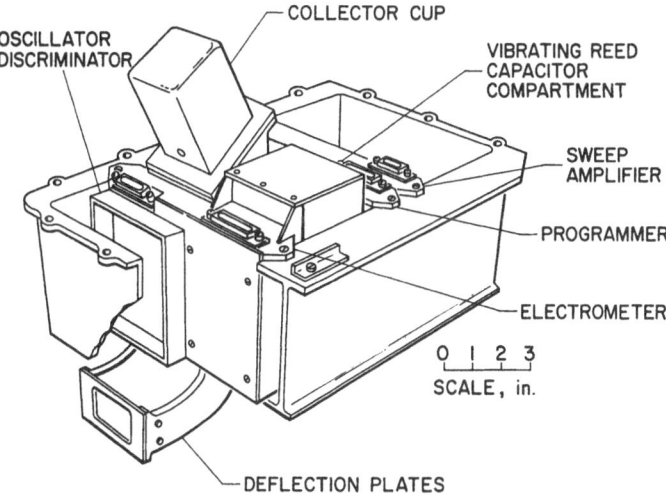

Fig. 1.   Plasma spectrometer, mounted in spacecraft instrument case (shown partially cut away).

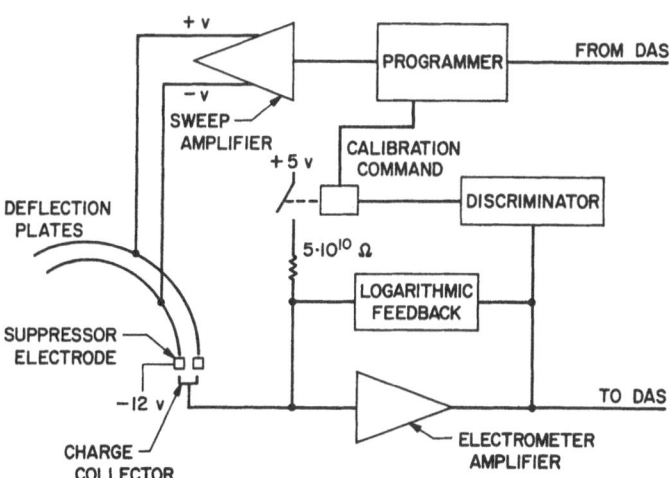

Fig. 2.   Block diagram of plasma spectrometer. (Data automation system (DAS) supplied stepping and reset pulses to control measurement sequence and accepted analog output signal, which it encoded for telemetry.)

in shape, such that the angular acceptance for charged particles in two perpendicular planes was approximately equal. Its theoretical angular and energy response functions in the plane of analysis are shown in Figure 3.

CONWAY W. SNYDER AND MARCIA NEUGEBAUER

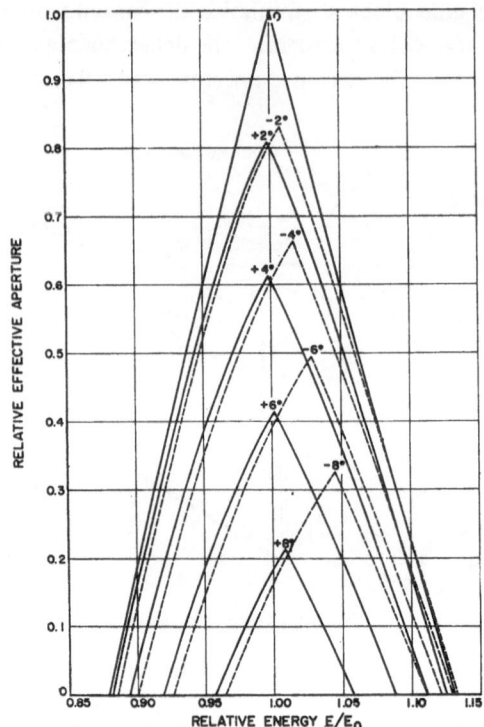

Fig. 3. Response functions of spectrometer in angle and in energy. (Curves give relative effective aperture for perfectly collimated beams of monoergic charged particles incident in various directions in plane of analysis; E_0 is energy of a particle which, entering along axis of input aperture, would traverse analyzer at a constant radius.)

TABLE I

ENERGY AND VELOCITY OF PROTONS ACCEPTED BY THE SPECTROMETER
IN ITS 10 MEASUREMENT STEPS*

Step No.	Proton energy (eV)	Proton velocity (km/sec)
1	231	210
2	346	257
3	516	314
4	751	379
5	1124	464
6	1664	565
7	2476	689
8	3688	841
9	5408	1018
10	8224	1255

* Numbers correspond to the peak of the response function for protons at normal incidence.

In order to obtain an energy spectrum of the solar plasma ions, the deflecting voltage on the analyzer plates was changed by the programmer at intervals of about 18 sec in an ascending sequence of ten values, as shown in Table I. Then, with the plates connected together, a zero reading and a calibration reading (using a standard current of 10^{-10} A) were taken. The calibration permitted us to correct the data for long-term drifts resulting from changes in temperature or in the analog-to-digital converter of the data-automation system, which encoded our output for the telemetry system.

The electrometer, which utilized a dynamic-capacitor modulator to assure long-term stability, was capable of measuring currents from 10^{-6} to 10^{-13} A, and its output was an accurately linear function of the logarithm of the input current over all but the lowest $\frac{1}{2}$ decade of this 6-decade range. The output was a $d-c$ voltage, which was converted by the spacecraft data-automation system into a digital reading with a quantization interval of 0.1 decade.

One feature of the operation of the electrometer is important to the data analysis. The required large dynamic range for current measurement was obtained by the use of a thermionic diode (Raytheon CK 5886) in the feedback loop of the electrometer amplifier. The action of this circuit element provided the electrometer with two distinct modes of operation: a "static" or current-measuring mode and an "integrating" or charge-measuring mode. In the static mode, when the diode was conducting, the potential drop across it was proportional to the logarithm of the input current, and the output remained constant for a constant input current.

In the integrating mode, when the diode was cut off, changes in the output voltage reflected the charging or discharging of the 5-pf capacitance of the diode, so that the mean input current could be determined from the change in output voltage between readings. A net negative input current would push the electrometer even further "off scale", i.e., away from the static mode. To prevent this occurrence, a voltage discriminator operated to inject a short pulse of positive current, thus bringing the electrometer back into its static range if it drifted off too far.

A net positive current of 5×10^{-13} A or more was sufficient to bring the electrometer from the discriminator threshold back to the proper point on its static characteristic in less time than the 18 sec between readings. Thus, currents larger than this are always read correctly. Smaller currents may be read correctly, depending upon the previous reading. Currents an order of magnitude smaller, and of either polarity, can be inferred by analysis of the two successive readings. To date, we have confined our attention mainly to the larger currents.

The existence of these two electrometer modes had both desirable and undesirable consequences. It enabled us to measure the electron current in energy steps which contained no plasma current, thus assuring ourselves that the effect of photoelectrons upon our results was indeed very small. The electron currents measured were always below 10^{-13} A, and frequently five to ten times smaller. They varied with time, but the significance of the variations has not been investigated as yet.

At the beginning of each measurement cycle, the electrometer was in the integrating mode because of a switch transient at the end of the preceding cycle. This fact introduces an ambiguity into some of our spectra, as it may not be clear whether the electrometer returned to the static mode in a particular step because a large positive current was detected or because the discriminator had injected the current pulse. Such uncertain cases are relatively rare, however.

4. Limitations and Assumptions

Before proceeding to discuss the data, we should like to call attention to certain important limitations of the experiment which permit the experimental data to be interpreted in more than one way. In each case, we shall indicate what assumption we have used in our analysis; thereafter, in order to simplify the discussion, we shall ignore other possible interpretations, realizing, however, that they might indeed be the correct ones.

First, our measurements were taken serially, so that we never obtained an energy spectrum in a time known to be short relative to the characteristic fluctuation period of the plasma. We believe that this is not a serious limitation. Undoubtedly, there are times when distorted spectra were obtained because the plasma was changing rapidly. Usually, however, it is possible to find several consecutive readings that are essentially (or even exactly) in agreement, which tends to confirm their validity.

A second limitation is our relatively small solid angle for accepting particles and our invariable orientation. We have chosen to assume that the bulk velocity of the plasma is always radially outward from the Sun, since this seems to be the most reasonable and the only tractable assumption. Nevertheless, it must be understood that changes in collected current, which we interpret as flux changes in the plasma, may instead reflect changes in the direction of the flow. Another facet of the same problem is the aberration effect, which can distort our energy spectra. We attempt to correct for this effect in our analysis, but there are limitations to this correction, as can be seen from the following consideration. Assume that the azimuthal velocity of the spacecraft around the Sun is 40 km/sec (which was its highest value during the mission). Then, particles travelling radially outward at 210 km/sec (which should be detected in our first energy step) will encounter the spacecraft at an aberration angle of arctan 40/210 = 11 deg. But this angle puts the particles outside the acceptance cone of the analyzer, and hence, they will not be detected at all. We do not believe that such low velocities ever occurred, and therefore do not consider the aberration effect to have been serious.

Third, having no way to measure the absolute electric potential of the spacecraft, we have assumed it to be neutral – whatever neutrality may mean in such a context. We observe that the peak current reading occurs anywhere from step 3 to step 8 in the sequence at various times. We interpret this observation in terms of a change in the bulk velocity of the plasma. The same readings could be produced by changes in the spacecraft potential, which accelerates the ions. We do not consider this interpretation to be reasonable, as negative potentials of nearly 4 kV would be required; but further study of this rather difficult problem in plasma physics is desirable.

The fourth, and probably most important, limitation was the fact that our energy windows were narrow, too widely spaced, and too few in number to give an ideal characterization of the plasma energy spectrum. This limitation, when compounded with our narrow acceptance cone and with certain fundamental uncertainties about the nature of the plasma (such as its isotropy, its velocity distribution function, and the direction of the magnetic field in it) makes it difficult, if not impossible, to calculate values of the plasma composition, density, and temperature with an accuracy comparable to that with which we measure the currents and the bulk velocity. Better than order-of-magnitude estimates can be made, however, and no better accuracy was hoped for in a pioneering experiment of this type.

5. General Results of the Plasma Experiment

We have available approximately 40000 current-vs.-energy spectra of the interplanetary plasma. At present, we have almost (but not quite) completed the task of getting all these data listed on a single master magnetic tape which is free from noise, transmission errors, and misplaced data, so that we can commence more detailed analysis of this mass of data by electronic computing techniques. All the conclusions that we have been able to obtain up to now have come either from inspection of the raw data or from relatively simple calculations on small samples of the data.

A preliminary report of the experimental results was published before *Mariner* had reached Venus (NEUGEBAUER and SNYDER, 1962). The general conclusions about the plasma which were listed there have been confirmed by the later data; namely:

(1) A measurable flow of plasma from the direction of the Sun existed at all times during the *Mariner* mission.

(2) The plasma appears to contain two components which differ by a factor of approximately 2 in energy per unit charge; presumably, these are protons and alpha-particles having the same mean velocity.

(3) The flux is in agreement with the values found by *Explorer X* and by the ion traps on the *Lunik* satellites.

(4) The plasma bulk velocity is high, in rough agreement with the observations of *Explorer X* and the predictions of the "solar-wind" theory of Eugene PARKER (1958).

(5) The energy density of the plasma is dominant over that of the magnetic field, indicating that the interplanetary field is trapped in and carried along with the plasma. We believe that this fact strongly supports our assumption that direction of flow is nearly radial at all times.

(6) The plasma velocity is greater than the Alfvén velocity, so that magneto-hydrodynamic shock phenomena may be expected.

A few samples of our 40000 spectra are shown in Figure 4. The vertical scale is the logarithm of the collected current, and since the interval between energy steps is a constant ratio, a log-log representation of the spectrum is given. The spectrum

CONWAY W. SNYDER AND MARCIA NEUGEBAUER

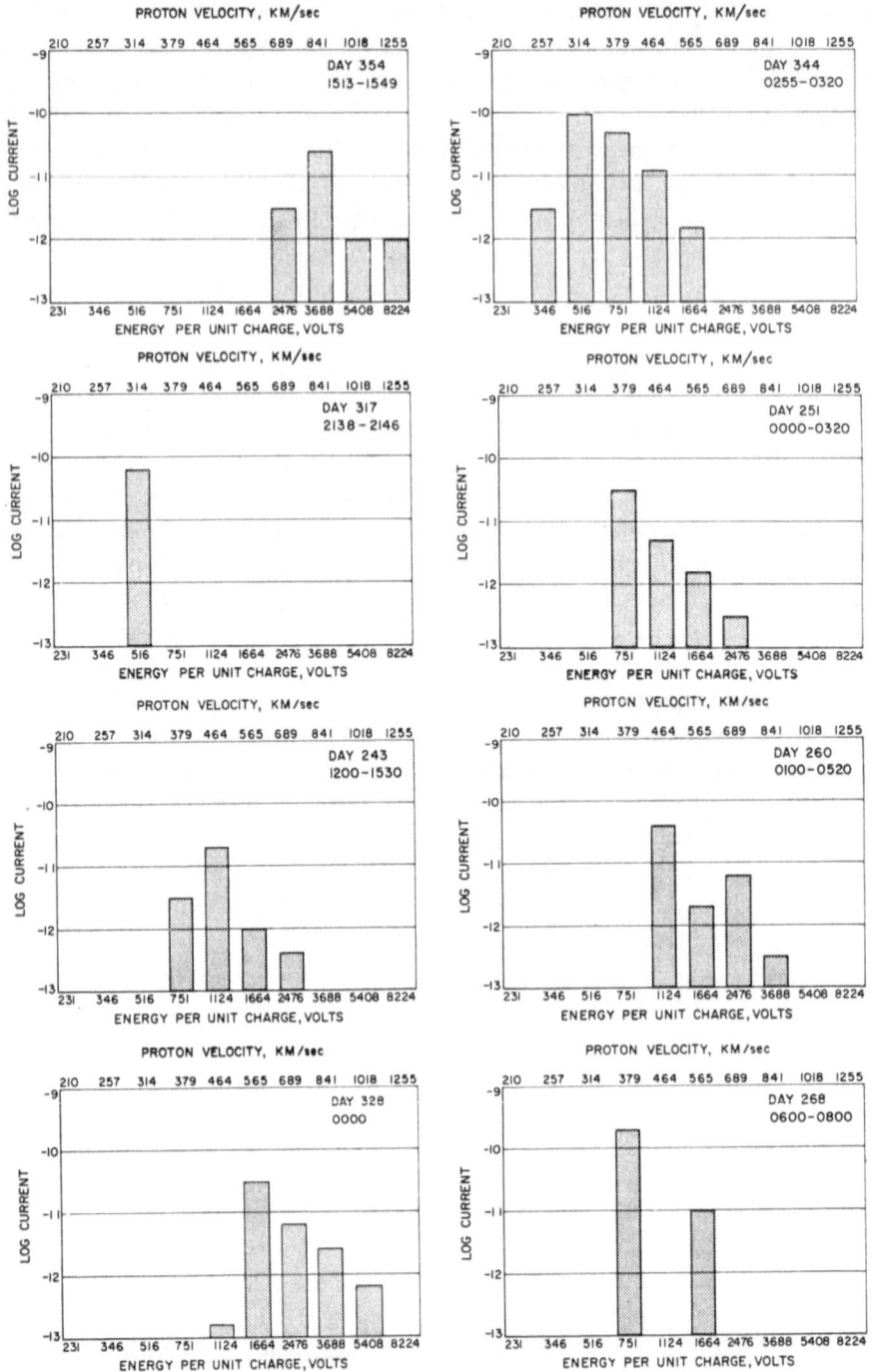

Fig. 4. Sample current-vs-energy spectra. Times in hours and minutes, Universal Time. Day 1 was January 1, 1962.

of day 354 is an average over 36 min in which the readings were steady; it represents the highest sustained solar-wind velocity that we observed – approximately 830 km/sec. Note that no measurable current was observed in the first six steps. In contrast to this spectrum is that of day 344, a 25-min average; it represents one of the very few times when a large current appeared in step 2. The solar-wind velocity at this time must have been near 330 or 340 km/sec. This is not the lowest velocity observed, however. This distinction is claimed by the spectrum of day 317, observed in three consecutive cycles; the velocity at this time was 315 km/sec.

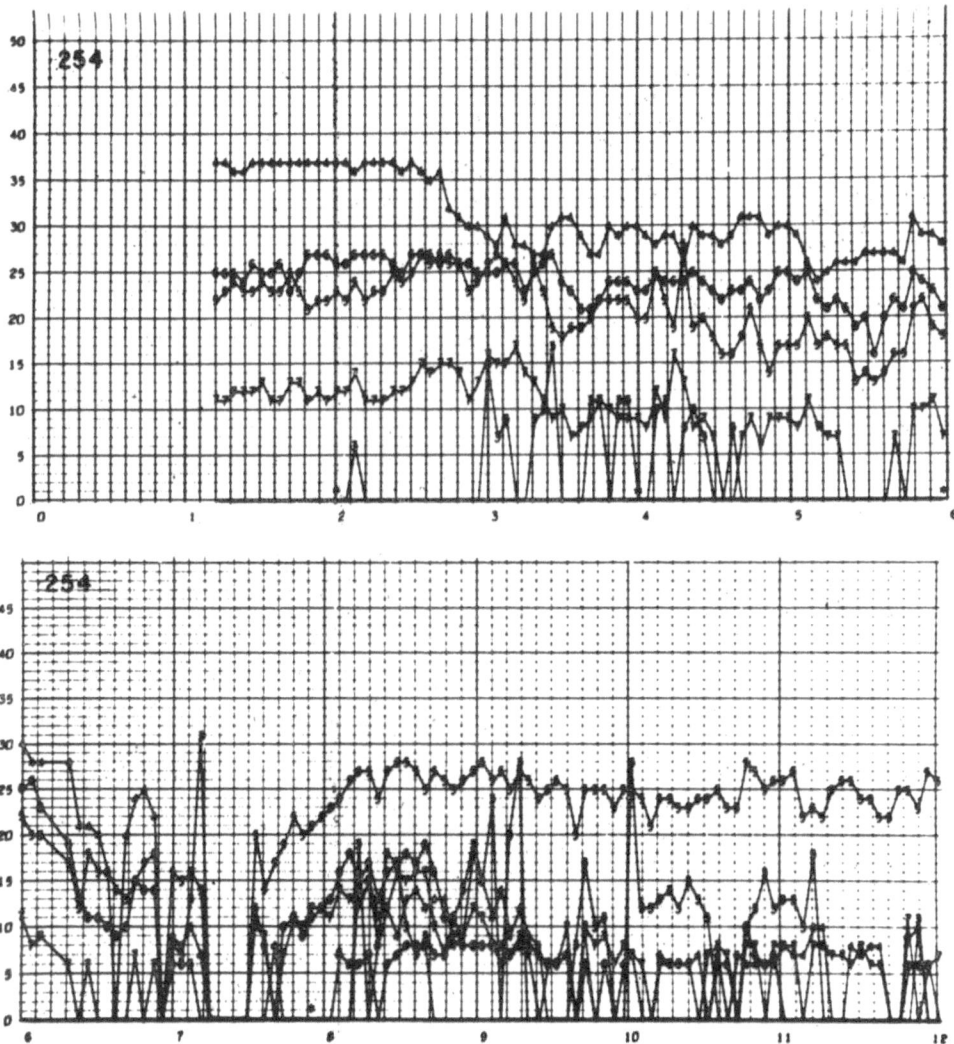

Fig. 5. Plasma record for first half of day 254 (Sept. 11), showing period of smallest plasma flux during mission (numbers on graph represent energy steps as listed in Table I).

Approximately half the spectra are generally similar either to that of day 251 or of day 243, where the peak may be anywhere from step 3 to step 8. Somewhat more than one third of them are similar to the spectra of day 260, or a more extreme variant, that of day 268. The high frequency of these double-humped spectra is taken to be proof of the presence of an appreciable fraction of alpha-particles in the plasma, but we cannot yet make a quantitative estimate of their abundance. The difficulty in making such an estimate comes from the fact that protons and alpha-particles of the same velocity differ by a factor 2.0 in E/Q, whereas two of our energy-step intervals amount to a factor of 2.2. Thus, if we happen to sit exactly on the center of the proton peak, we catch the alpha-particle peak somewhere down from the top, and vice versa.

Several examples of the data are shown in Figures 5 to 10. These are graphs generated by an electronic computer from the telemetry records, showing the logarithm of the measured currents for all energy steps in which the apparent current was greater than about 3×10^{-14} A. The correction for the nonlinearity of the electrometer characteristic at very low currents has not been applied, so that currents in the bottom decade of the graph are not accurately represented. The units on the ordinate, when divided by ten, represent 14 plus the logarithm of the current, so that 20 represents 10^{-12} and 50 represents 10^{-9} A. Each graph contains all the data obtained in 6 hour.

It will be noted that the maximum current usually was in the range 10^{-11} to 10^{-10} A, with occasional times when it was somewhat outside this range. A notable exception occurred between 0700 and 0730 UT on September 11 (day 254), as shown in Figure 5. At this time, the plasma seemed to disappear; what actually happened was that the currents were so small that the electrometer remained in the integrating mode throughout five consecutive measurement sequences. Analysis shows that the current in step 5 was close to 3×10^{-13} A during this period, while currents in nearby steps were at least five times less. This was the only period of such low measured current encountered during the mission. It is interesting to note that, prior to this time, the peak current was in step 4, whereas subsequently it was in step 5. Probably, therefore, during this half hour, the peak fell between our measurement levels.

Note the prevalence of flux variations of approximately a factor of 2 with time constants of about a quarter of an hour to an hour. Such variations are a rather common feature of the data.

Figure 6 shows an exceptionally turbulent period, with the current peak shifting 24 times in 6 hour between step 5 and step 6. We interpret this as velocity fluctuations over the range from about 470 to 560 km/sec. Note that dips in the current in step 7, when they occur, are in phase with dips in step 5. This very common phenomenon we attribute to the simultaneous velocity changes for the protons and the alpha-particles.

A contrasting record is that of Figure 7, showing one of the quietest times during the mission. Note, however, a sudden increase in velocity at 4 o'clock, with step 4 disappearing and step 5 increasing suddenly by a factor of 5.

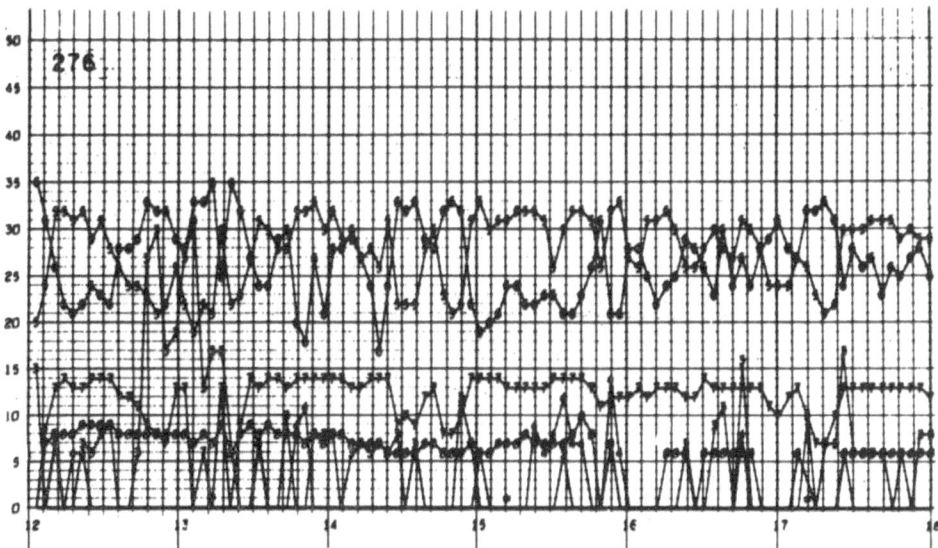

Fig. 6.   Plasma record for third quarter of day 276 (Oct. 3), showing frequent fluctuations in velocity (numbers on graph represent energy steps as listed in Table I).

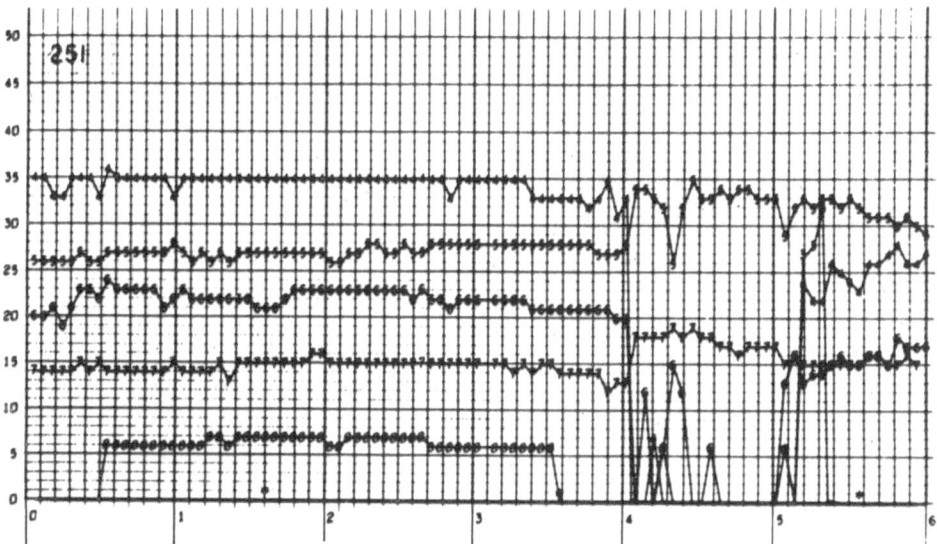

Fig. 7.   Plasma record for first quarter of day 251 (Sept. 8), showing period of very calm conditions and sudden increase in plasma velocity (numbers on graph represent energy steps as listed in Table I).

Figure 8 shows a period of 12 hour when the plasma was fairly quiet, but its velocity was gradually dropping. This is indicated by the current in step 4, which starts in the cellar, rises gradually, and at the end of the record takes over from step 5, which had previously contained most of the current.

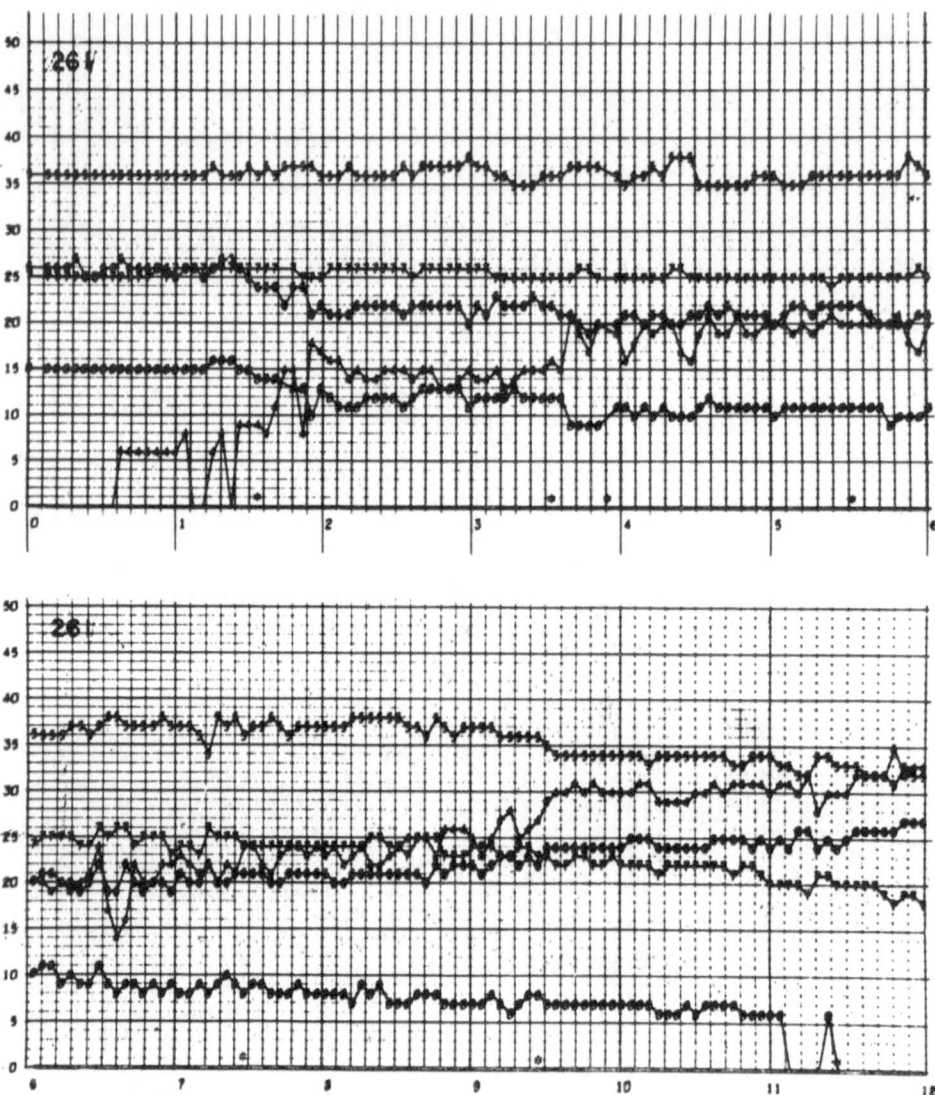

Fig. 8.   Plasma record for first half of day 261 (Sept. 18), showing moderately calm conditions, with gradual decrease in plasma velocity (numbers on graph represent energy steps as listed in Table I).

A very low-temperature plasma is shown in Figure 9. All the proton current is in step 4, all the alpha-particle current is in step 6, and that is all until step 3 begins to become appreciable near the end.

Figure 10 shows the prelude to and the beginning of one of the most interesting events of the mission, which occurred on October 7. Note that the proton peak is centered fairly accurately on step 4 for the first 10 hour, giving a bulk velocity near 380 km/sec. After 11 o'clock, the currents in steps 3 and 5 become equal and 40

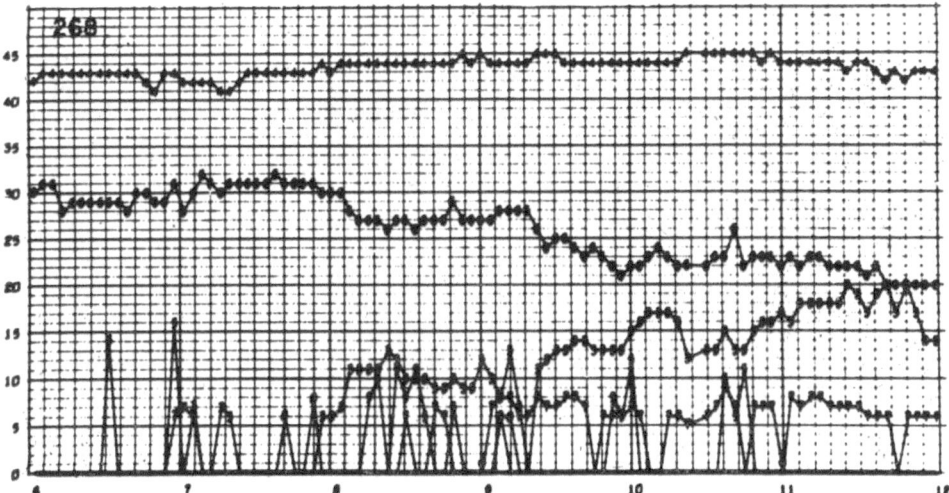

Fig. 9. Plasma record for second quarter of day 268 (Sept. 25), showing fairly calm plasma with low velocity and very low temperature (numbers on graph represent energy steps as listed in Table I).

times less than the current in step 4. This corresponds to a plasma temperature of about 0.8×10^5 °K. The temperature drops slightly (15 or 20%) and then, at about 1440, increases suddenly to approximately 1.2×10^5 °K. Note that, during this temperature jump, the currents in the proton peak (step 4) and the alpha-particle peak (step 6) remain relatively constant. The magnetic field is also constant, as we shall see.

At 1547, the picture changed abruptly. Subsequent events can be followed in Figure 11. It is tempting to call this event a shock, because of its suddenness and its superficial resemblance to the graphical record of a shock-tube experiment. We hope that detailed theoretical analysis may indicate whether indeed it is. It is apparent from Figure 11 that the velocity and the density both increase abruptly and simultaneously, but whereas the density peaks at 1630 and falls below its "preshock" value at 2030, the velocity continues to rise. Its peak value, 840 km/sec, was not reached until nearly 33 hour after the passage of the front. This short-time increase in density, followed by gradual rise in velocity, appears in other events as well.

The correlation between plasma and magnetic field is shown in Figure 12. It is typical of all the data that abrupt changes in plasma correlate in time with abrupt magnetic field changes, but we have not yet been able to discern any consistent pattern in these simultaneous changes.

6. Preliminary Analysis of Plasma Properties

The properties of the interplanetary plasma which we would most like to know are probably its particle density, energy density, and composition. We covet also its distribution function in velocity space, or at least some simple approximation to it, which we might call a "temperature". The investigation of all these properties is

certainly going to require many interplanetary spacecraft missions of greater so-
phistication than Mariner II.

To learn as much as we can about these properties from our data is a task which
we have barely commenced. If we consider only currents above 5×10^{-13} A, then
about 10% of our measured spectra contain only one or two points, 25% contain
three points, 55% contain four, and the fraction of five-point spectra is 10% or less.
Yet, even if we assume isotropy for the plasma, the independent parameters in the

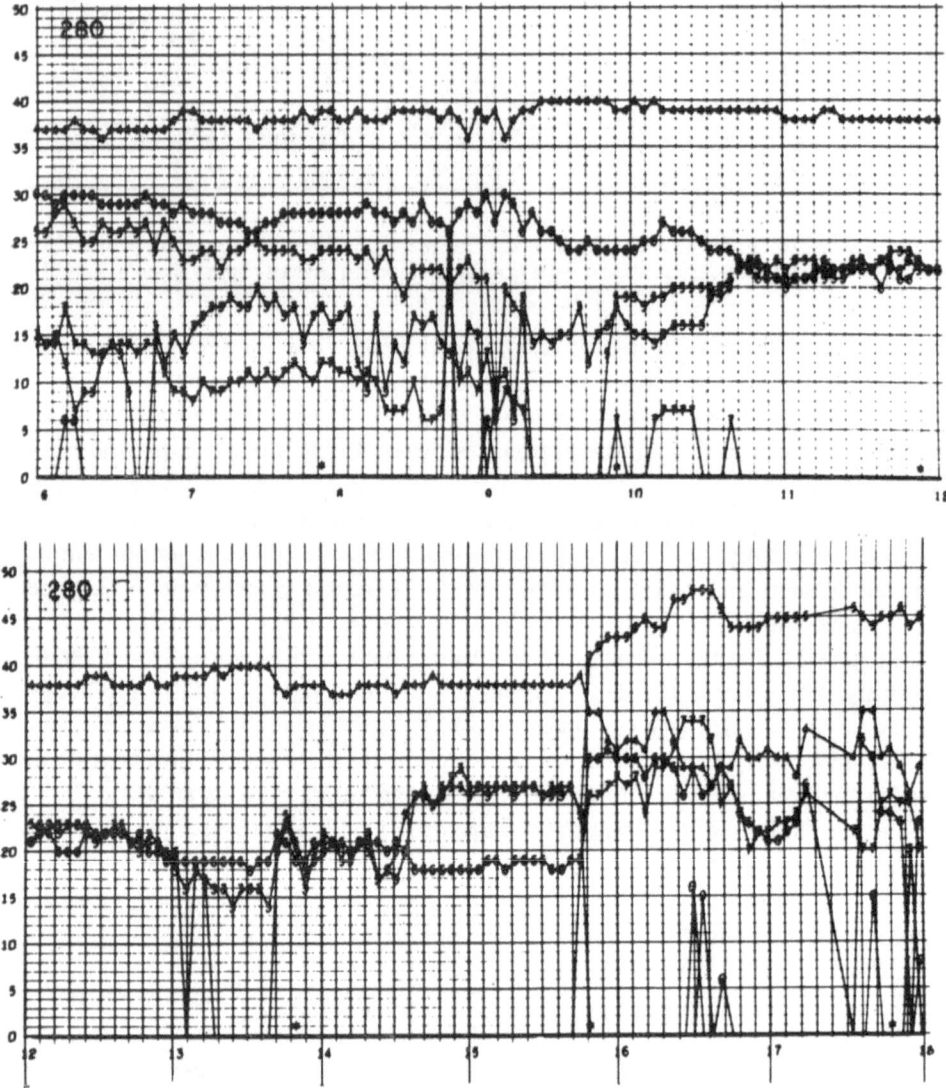

Fig. 10.   Plasma record for second and third quarters of day 280 (Oct. 7), showing prelude to and
early part of "sudden-commencement magnetic storm" (numbers on graph represent energy steps
as listed in Table I).

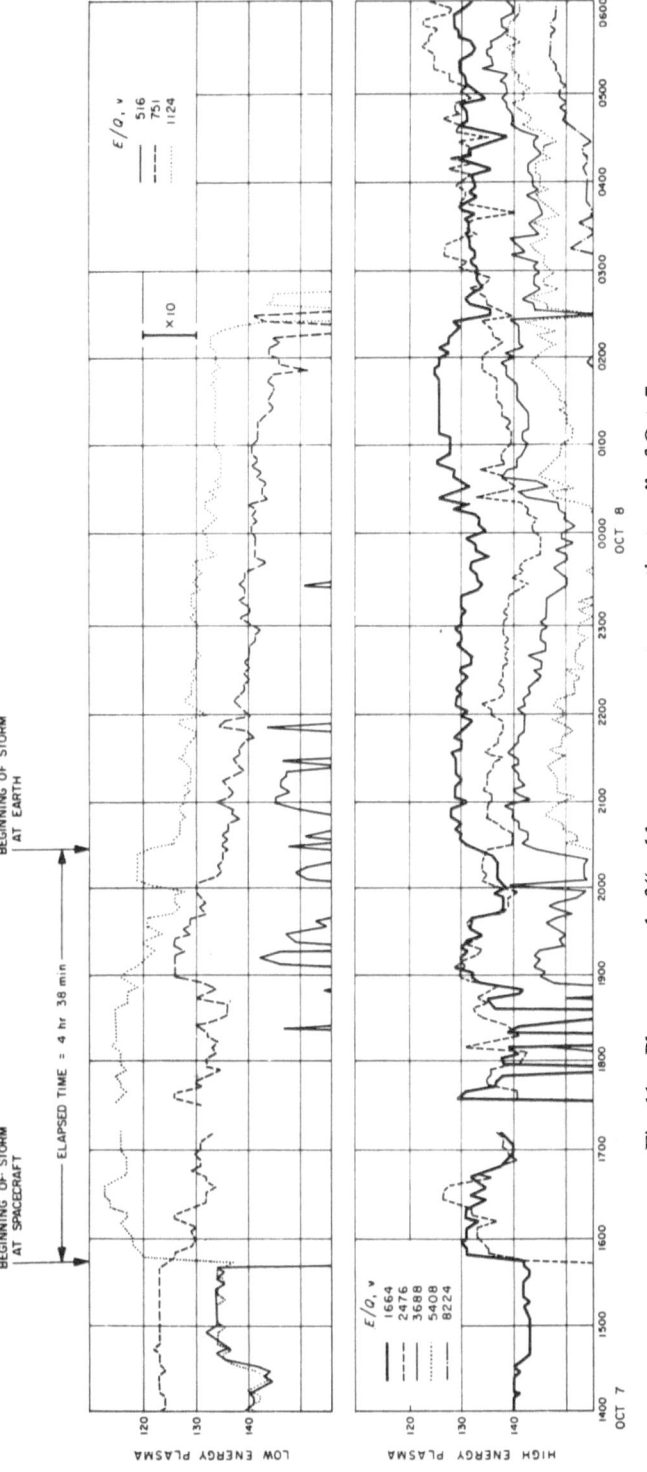

Fig. 11. Plasma record of "sudden-commencement magnetic storm" of Oct. 7.

problem include density, bulk velocity, proton-alpha-particle ratio, proton temperature, and alpha-particle temperature – five parameters at least. Complete characterization of the plasma is apparently unattainable, but the application of sophisticated function-fitting schemes to our large volume of data should yield reasonably
accure estimates of most of the interesting plasma properties, limited, of course, by
our inescapable assumption of strictly radial bulk velocity.

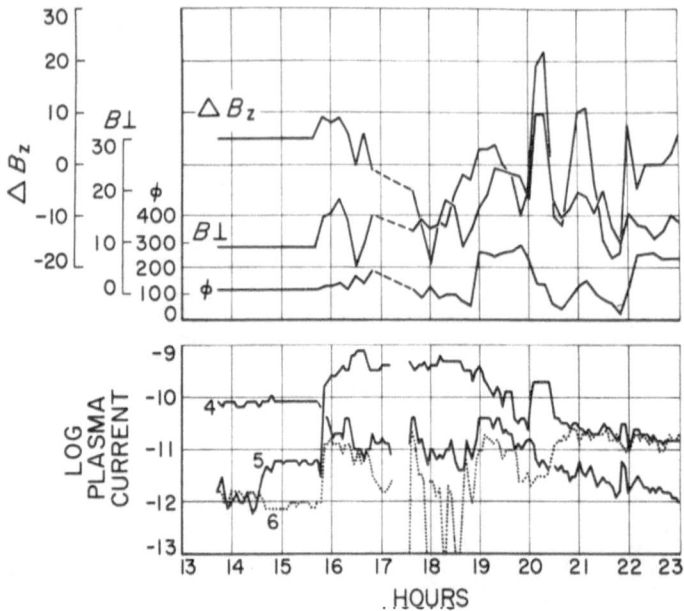

Fig. 12. Simplified record of plasma and magnetic field during early part of "sudden-commencement
magnetic storm" of Oct. 7. (ΔB_z is radial component of interplanetary field, measured from arbitrary
zero level; $B_\perp$ is field component perpendicular to radial direction; and ϕ gives orientation of this
transverse component. $\phi = 0$ represents magnetic vector pointing toward north ecliptic pole and
$\phi = 90$ deg vector in ecliptic plane pointing in direction opposite to motion of the planets.)

Our study of plasma parameters has, up to the present time, been limited mainly
to the analysis of about 100 spectra, one per day, spread throughout the mission.
We choose a spectrum, such as the one for day 328 in Figure 4, in which measurable
currents are observed in the steps on either side of the proton peak. Assuming a one-
dimensional Boltzmann distribution function, we fit to these three points of the
spectrum, thus determining three parameters, which we ordinarily choose to be the
velocity V, the proton density n, and the temperature T. In fact, the highest one of
these steps may contain a contribution from alpha-particles as well as protons, but
as we have no way to determine this, we ignore it. The effect of this assumption is to
make the calculated values of V, n, and T all too high.

The plasma model assumed is a one-dimensional one which is artificial but analytically tractable. Transverse thermal motions are not considered (i.e., the "temperature" perpendicular to the radial direction is assumed zero), and only a "radial

temperature" is calculated, it being a measure of the observed spread in proton velocity. The proton velocity is considered to be the sum of the bulk velocity V and thermal velocity v (both radial), and the currents I_p for the three successive steps are fitted to the expression

$$I_p(n, V, T) = QA \int_{-\infty}^{\infty} (V + v)\varepsilon \, \mathrm{d}n_v$$

where

$$\mathrm{d}n_v = n \sqrt{\frac{m}{2\pi kT}} \exp \frac{-mv^2}{2kT} \, \mathrm{d}v$$

and ε is the resolution function of the spectrometer, a complicated function which takes into account the instrument response as shown in Figure 3 and the aberration effect introduced by the transverse velocity of the spacecraft. Q and A are, respectively, the proton charge and the aperture area. As an example, the spectrum of day 328 in Figure 4 has the following parameters according to the one-dimensional analysis: $V = 600$ km/sec, $n = 2.9$ cm^{-3}, and $T = 1.9 \times 10^5$ °K.

A small number of analyses which have been made using an isotropic model of the plasma seem to indicate that the densities calculated from the one-dimensional model are about a factor of 2 too small. The temperatures we believe to be correct to within a factor of 2. The velocities can be in error by more than about 10% only if the fraction of alpha-particles in the plasma is relatively large.

The results of this analysis, which must be considered preliminary, as it encompasses only about one five-hundredth of our data, are shown as scatter diagrams in Figure 13, 14, and 15. Ninety percent of the velocities lie between 360 and 700 km/sec. Ninety percent of the temperatures lie between 6×10^4 and 5×10^5 °K. Ninety percent

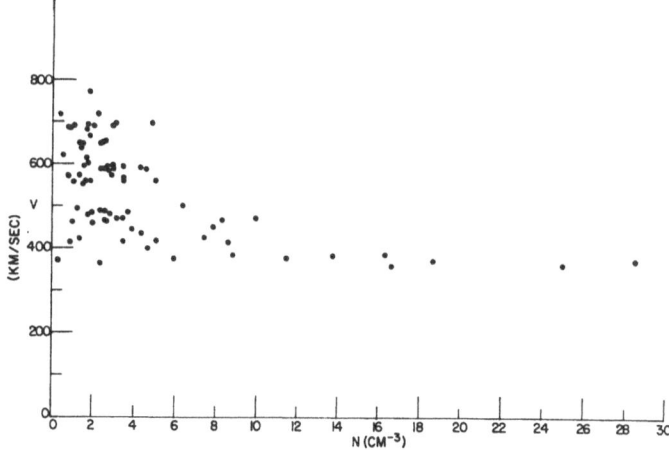

Fig. 13. Scatter diagram of plasma velocity V and proton density n, calculated by one-dimensional model for random sample of spectra.

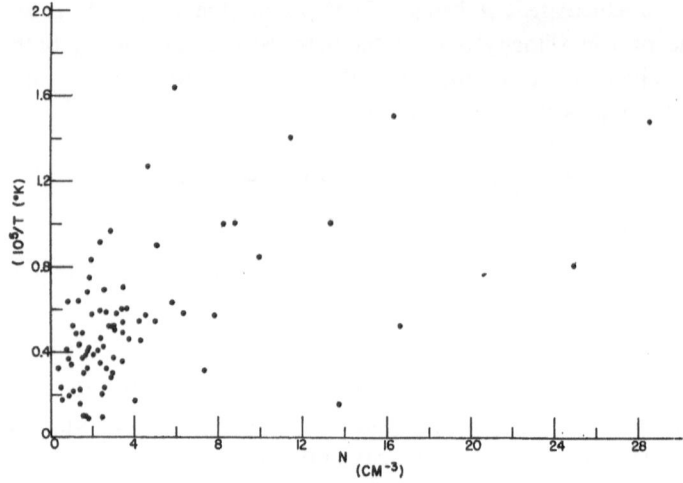

Fig. 14.   Scatter diagram of reciprocal of plasma temperature T and proton density n, calculated by one-dimensional model for random sample of spectra.

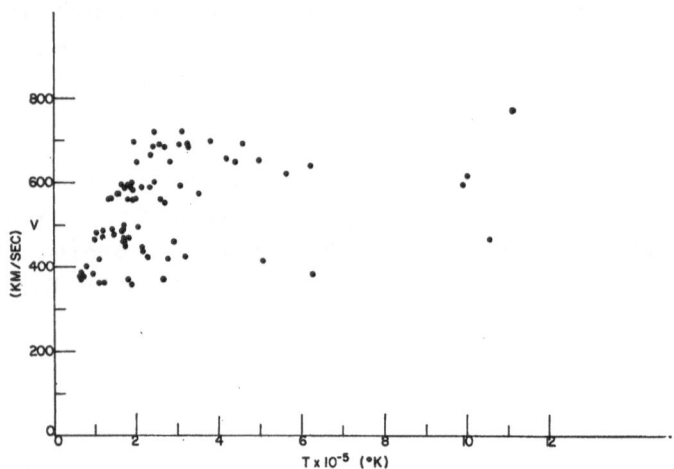

Fig. 15.   Scatter diagram of plasma velocity V and temperature T, calculated by one-dimensional model for random sample of spectra.

of the number densities are between 0.3 and 10 protons/cm^3. Ninety percent of the energy densities are between 2×10^{-7} and 2×10^{-8} erg/cm^3. From inspection of the plots, one gets the impression that V and T are inversely correlated with n and, hence, directly correlated with one another. We hope shortly to be able to test these conclusions with vastly more statistical validity.

7. The Solar-Wind Velocity

The one solar-wind parameter which we measure with greatest precision is the bulk velocity. Since our measurement steps are only 22% apart in velocity, it is almost impossible to make a mistake of more than 10%, even without any analysis. Hence, we have been particularly interested in seeing what can be learned by studying the velocity alone.*

A convenient way to summarize the plasma velocity data for the entire mission is by daily averages. A simple and reasonably accurate way to obtain such an average is to take as the velocity of a particular plasma spectrum the velocity number from Table 1 which corresponds to the measurement step containing the highest current. Daily averages of these velocity numbers are plotted in Figure 16.

It is instantly apparent from this graph that the steady, symmetrical solar wind, which many of us have pictured in our minds and associated with the smooth Archimedean spiral of the garden-hose model of the interplanetary magnetic field, does not exist. The graph shows nineteen peaks in plasma velocity, each of which is preceded or followed by a companion peak approximately 27 days later. In only three cases is the interval between peaks as much as 29 or as little as 24 days. These peaks fall naturally into five series, one with five members, three with four members, and one with two members. Two of these series have been marked with *A*'s and *C*'s, respectively, because they appear to have particular significance. Each peak in the

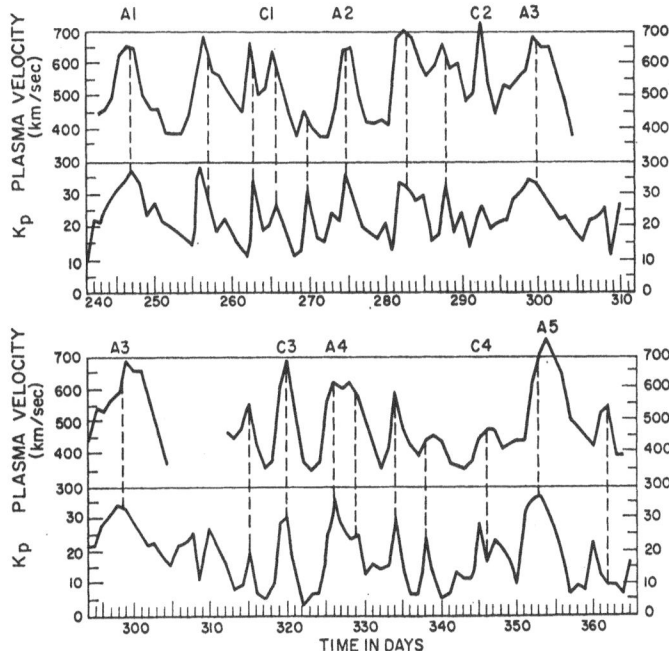

Fig. 16. Daily "average" plasma velocity compared with planetary magnetic activity index K_p. (See text for definition of average velocity.)

* We are indebted to U. R. Rao for valuable suggestions in this regard (SNYDER, NEUGEBAUER and RAO, 1963).

C series follows by 1 or 2 days the central meridian passage of a calcium plage region just north of the solar equator, which was observed in five consecutive solar rotations. Each peak in the A series follows by 1 or 2 days the central meridian passage of a calcium plage region about 10 deg north of the solar equator, which was observed in seven consecutive solar rotations. The interval between peak A2 and peak A3 is abnormally short, and the corresponding interval between plage passages was the shortest in the series.

The conclusion appears inescapable that most of the solar plasma that we observed was not produced by evaporation or by hydrodynamic expansion of a *homogeneous* solar corona but came instead from long-lived local regions in the corona which were abnormal in some respect, perhaps having higher than average temperature.

The correlation of plasma velocity peaks with plage regions is interesting, but we do not know what it means. Attempts to use the measured velocity to extrapolate back to the point on the Sun whence these high-velocity streams originated have not yielded useful results. Some of the extrapolated source points coincide with the plage regions, and others fall far away. This study is being continued.

It has been known for several decades that geomagnetic storms of moderate size tend to recur at 27-day intervals, and the hypothetical sources of plasma streams assumed to produce the storms have been called M-regions (BARTELS, 1934). These M-regions have never been found to correlate with any observable features on the Sun, except perhaps the unipolar magnetic regions observed with the Mount Wilson magnetograph by BABCOCK and BABCOCK (1954, 1955).

Our results appear to confirm the hypothesis that recurrent geomagnetic storms are produced by the impact on the Earth's magnetosphere of plasma streams of particularly high velocity which flow outward from M-regions. Our A series of peaks coincides with a long sequence of recurrent M-region storms. More striking evidence is provided by the comparison of our daily average plasma velocity graph with the daily planetary K index of geomagnetic disturbance, which is also plotted in Figure 16.

These graphs match almost peak for peak and valley for valley. The displacement in time which becomes noticeable toward the end of the record is explained by the fact that the spacecraft was far ahead of the Earth, so that a point on the solar surface passed beneath the spacecraft as much as 2 days later than beneath the Earth. This correlation between plasma velocity and K_p can be extended to shorter time periods than 1 day by making appropriate corrections for the relative positions of Earth and Mariner.

It has long been assumed, as suggested by Bartels (BARTELS and VELDKAMP, 1949) that K_p is a measure of the intensity of the solar plasma flowing past the Earth. Our data would appear to indicate that this is the case, and that K_p has a high correlation with the plasma velocity. This interpretation has recently been disputed by DESSLER and FEJER (1963), who believe that it is correlated instead with turbulence or irregularities in the solar wind. It is, of course, possible that both interpretations could be correct, if a high velocity is always associated with a large amount of turbulence. Our data analysis has not progressed sufficiently so that we can state positively that K_p does not correlate with the flux, the energy density, or the turbulence of the plasma as closely as with its velocity, but the preliminary indications are that it does not.

One pertinent bit of new evidence is available, however, as shown in Figure 17.

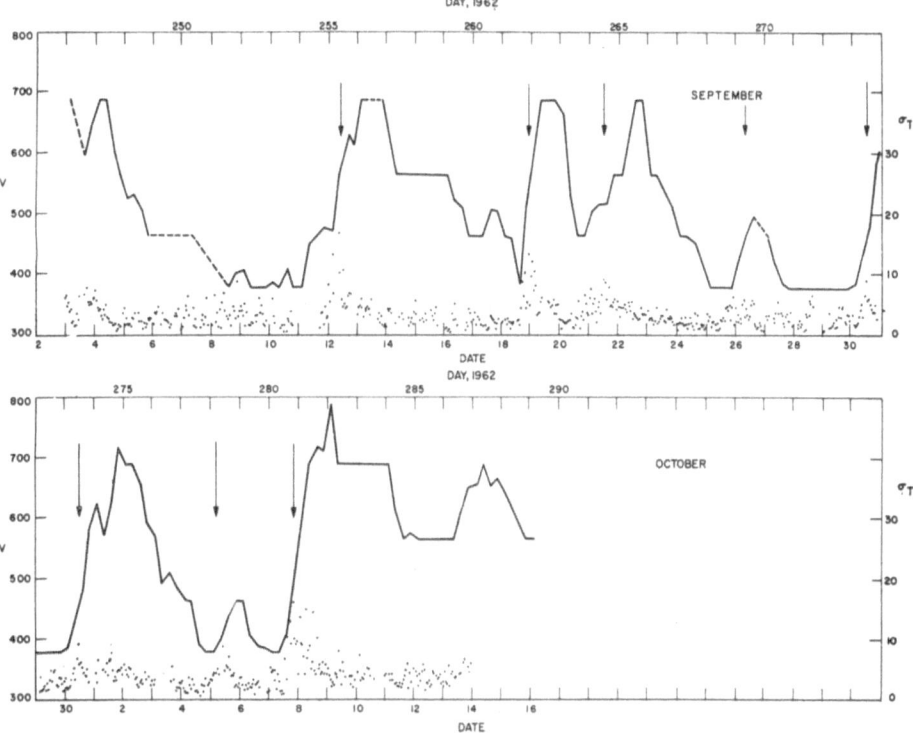

Fig. 17.   Comparison of variance of total magnetic field with 6-hour "average" plasma velocity.
(See text for definition of average velocity.)

Here the hourly variance of the total magnetic field as measured by Mariner II is compared with 6-hour averages of the plasma velocity (calculated as explained above). Clearly, the peaks in magnetic variance do not coincide with the velocity peaks, but rather occur at the times when the plasma velocity is just beginning to rise from a low value. This fact would appear to contradict the theory of Dessler and Fejer, but a more detailed comparison of plasma and magnetometer data should be completed before firm conclusions are drawn.

Acknowledgements

The authors desire to express their profound gratitude to Conrad Josias, James L. Lawrence, Jr., and Henry R. Mertz, whose outstanding competence in electronic design and diligence in assembly and testing produced a plasma spectrometer with such exceptional stability and reliability. We desire also to thank Kurt Heftman who, as data engineer, was so helpful in supplying us with our data quickly and in convenient form. Manifestly, it is impossible to acknowledge the small army of engineers, at the Jet Propulsion Laboratory and elsewhere, whose skill and effort were responsible for the successful operation of all the complex spacecraft systems which had to operate correctly for more than four months in space to enable our experiment to fulfil its mission.

Mariner II was designed and built by the Jet Propulsion Laboratory of the California Institute of Technology under contract with the National Aeronautics and Space Administration.

References

BABCOCK, H. W. and BABCOCK, H. D.: 1955, 'The Sun's Magnetic Field 1952–1955', *Astrophys. J.* **121**, 349–366.

BARTELS, J.: 1934, 'Twenty-seven Day Recurrences in Terrestrial Magnetic and Solar Activity 1923–1933', *Terr. Magn.* **39**, 201–202.

BARTELS, J. and VELDKAMP, J.: 1949, 'Geomagnetic and Solar Data', *J. Geophys. Res.* **54**, 295–299.

DESSLER, A. J. and FEJER, J. A.: 1963, 'Interpretation of K_p Index and M-Region Geomagnetic Storms', *Plan. Space Sci.* to be published.

JOSIAS, C. and LAWRENCE, J. L.: JR.: 1963, 'An Instrument for the Measurement of Interplanetary Solar Plasma', to be published.

'The Mission of Mariner II: Preliminary Observations': 1962, *Science* **138**.

MUSTEL', E. R.: 1963, 'The Sun and the Interplanetary Plasma', paper presented at the Fourth International Space Science Symposium, Warsaw (COSPAR).

NEUGEBAUER, M. and SNYDER, C. W.: 1962, 'The Mission of Mariner II, Preliminary Observations, Solar Plasma Experiment', *Science* **138**, 1095–1100.

PARKER, E. N.: 1958, 'Dynamics of the Interplanetary Gas and Magnetic Fields', *Astrophys. J.* **128**, 667–676.

ROSSI, B.: 1962, 'Interplanetary Plasma', *Space Research* **3**, North-Holland Publishing Company, Amsterdam.

SNYDER, C. W.: NEUGEBAUER, M., and RAO, U. R.: 1963, 'The Solar-Wind Velocity and its Correlation with Cosmic-Ray Variations and with Solar and Geomagnetic Activity', to be published.

Discussion and Questions

L. Tonks: In view of the distance between the sun and the solar probe and the drift time required, one would expect a good deal of evening out of radial velocities. I wonder whether the variations of the radial velocity are consistent with the evening out which must occur due to the fast particles from the sun catching up with the slow ones and tending to equalize the velocity.

C. Snyder: All I can say about that one is that I'm sure nothing has been done about it. I'm sure that this problem is going to be a very difficult one to unravel. We simply have recognized it as a problem, but that is as far as we have gone.

H. Zirin: I am interested in your identification of the plage regions crossing the central meridian at the time of the peak. Was this one of the many regions on the sun that you picked out as always turning up on the meredian at the time of the peak? Or did each plage region appearing on the meridian give you a series of peaks?

C. Snyder: There are of course far more plage regions on the sun than we see peaks. We only see 19 peaks. Now the reason I picked out these two active regions in particular is that these are listed on the McMath-Hulbert observatory record as being long life plage regions on the basis of completely independent evidence. I don't know what their evidence is, but the observers go through all these plages and call out the constant ones. Those do correspond very well to one series of our peaks. This is true of both the A series and the B series. Now there are other ones of our 19 peaks that appear also with about the same delay time after other plage regions that are again near the sub probe point on the sun. But in these cases I think one has to be a little careful because the plages weren't observed for such a long time. So the identification here is not obvious. I would like to say one more thing about it than I have said. All the plage regions that appear to be possible sources of high velocity streams do lie on a region of the sun that is under the spacecraft. In other words ones that are at latitudes of 20 or 30 degrees never appear to be good candidates for sources. But let me reiterate that as far as we have gone up till now is to try to take the velocity that we see and extrapolate back to a point on the sun. About as often as not it turns out to be a completely featureless region on the sun having nothing to do with a plage region. In fact our delay times of one or two days which correspond to 800 and 1600 km/sec don't appear to fit at all with the velocities we see. So I really don't know what the connection is to these plage regions except that I think we can't ignore it because it is so close.

H. Zirin: I think perhaps this closeness is the strongest argument. You have the advantage that you were able to pick out distinct plages.

C. Snyder: I hope that Dr. Biermann will have something to say about this, it is a very interesting point.

N. Ness: The last slide you showed illustrated a very striking correlation between the bulk velocity of the plasma to the variation of the magnetic field. Could you tell us how this variance of the magnetic field is determined?

C. Snyder: The variance of the magnetic field was determined in the following way. First of all, one makes an assumption as to where the zero of the Z axis field is. This then gives you the three magnetic field components. Then the resultant magnetic field is calculated. The average of all the reading taken in one hour (about a hundred of them) is calculated, and the root mean square deviation from the mean is taken.

N. Ness: The mean is then assumed to remain constant around the spacecraft during the entire flight?

C. Snyder: No, just during the hourly interval.

N. Ness: So that is the variance from the mean during the hour.

C. Snyder: Well, the zero point of the Z axis is assumed to be always the same but not the mean field.

S. Bowyer (Catholic University): You indicated that there were three times when you had the same reading in one channel. Could you infer from that since the Mariner was travelling at 40 km/sec each stream would be less than 80 km cross section?

C. Snyder: I'm sorry, I don't think I understand the question.

S. Bowyer: I thought that you said on three occasions there were three readings which were identical.

C. Snyder: Successive repetitions of the same spectrum were very common. Sometimes it was identical for six hours.

THE PLASMA IN INTERPLANETARY SPACE

L. BIERMANN

Max Planck Institut für Physik und Astrophysik, München

Abstract. After a review of relevant observations the state of motion of the interplanetary plasma and the properties of the interplanetary magnetic fields including their relation to those of the sun are discussed with reference to published theoretical models and to a model recently discussed by R. Lüst and the author. Finally, a brief account is given of some recent theoretical work on the interaction between solar plasma and plasma of different origin (cometary or artificially produced).

The most general observation, which can be made with respect to the interplanetary plasma, is that it constitutes the nearest example of a dilute cosmic plasma. As seems to be the rule in cosmic plasma clouds, magnetic fields are found to be present, which are effectively "frozen" in the interplanetary plasma; by these magnetic fields also the cosmic ray particles coming from interstellar space, and the solar particles of similar energy, are affected in their motion and partially coupled to the plasma. The magnetic fields are furthermore believed to play an important role in the interaction of the several kinds of plasma which are or may be present in interplanetary space: that is the plasma ordinarily emitted by the quiet or the active sun, the plasma originating from a comet or possibly a planet, and finally the plasma clouds, which can now be artificially produced from a space craft.

For these reasons, the physical behaviour of the interplanetary plasma is of far more general interest than would seem from the point of view of solar physics alone or of the physics of the comets and planets, how important these aspects may be (cf. e.g. BIERMANN, 1960).

Having this in mind, we shall in the following first group the observational evidence in perhaps a slightly different way than is usually done. I shall then present certain considerations pertaining to the interplanetary magnetic fields, and some new results on the comets as natural plasma probes. In conclusion I shall say a few words about the state of our own experiment for producing plasma clouds in interplanetary space.

Sources of Information on the Interplanetary Plasma, and Results

As is well known, we have in addition to the direct measurements (cf Dr. Snyder's report, see page 67; also BRIDGE *et al.*, 1962) some indirect methods for obtaining information on the interplanetary plasma. Some of these reveal only the presence of plasma and its distribution in space, while others yield information mainly on its state of motion.

The oldest indirect method is provided by the geomagnetic records from which the existence of long-lived M-regions on the sun was deduced, long before their connection with the chromospheric and magnetic surface features could be studied. The basis for the conclusion that magnetic disturbances are due to plasma clouds or

streams coming from the sun is of course Dr. Chapman's work done jointly with
Dr. Ferraro a long time ago. Other indirect lines of evidence are provided by changes
in the appearance of radio point sources when they are within an angular distance of
30 degrees or so from the sun, and – for the interplanetary magnetic fields – the
properties of the energetic particles generated on the sun during flares and observed
later on the earth.

However, by far the largest volume in interplanetary space and the largest span
of time is covered (though not continuously) by the comets with plasma tails. The
observed kinematical properties of the cometary plasma can be understood, it seems,
only as a consequence of its interaction with the solar plasma – the pressure of solar
light being entirely inadequate for producing the observed accelerations. Hence the
plasma tails of comets may be looked upon as natural probes indicating the presence
and the direction of motion of the solar plasma as well as the order of magnitude of
its velocity. Furthermore, their structural features usually indicate the presence of
magnetic fields, and it is very likely that the interplanetary magnetic fields play an
important role in coupling together the solar plasma and that of cometary origin.

Since more than 130 comets with plasma tails are known, with a frequency of new
discoveries of 1 or even 2 per year, the weight of this body of evidence is obvious.
In order to make this point still clearer, I shall use the following diagram, in which
the positions of the sun and the planets out to 5 AU, the approximate distance of
Jupiter, are indicated:

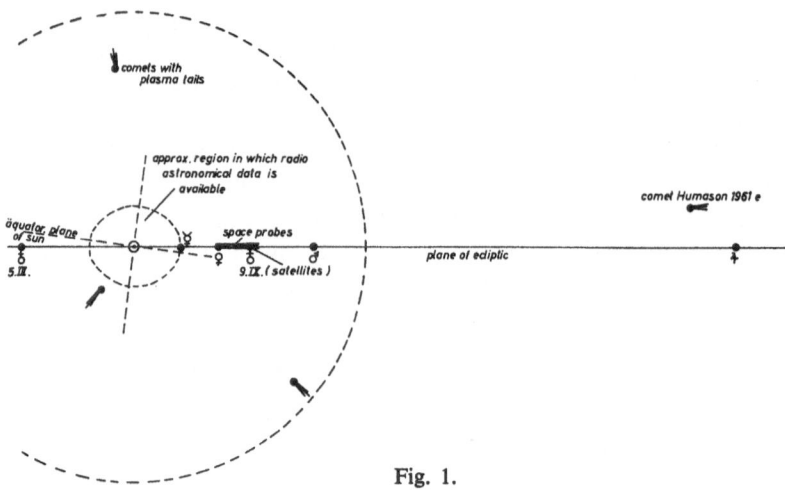

Fig. 1.

The evidence from the observations of radio point sources, specifically the fact
that the diameter of these point sources is increased near the sun, because the radio
waves are scattered by interplanetary electrons owing to the inhomogeneities of their
distributions, gives information concerning the density distribution of the latter out
to say about 0.3 or 0.4 AU. According to the usual interpretation of these observations

the surfaces of equal density of the interplanetary electrons seem to be flattened but not very much so. Thus out to this distance from the sun we have a certain body of evidence which indicates the presence of plasma, but nothing about its state of motion.

Next we have the evidence from geomagnetism; here it is important to note that the equatorial plane of the sun is inclined by an angle of 7.2 degrees to the ecliptic plane. This means that observations from the earth or any space craft close to the ecliptic plane, e.g. near Venus, which is approximately in the same plane, will be off from the equatorial plane of the sun by up to 7 degrees. Unfortunately that is not very much because the activity belt of the sun reaches to about 30 to 35 degrees North and South heliographic latitude.

The evidence from direct measurements of the solar plasma pertains thus to a region with heliographic latitude less than $\approx 7°$ (north or south) and solar distances r between 0.7 and 1, with some information pertaining to larger distances from the Russian Mars mission. Apart from the vicinity of the earth, most of this evidence comes from the 100 days of observations by Mariner II.

By contrast comets with plasma tails are observed at all heliographic latitudes and out to solar distances of usually 1.5–2 AU. A few comets with plasma tails have been found at larger distances, a noteworthy example being comet Humason 1961e, which became visible at the distance of Jupiter approximately 5 AU and a heliographic latitude of 25°. The general properties of its tail, e.g. direction and color, indicated a plasma tail; this conclusion was subsequently confirmed, when GREENSTEIN (1962) took a spectrum, which clearly showed the emissions of ionized molecules.

Pictures of this comet have been published in various journals (e.g. by ROEMER, 1962; VAN BIESBROUCK, 1962; MILLER, 1962). On one day the tail showed a huge mass of gas released by the nucleus with a velocity of 9 km/sec. The general behaviour of the tails shows that the effective acceleration is always large compared to that of gravity. Thus the main phenomenon indicating the presence of solar corpuscular radiation is found even at that distance and at a period during which solar activity had already markedly declined from the maximum reached in 1957.

The main point which I wanted to bring out is this. The information derived from the comets pertains to a volume in interplanetary space larger by orders of magnitude than that covered by direct measurements. Furthermore, latitudes outside the ones of the active regions on the sun are being reached, the phenomena in which could be entirely different from the ones observed at low solar latitudes, e.g. from the vicinity of the earth. In view of the very great technical difficulties of sending space probes to high latitudes and large distances, it is obvious that a combination of the cheap, but indirect method of observing the plasma tails of comets from the earth with that of direct measurements of the plasma flux from a space vehicle, is the most efficient way of obtaining more complete information on the interplanetary plasma.

For the vicinity of the ecliptic plane the direct measurements as discussed by Dr. Snyder in conjunction with the indirect evidence on the prevailing direction of motion of the solar plasma, reveal the main properties of the interplanetary plasma and its nature; that is outside the periods of major disturbances of the solar wind.

For larger distances from the ecliptic plane we have until now only the indirect evidence from the comets. To a first rough approximation this does not indicate any great change of conditions in interplanetary space with increasing heliographic latitude. STUMPF (1961), who studied a limited number of comets of mainly before 1900, found some evidence for a decreasing action of the solar wind at high heliographic latitudes. This result, however, needs further study and confirmation.

Another point of great interest is the dependence on average solar activity as measured, e.g. by the monthly averages of the sunspot number. There we might expect, e.g. a decline to zero in analogy to the sunspots, or, as an alternative, only a moderate decline (cf. geomagnetic activity) or even less variation. This question has, after earlier work by BIERMANN (1951, 1952), been investigated in much greater detail by Rhea LÜST (1961, 1963), who studied five comets which appeared during periods of minimum solar activity. Her results show that these comets do not differ in their general properties from those observed at medium or high solar activity.

We conclude then, that also around the minima of solar activity there is always plasma emitted by the sun. For the vicinity of the ecliptic plane this is in line with the geomagnetic evidence, that days of complete geomagnetic quietness are extremely rare even at minimum periods.

Theoretical Considerations

Next we turn to the mechanism, by which momentum is transferred from the solar wind, or from the storms in interplanetary space to the plasma tails of comets. With temperatures of $\approx 10^5$ degrees and the densities in the solar wind which are now usually accepted, the old proposal of mine of a transfer by the thermal motion of the electrons would not seem to be efficient anymore. An alternative, also proposed in 1951 (BIERMANN, 1951; ALFVÉN, 1957) is the coupling by magnetic fields; with the present observational evidence from the Mariner II measurements this becomes much more likely than it was at that time.

This leads us to the question of the characteristics of the interplanetary magnetic fields. Here we have obviously the main alternatives: either the fields are always linked to the sun in such a way that the configuration in interplanetary space is consistent both with solar rotation and with the radial motion of the plasma, and consequently forms spirals as has been proposed by PARKER (1958); or else the fields are disordered and we have a magnetized moving plasma (R. LÜST, 1962). Observation shows that the magnetic energy density is usually small (10^{-2}) as compared to the kinetic energy of the moving plasma. Hence in the second case the fields could be regarded as effectively disconnected from the sun. Up to now the discussion of the Mariner II results does not seem to have answered the question of the relative merits of these two models.

The second model has recently been rediscussed by BIERMANN and Reimar LÜST (1964). The question of how magnetic lines of force could be disconnected from the sun, may be discussed by means of the appropriate formulae of magneto-hydrodynamics. Assuming for the transport of magnetic flux from the sun into interplanetary space a figure given by H. Babcock in connection with the theory of the sun's magnetic cycle (H. W. BABCOCK, 1961), namely 10^{15}–10^{16} Mw/sec, we find, that under thermal conditions this rate of disconnection would not be possible. This result may be related to the recent observation – which was discussed yesterday – that acceleration of charged particles to energies in the MeV range, which again should require non-thermal conditions, is apparently a normal feature on the sun.

The figure of 10^{15}–10^{16} Mw/sec is now used to estimate the magnetic fields in interplanetary space. For this we have to make certain assumptions about the geometry of the fields. Let us assume closed loops, say loops which have originally been part of bipolar magnetic regions, which have been blown up into the corona, as described by Babcock, and finally after having become disconnected from the surface been carried into interplanetary space with the solar wind. Taking dimensions which seem plausible on the basis of the number of active regions on the sun, which are seen at a given time, we can compute the magnetic intensity, which should occur in interplanetary space; it is found, that fields of $\approx 3\gamma$ combined with a radial velocity of 500 km/sec and a linear dimension of $\approx 10^{13}$ cm ($\frac{2}{3}$ AU) near the earth's orbit are indeed consistent with the transport of magnetic flux as given above.

Another point of interest here is the fact that the magnetic flux carried away from the sun must eventually reach true interstellar space and might conceivably add to the interstellar magnetic fields. An estimate, which may be found in our paper appears to show that this flux might contribute significantly to the interstellar fields.

Let me come back to the question of the coupling between the solar plasma and the plasma of a comet by the interplanetary magnetic fields; this question has also a practical interest in connection with an experiment to be described below. From the theory of plasma, it may be shown that two kinds of ions, with different initial mass velocities, will, in the presence of a perpendicular magnetic field, equalize their mass velocities within a time scale given by their times of gyration (e.g. 4π seconds for protons in field of 5γ). The presence of a comet in interplanetary space means that there is a source of heavier molecules and atoms, which become ionized by charge transfer or by photoionization, while reaching a distance of the order of 10^6 km from the cometary nucleus. Before ionization, the cometary particles practically do not interact with the solar plasma, but after ionization they are rapidly coupled to the solar plasma, thus increasing its mean molecular weight $\bar{m}$ from 0.6 to possibly ≈ 15. The balance of mass, momentum and energy under these circumstances has been studied in a simplified model by BIERMANN, BROSOWSKI and SCHMIDT (1962). They found that initially the plasma mass velocity decreases with increasing $\bar{m}$ still faster than would correspond to constant momentum; the problem of the total flow pattern of the plasma in the vicinity of a comet, is as yet only qualitatively understood.

Another study by BIERMANN and TREFFTZ (1964) was initiated for the purpose of understanding the appearance of the forbidden lines of atomic oxygen in a number of brighter comets; it led subsequently to a new model of the production of the observed CO^+ structures. In regard to the latter, the main problem is to account for the fact, that the observed time scales, with which CO^+ structures appear in a comet, are considerably shorter than can be understood on the basis of photoionization or charge transfer ($\lesssim 1$ day, occasionally much less, as compared to at least 10 days, which should be expected). The model proposed in this investigation demands the production under the influence of solar heat of much more neutral gas from the cometary nucleus than becomes visible in the form of C_2, CN and CO^+, the main constituents contributing to the optical picture. In the vicinity of the nucleus, out to a distance of the order of 10^9 cm, the resulting stationary densities are then high enough to allow exothermic chemical reactions between ions produced by the solar far ultraviolet radiation and neutral molecules. The CO^+ would be a natural result of such reactions, if both O and C would be preferentially present in the form of hydrides, because the binding energy between C and O is relatively much larger. The shorter time scale would come in by the fact, that the zone of sufficiently high densities would be traversed with the thermal velocity of about 1 km/sec of the non-ionized molecules in a few hours or less; hence fluctuations in the gas production would be reflected in the production of CO^+. A number of models were computed in order to establish this last point.

If this model is correct, the total contribution of a comet to the interplanetary plasma could be of the order of 10^{31} molecules/sec for a rather bright comet, that is 100 to 1000 times more than is usually thought. Whether or not this is so, can be checked by spectroscopic observation in the far ultraviolet.

Experiments with Artificial Plasma Clouds

In closing this discussion I would like to say a few words on the status of a rocket experiment, which has suggested itself to us by our work on comets. The idea of using an artificial plasma cloud, say of Ba^+ or Sr^+ ions, for getting information on the solar plasma, has been described repeatedly (BIERMANN et al., 1961; BIERMANN, 1963); quantities of the order of some kg should be visible from the ground at distances of say, 200000 km, without great difficulty. Apparatus for vaporizing Ba, which should then become ionized by solar light, by means of chemical reactions has been developed by Dr. Lüst's group; after extended tests in the laboratory the first successful experiments have been carried out at ionospheric heights (≈ 160 km) in cooperation with Dr. Blamont's group near Colomb Bechar in the French Sahara. This experiment has originally been planned as one of the tests to be carried out, before a similar experiment outside the earth's magnetosphere, in true interplanetary space, is made. However, it should have independent value as an ionospheric experiment for giving information on the flow patterns of the ionospheric plasma, as soon as the artificial plasma is produced in sufficient quantity.

References

ALFVÉN, H.: 1957, *Tellus* **9**, 92.
BABCOCK, H. W.: 1961, *Astrophys. J.* **133**, 572.
BIERMANN, L.: 1951, *Z. Astrophys.* **29**, 274.

BIERMANN, L.: 1960, *Rev. Mod. Phys.* **32**, 1008.

BIERMANN, L.: 1963, *NASA Technical Note* TN D-1901.

BIERMANN, L., BROSOWSKI, B., and SCHMIDT, H. U.: 1962, *Sitzber. Bayer. Akad. Wiss.* July, 1963.

BIERMANN, L. and LÜST, R.: 1964, *Astrophysica Norwegica* **9**, 61.

BIERMANN, L., LÜST, R., LÜST, RH., and SCHMIDT, H. U.: 1961, *Z. Astrophys.* **53**, 226.

BIERMANN, L. and LÜST, RH.: 1963, *The Solar System* (ed. by B. M. MIDDLEHURST and G. P. KUIPER),
    Vol. IV, Chapter 18, p. 618.

BIERMANN, L. and TREFFTZ, E.: 1964, *Z. Astrophys.* **53**, 226.

BIESBROUCK, G. VAN: 1962, *Astrophys. J.* **136**, 1155.

BRIDGE, H. S., DILWORTH, C., LAZARUS, A. J., LYON, E. F., ROSSI, B., and SCHERB, F.: 1962, *J. Phys.
    Soc. of Japan* **17**, Suppl. A–II.

GREENSTEIN, J. L.: 1962, *Astrophys. J.* **136**, 688.

LÜST, R.: 1963, *Space Sci. Rev.* **1**, 522.

LÜST, RH.: 1961, *Z. Astrophys.* **51**, 163.

LÜST, RH.: 1963, *Z. Astrophys.* **57**, 192.

MILLER, F. D.: 1962, *Publ. Astron. Soc. Pacific* **74**, 528.

PARKER, E. N.: 1958, *Astrophys. J.* **128**, 664.

PARKER, E. N.: 1963, *Interplanetary dynamical processes*, Interscience Division of John Wiley and
    Sons, New York.

ROEMER, E.: 1962, *Publ. Astron. Soc. Pacific* **74**, 351.

STUMPFF, P.: 1961, *AN* **286**, 87.

Discussion and Questions

E. Hones (IDA): I'd like to ask a question about your barium experiment. What fraction of ionization can you get in the initial chemical bomb burst?

L. Biermann: Well, the ionization is supposed to be made by solar ultraviolet light. As far as we know, using quantum mechanical estimates, for the cross-sections involved it would take about five minutes or so to get it ionized. But that is not very reliable and I hope that our plates already give this information.

H. Alfvén: In the theory of the interaction between an ionized magnetized interplanetary plasma and the gas of a comet, it may be also important to point out that there is a special phenomenon which gives rise to a very rapid ionization. It is known from a series of plasma experiments one of which was done by Dr. Wilcox here that as soon as the relative velocity between non-ionized gas and a magnetized plasma reaches a certain critical velocity the ionization comes very rapidly. This critical velocity depends upon the chemical constitution of the gas and is usually in the region of 10 or 50 km/sec and in your case it may be possible that this phenomenon is of importance. It has been confirmed experimentally but the theory of it is rather doubtful and consequently it may not be possible to introduce it into your theory. I just mention it as having a possible significance.

E. Fontheim (University of Michigan): Would you care to tell us what measurements were taken when the barium cloud was injected into the ionosphere?

L. Biermann: Both photographs and spectra were taken, the latter ones for determining the relative amounts of molecular and ionized gas present.

A. Kantrowitz (AVCO-Everett Research Lab.): Could you say something about how many mean free paths there are across a comet tail? I'm particularly wondering whether it is possible for the small fraction of ionization to be enough to result in mass motion of the comet tail.

L. Biermann: I didn't quite get the question.

A. Kantrowitz: Is the density or mean free path such that a small fraction of ions could move the rest of the gas like fluid? In other words you don't have to ionize the whole thing if you have many mean free paths.

L. Biermann: The densities are extremely small. They are comparable to the densities in interplanetary space but it seems that the materials in the comet tails are very unevenly distributed, the pressure differences being balanced by the magnetic forces. But even so I think the non-ionized material is practically not coupled with the ionized material. It becomes coupled only after the processes of ionization.

CORONAL EXPANSION AND SOLAR CORPUSCULAR
RADIATION*

E. N. PARKER

*Enrico Fermi Institute for Nuclear Studies and Department of Physics,
University of Chicago, Chicago, Ill.*

The streaming of particles from the sun (solar corpuscular radiation) leads to a variety
of observable geophysical and astrophysical effects. The aurora, geomagnetic activity,
cosmic ray variations, comet tail behavior, and perhaps even certain controversial
aspects of the over-all terrestrial weather pattern (see for instance MacDonald and
Roberts, 1960) are induced by the solar corpuscular radiation. The history of development of knowledge of solar corpuscular radiation is long and it is only in the
past few years that an idea of the mechanism of its origin at the sun has developed.
The manner in which the corpuscular radiation brings about the various effects
mentioned above is understood qualitatively, and even semi-quantitatively, in some
cases, and very little in others such as the aurora. The discussion presented here
deals principally with the dynamics of the origin at the sun and behavior of the solar
corpuscular radiation in interplanetary space. It pays but little attention to the details
of the explanation of its effects, except insofar as the effects permit estimates of the
nature of the corpuscular radiation.

1. The Existence of Solar Corpuscular Radiation

The elusive phenomenon of solar corpuscular radiation has been under study since
the beginning of the century. But until space vehicles became available a few years
ago, the solar corpuscular radiation could be known only through its indirect effects,
with the result that a wide diversity of opinion existed concerning its nature and origin.
Evidence for the existence of solar corpuscular radiation had its first substantial
beginnings with the pioneering work of Störmer (see STÖRMER, 1955 for an extended
discussion) on the aurora. Störmer showed that much of the latitude behavior and
curtain-like forms of the aurora could be accounted for if it were assumed that sprays
of mono-energetic particles with energies of the order of several MeV were emitted
from the sun and propagated in straight lines to the magnetic dipole field of Earth.
Störmer's well-known calculations of the charged particle motion in the geomagnetic
field illustrated how the auroral forms might come about. It has become evident
in the last couple of decades that the aurora cannot be produced as Störmer imagined
because charged particles with enough energy to behave in the geomagnetic field as

---

* This paper was already published in *Amer. Inst. Aeronaut. Astronaut. J.*

Chang & Huang (eds.), Proc. Plasma Space Sci. Symp. All rights reserved.

imagined by Störmer would penetrate much more deeply into the terrestrial atmosphere than is observed. The present value of Störmer's work, so far as solar corpuscular radiation is concerned, lies in his calling attention to the particle nature of the aurora, and to the possibility of particle emission from the sun.*

The next major step in building toward the present concepts of solar corpuscular radiation had to do with geomagnetic activity and the magnetic storm phenomenon. It was suggested by Lindemann that the geomagnetic storm was the result of neutral clouds or streams of ionized hydrogen shot out of the sun at the time of a solar flare. Following this CHAPMAN and FERRARO (1931, 1932, 1933, 1930) worked out a number of examples of the dynamical interaction of ionized hydrogen with the geomagnetic field and demonstrated the soundness of the idea. Their pioneering work has served as the foundation for all subsequent developments in the theory of the geomagnetic storm, which are centered upon the idea that the geomagnetic storm, and geomagnetic activity in general, are the result of the impact of interplanetary plasma clouds against the geomagnetic field (see discussion and references in PARKER, 1962).

It was shown by Forbush (see review in FORBUSH, 1954) that the intensity of the galactic cosmic rays varies over the 11-year (or 22-year) cycle of solar activity. The cosmic ray intensity tends to be higher when solar activity is low than when solar activity is high. Forbush also pointed out the tendency for the galactic cosmic ray intensity to decline in general association with the geomagnetic storm. With the development of the neutron monitor (SIMPSON, 1951) enabling one to look at the cosmic ray protons in the low energy range of a few GeV, these variations were large enough to study in some detail. SIMPSON (1954) was able to show that the variations were of extraterrestrial origin and evidently the work of the solar corpuscular radiation. It became clear that the active sun and the associated corpuscular radiation was in some way able partially to exclude the galactic cosmic rays from the vicinity of Earth. Variations in the cosmic ray intensity became then a means for studying interplanetary conditions (see discussion and review in SIMPSON, 1960). A great variety of models, such as a plasma cloud carrying a small scale disordered magnetic field (MORRISON, 1956) were considered (see references and discussion in PARKER, 1963). The important point here is that solar corpuscular radiation turns up again as the perpetrator of an observable effect in the terrestrial environment.

For several centuries it has been known that the gaseous tails of comets tend to point straight away from the sun irrespective of the direction of orbital motion of the comet around the sun. The classical explanation was always radiation pressure. However about twenty years ago the analysis of the acceleration of the visible inhomogeneities in the tail of the comet (WURM, 1943) began to make it clear that the outward pressure on the gas in the comet tail must be enormous. BIERMANN (1951, 1952, 1957) pointed out that radiation pressure was inadequate by a large factor and

---

* It has become evident in the past few years that the outer boundary of the geomagnetic field lies so far out (SONETT et al., 1959; DOLGINOV and PUŠKIN, 1959; HEPPNER et al., 1960; CAHILL and AMAZEEN, 1962) that the energetic particles in the aurora are probably not the same particles as are emitted by the sun but are only excited by the solar particles in some indirect way.

that the only tenable assumption was solar corpuscular radiation. Biermann's realization of the role of solar corpuscular radiation in comet tail behavior was of particular importance because it made clear in a striking manner that solar corpuscular
radiation is evidently emitted in *all directions* from the sun at *all times*. (See further
discussion in STUMPFF, 1956; LÜST, 1961.) It made it clear that solar corpuscular
radiation is not a special property of active regions and flares on the sun, but is a
general and continuing property of the quiet, as well as the active, sun.

There are a number of means by which the velocity of propagation of the solar
corpuscular radiation has been estimated. (a) The one or two day delay of the geomagnetic storm and the cosmic ray decrease after a large flare on the sun shows clearly
that the velocity of propagation is something of the order of 1000–2000 km/sec from
the very active sun. (b) If it is assumed that plages on the central meridian of the
sun are associated with slight enhancements in the quiet-day corpuscular radiation
emitted in the direction of Earth, then the delay of the small increases in geomagnetic
activity suggest quiet-day propagation velocities of 150–400 km/sec (MUSTEL', 1959;
MITROPOLSKAJA, 1959; MUSTEL' and MITROPOLSKAJA, 1959), with the lower velocities
being more prevalent during years of sunspot minimum. (c) Comet tail acceleration
reflects clearly the variations in the strength of the solar corpuscular radiation with
the level of solar activity. Biermann has pointed out that the gaseous comet tails
seems always to lie within about 3° of the radial direction from the sun, which when
combined with a typical transverse orbital velocity of, say, 30 km/sec gives a corpuscular velocity which is about 500 km/sec.

So far as we are aware, there are no methods for estimating the density of the
solar corpuscular radiation indirectly. Early attempts to deduce the interplanetary
electron density from observations of the polarized component of the zodiacal light
(BEHR and SIEDENTOPF, 1953; ELSÄSSER, 1954) and from free particle models of comet
tail interaction (BIERMANN, 1957) gave quiet-day values of the order of $500/cm^3$ at the
orbit of Earth. BLACKWELL (1960abc) pointed out however that the density must be
much lower, less than $100/cm^3$, and that the polarized component of the zodiacal
light was scattered principally by dust grains rather than electrons. Then BLACKWELL
and INGHAM (1961) suggested that the density was of the order of $300/cm^3$ at the peak
of a strong outburst from the sun, based on a supposed brightening of the zodiacal
light at that time. It was pointed out by PARKER (1962) that the effect of the outburst
on the geomagnetic field suggested a maximum density of more like $30/cm^3$.

Recent direct observation in space shows the continuing presence of solar corpuscular radiation (ŠKLOVSKIJ *et al.*, 1960; GRINGAUZ *et al.*, 1960). The quiet-day velocity
is 300–600 km/sec (about 1 keV per proton) and the density is 2–20 protons/cm^3
(BRIDGE *et al.*, 1962; BONETTI *et al.*, 1962; NEUGEBAUER and SNYDER, 1962). The
observations show clearly that the intensity of the solar corpuscular radiation may
fluctuate rapidly at times, even on what would otherwise be considered a quiet day.
There are as yet no certain direct quantitative measurements of the typical enhanced
solar corpuscular radiation, so that we can only speculate that it may reach a maximum
of $30/cm^3$.

2. The Origin of Solar Corpuscular Radiation

There have been proposed over the years a variety of interesting ideas concerning the origin of solar corpuscular radiation at the sun. Many authors have looked to the plage as the origin of much of the corpuscular emission and others have favored the more violent solar flare. Some have considered the sunspot with its intense magnetic fields and great store of energy and constant change. In the minds of many workers in the field there is a close association of the origin of kilovolt solar corpuscular radiation, with the energetic (1 MeV–10 GeV) particles emitted from so many solar flares. Very little has been worked out quantitatively, but it is evident that most of the ideas have centered around the magnetic field as a major factor in catapulting the corpuscular radiation into space. Perhaps the most clear cut of these ideas is the well-known "melon seed" model put forth by SCHLÜTER (1950).

It has been our own view that the solar corpuscular radiation may have a somewhat less special origin than has generally been imagined. This view was impressed upon us by Biemann's demonstration of the continuing nature of the solar corpuscular radiation. A suggestion as to where to start looking for the origin of the solar corpuscular radiation was provided to us by Chapman's theoretical studies of the thermal conductivity of the solar corona. CHAPMAN (1957, 1959) pointed out that the thermal conductivity of the ionized hydrogen, of which the corona is composed, is extremely large ($6 \times 10^{-7} T^{5/2}$ ergs/cm sec °K) with the result that the coronal temperature, of some 10^6 °K, must extend far into space. He pointed out that, except perhaps in the lowest portions of the corona, radiation losses may be neglected compared to the energy transported by thermal conduction. Then assuming that the corona is static, and more or less symmetric about the sun, he showed that the temperature declines outward only very slowly, like $1/r^{2/7}$. This would mean that the temperature of the corona at the orbit of Earth is about 0.2 what it is at the sun: 10^6 °K at the sun would imply 2×10^5 °K at the orbit of Earth. The density of the corona at the orbit of Earth proved to be 10–100 atoms/cm^3 on the basis of this model. Now there may be other means of energy transport besides thermal conduction available, so that we do not take seriously the $1/r^{2/7}$ temperature dependence of this particular model. The great importance of Chapman's calculation lies in its demonstration that, unless some powerful mechanism inhibits the outward extension of coronal temperature, the temperature must extend far into interplanetary space.*

The continuing nature of the solar corpuscular radiation and enormous extent of the coronal temperature and density into space suggested to us that the ordinary hydrodynamic properties of the solar corona might be a fruitful field of investigation. The first step was to show that the solar gravitational field, no matter how tightly it

---

* There has been some concern that transverse magnetic fields might seriously inhibit the conduction of thermal energy in the radial direction outward from the sun. The re-entrant nature of some of the filaments over active regions strongly suggest the existence of strong inhibition there. However, apart from some active regions, the general radial configuration of corona streamers and striations suggests that there may be little or no inhibition to an outward conduction of heat throughout most of the solar corona.

may bind the lower corona to the sun, is not able to contain the more distant portions of the corona. One way to see this is to note that the gravitational potential decreases outward like $1/r$, but there is no evident way in which the temperature can be made to decline so rapidly. Hence, sufficiently far from the sun the thermal energy of the gas exceeds the gravitational energy and the gas should be free to escape. Consequently, it is necessary to abandon the traditional idea of a static corona and admit the possibility of continual expansion. We must turn from the hydrostatic barometric equation to the hydrodynamic equation. It is sufficient for the present purposes of discussion to consider the simple case of a solar corona with spherical symmetry about the center of the sun and composed solely of ionized hydrogen. Then for steady conditions the outward velocity $v(r)$, the numbers of ions/cm^3 $N(r)$, and the temperature $T(r)$ are related by the familiar equation*

$$v\frac{dv}{dr} = -\frac{1}{NM}\frac{d}{dr}2NkT - \frac{GM_\odot}{r^2} \tag{1}$$

where M is the mass of a hydrogen atom. Conservation of matter requires that

$$N(r)v(r)r^2 = N_0 v_0 a^2 \tag{2}$$

where $r = a$ is the reference level, taken for convenience to lie in the lower corona where $N(a) = N_0$, $v(a) = v_0$. Solution of the hydrodynamic equation for the simple case of an isothermal corona $T = T_0$ showed immediately that the corona not only expands, but that it expands with supersonic velocities of several hundred km/sec (PARKER, 1958a). Subsequent solution with an outward decline of $T(r)$ given by the polytrope relation

$$T(r) = T_0\left\{\frac{N(r)}{N_0}\right\}^{\alpha-1}$$

showed that the supersonic expansion is not critically dependent on the form of $T(r)$ nor on the value of T_0 (PARKER, 1960). Recent investigations (PARKER, 1964) of the solar corona with the assumption that $T(r)$ is determined by the heat flow equation** shows that, within the idealizations contained in the equations, the corona must

---

* There has been concern on the part of some individuals as to whether the hydrodynamic equation is applicable to the solar corona. The traditional treatment of the corona has centered on the idea of individually moving particles which may evaporate freely into space. We point out that the mean free path in the corona is small compared to the scale height. Far out from the sun in interplanetary space there is the interesting possibility, if the temperature is high enough, that the mean free path may be relatively long (PARKER, 1958b). However, the weak magnetic fields, and their associated hose and mirror instabilities, maintain the pressure in an approximately isotropic state and rather closely confine the motions of each individual ion and electron; the effect is much the same as a high collision rate, so that the usual hydrodynamic equation is a valid approximation for the overall large-scale motion of the gas.
** It is an open question at present as to what extent there may be an energy supply to the outer corona in addition to thermal conduction (NOBLE and SCARF, 1963) as a consequence of the dissipation of hydromagnetic waves, etc. Hence the heat flow equation probably represents a minimum energy supply.

expand supersonically for all coronal T_0. Finally, since the integration of the hydro-dynamic equations proceeds along the stream lines, the results are essentially inde-pendent of whether the corona is composed of many radial striations and filaments or whether it is more or less homogeneous. The basic properties of the equations are the same in all cases, predicting several hundred km/sec at the orbit of Earth when the coronal temperature is of the order of 10^6 °K, so in the present discussion we may as well consider the original isothermal coronal model because the computations are simpler than the others.

The investigation showed that the gravitationally bound stellar corona has the general property of being nearly static at its base and expanding with a velocity that increases monotonically outward to values of several hundred km/sec, i.e. con-siderably in excess of the rms thermal velocity at the base of the corona. It is evident at once that the "solar corpuscular radiation", postulated to account for so many geophysical effects, is probably the manifestation of this coronal expansion. The velocity requirements of 200–2000 km/sec are just those predicted from the observed coronal temperatures of 0.5–4 $\times 10^6$ °K. The hydrodynamic origin suggests that some such term as the "solar wind" might be more appropriate than "solar corpuscular radiation."

3. Physics of Coronal Expansion

The physics of the expansion of the solar corona can be understood on the basis of an analogy pointed out to the author by CLAUSER (1960), that the gravitational field of the sun plays the same role in coronal expansion as the throat in a de Laval nozzle. In both cases the gas attains supersonic velocity while passing from a high pressure into a vacuum as a consequence of a constriction in the flow. To see this directly note that the equation for conservation of mass for steady flow through a nozzle of cross section $A(r)$ is

$$N(r)A(r)v(r) = N_0 v_0 A_0. \tag{3}$$

The decrease in $A(r)$ as the gas moves into the throat requires a corresponding increase in the flow velocity $v(r)$, and in this way the throat causes $v(r)$ to increase up to the speed of sound. Beyond the throat the nozzle flares out and the velocity of the gas increases above the speed of sound as it expands into the vacuum. It is important to note that the gas can be made to decelerate, rather than accelerate, after reaching the speed of sound in the throat only if the nozzle does not open into a near vacuum. Only a suitably high back-pressure at the exhaust end will force the gas back down to subsonic velocities.

Equation (2) is the equation for conservation of mass in the solar corona. In this case $N(r)$ decreases rapidly with height in the corona because of the strong solar gravitational field. Consequently $v(r)$ must increase rapidly with height to maintain the net flow of matter, with the result that the velocity reaches the speed of sound at a distance of a few solar radii. Beyond that point the velocity increases further as a consequence of expansion into the vacuum of interstellar space, must as in the nozzle. It would be interesting to set up the analogous nozzle (PARKER, 1963) and observe the expansion in the laboratory. This analogue experiment might be one way to get at the otherwise difficult problem of the stability of the flow.

Now the hydrodynamic equations for the solar corona possess a critical point (PARKER, 1958a, 1960) across which passes the solution of physical interest for expansion into a vacuum. The critical point may be demonstrated by using (2) to eliminate $N(r)$ from (1). The result is

$$\frac{dv}{dr}\left(v - \frac{2kT}{Mv}\right) = -\frac{2kr^2}{M}\frac{d}{dr}\left(\frac{T}{r^2}\right) - \frac{GM_\odot}{r^2}. \qquad (4)$$

The critical point is the point (r_c, v_c) at which the right hand side of this equation and the coefficient of dv/dr both vanish. Equating the right hand side to zero leads to an expression that yields r_c as soon as the function $T(r)$ is specified. The temperature at the critical point $T(r_c)$ follows immediately. The velocity v_c at the critical point follows from the coefficient of dv/dr as $v_c^2 = 2kT(r_c)/M$. The critical point corresponds to the point at which the fluid velocity becomes supersonic.

For the simple case of an isothermal corona, which typifies the phenomenon of coronal expansion, the critical point has coordinates

$$r_c = \frac{GM_\odot M}{4kT_0}, \qquad v_c^2 = \frac{2kT_0}{M}. \qquad (5)$$

Requiring that the integral of (4) pass through the critical point leads to the solution

$$\frac{v^2}{v_c^2} - \ln\frac{v^2}{v_c^2} = -3 + 4\ln\frac{2v_c^2 r}{w^2 a} + \frac{2w^2 a}{v_c^2 r}$$

where $w^2 = GM_\odot/a$. It is readily seen that at large r, the velocity increases like $2v_c\left[\ln(r/a)\right]^{1/2}$. The velocity v_0 at the reference level $r = a$ is given by

$$\frac{v_0^2}{v_c^2} - \ln\frac{v_0^2}{v_c^2} = -3 + 4\ln\frac{2v_c^2}{w^2} + \frac{2w^2}{v_c^2}.$$

The mass loss to the sun is $4\pi N_0 v_0 a^2$, which is determined by the velocity v_0 in the low corona. Numerically we have $v_c = 1.3 \times 10^4 T_0^{1/2}$ cm/sec and $w^2 = 1.9 \times 10^{15}$ cm/sec. At the orbit of Earth we have very roughly that $v \simeq 5 \times 10^4 T_0^{1/2}$ cm/sec. As a specific example note that a coronal temperature of $T_0 = 1.0 \times 10^6$ °K yields $v_0 = 0.7$ km/sec at the base of the corona ($a = 7 \times 10^{10}$ cm, $N_0 = 2 \times 10^8$ atoms/cm³). The mass loss to the sun is 1.5×10^{12} gm/sec or about 10^{-4} $M_\odot$ in the 5×10^9 years that the sun has existed. The rms thermal velocity is 150 km/sec and the gravitational escape velocity is $2.5w \simeq 600$ km/sec. The critical point lies at $r_c = 4 \times 10^{11}$ cm $\simeq 6R_\odot$, where the expansion velocity is 120 km/sec. The velocity at the orbit of Earth is about 500 km/sec, and the density is about 7 atoms/cm^3. Velocity as a function of radial distance from the sun is plotted in Figure 1 for several values of T_0. Note that for the higher temperatures of 3 and 4×10^6 °K the expansion velocity exceeds 10^3 km/sec. The solar wind from this simple 1×10^6 °K model falls near the middle of the velocity and density ranges observed inspace. It is evident that the observed variation of coronal temperature with time and with position around the sun must lead to marked variations in the solar wind velocity and density with time and in different directions from the sun.

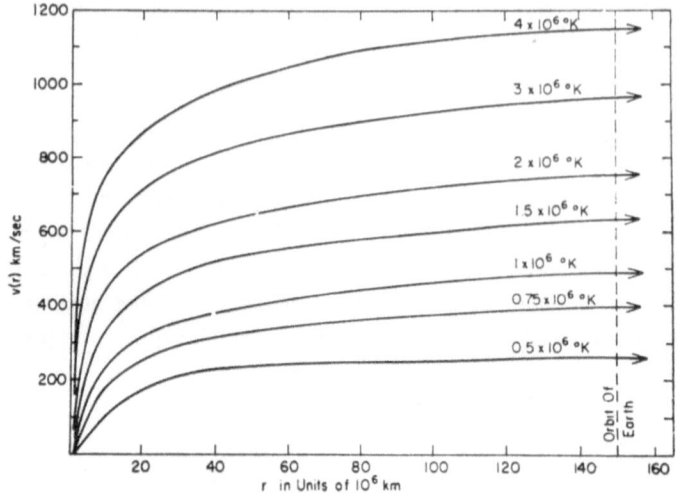

Fig. 1.   A plot of the radial expansion velocity of an isothermal corona as a function of distance from the sun (measured in units of 10^6 km) for various values of the coronal temperature T_0.

One interesting feature of the expanding corona is the enormous rate of decrease of the density of the solar wind with declining coronal temperature. The velocity of the solar wind varies approximately as $T_0^{1/2}$, as we would expect, but the density varies much more rapidly. For instance a uniform coronal temperature of 2×10^6 °K gives a wind velocity of 700 km/sec and a density of more than 10^2/cm^3 at the orbit of Earth. The density is up by about a factor of 14 from what it is for 1×10^6 °K. Obviously the density of the solar wind at the orbit of Earth cannot be predicted from the temperature $T(r)$ of the corona at the sun unless that temperature is known rather precisely. One can fit the old indirect density estimates of 10^2/cm^3 or more at the orbit of Earth, as well as the new direct measurements in space of 2–20/cm^3 with temperatures within the observational range of $0.5–3 \times 10^6$ °K.*

It is evident that the observed variation of coronal temperature with time and with position around the sun must lead to marked variations in the solar wind velocity and density with time and with direction from the sun. In this connection it is inter-

---

* The reader is referred to such reviews as VAN DE HULST (1953), BILLINGS (1959, 1963), LÜST et al. (1963) for a discussion of the observational determination of the temperature of the solar corona. The 10^6 °K employed in the present discussion has been chosen as a representative mean value suggested by the several methods of observational determination.

esting to note how the different effects produced by the solar corpuscular radiation tend to emphasize one or the other aspects of the solar wind. Studies of geomagnetic activity (MUSTEL', 1963; VSEHSVJATSKIJ, 1963; MOGILEVSKIJ, 1963) exhibit the effects of solar corpuscular radiation in such a manner as to suggest isolated streams of particles emitted from the sun, whereas comet tail behavior emphasizes the broad continuity of solar corpuscular radiation in space and time. Both aspects are real, of course. Broad streams of faster denser solar wind come from expansion of the hotter denser portions of the corona. It has been pointed out by SARABHAI (1963) that the appearance of narrow streams may be simulated by the shear and disorder in a narrow transition layer between much broader streams.

4. Quiet-Day Interplanetary Magnetic Fields

It is interesting to note that the expansion of the solar corona carries with it the general solar magnetic field of one gauss as a consequence of the high electrical conductivity of the coronal gas (PARKER, 1958). If the sun did not rotate, the result of the radial expansion of the corona would be to stretch out the magnetic lines of force in the radial direction, giving a magnetic field which declines outward like

$$B_r(r) = B_0 \left(\frac{a}{r}\right)^2.$$

The sense of the radial field in any direction from the sun would be determined by the sense of the field B_0 at the base of the corona in that direction. The rotation of the sun, with a period of about 25 days, has little or no effect on the radial motion of the coronal gas, but it introduces a pronounced spiral in the magnetic field configuration. In the equatorial plane of the sun there results the azimuthal component

$$B_\phi(r) \simeq B_0 \frac{a\Omega}{v} \frac{a}{r}$$

in addition to the radial component, where v is the solar wind velocity and Ω is the angular velocity of the sun. The lines of force form an Archimedes spiral, $r \simeq (v/\Omega)(\phi - \phi_0)$. The observed quiet-day solar wind velocities of 400–600 km/sec imply that the lines of force at the orbit of Earth are inclined about 45° to the radial direction. A field of one gauss at the sun leads to about 3×10^{-5} gauss at the orbit of Earth, which seems to be in agreement with the observations of Mariner II (COLEMAN et al., 1962). Except in the low corona the magnetic energy density appears to be small compared to the kinetic energy of the gas, so that the reaction of the magnetic field on the coronal gas can be neglected in a first rough approximation when considering the gross radial expansion of the solar corona. This is not to say that the solar magnetic field is negligible under every circumstance, of course. In particular it is evident from observations of coronal streamers that the field tends to channel the solar wind, or corpuscular radiation, toward the solar equatorial plane at times of sunspot minimum.

It is also evident that the magnetic field plays a role in isolating and preserving the observed filamentary structure of the corona for some distance into space.

When the observed variations of coronal temperature with solar latitude and longitude are considered, leading to a solar wind velocity v at large distance from the sun which varies considerably around the sun, it is evident that in the actual case the idealized Archimedes spirals discussed here must become considerably distorted and entangled.

5. Enhanced Solar Corpuscular Radiation

Thus far the discussion has dealt with the quiet-day solar corpuscular radiation. When the sun is particularly active, with one or more large flares, the velocity of the corpuscular radiation may rise to 1000–2000 km/sec, as noted earlier. The density may be estimated to rise to perhaps $30/cm^3$ or more at the orbit of Earth (see discussion in PARKER, 1962) for periods of a few hours. Calculations (see Figure 1) would seem to suggest that this enhanced solar corpuscular radiation, like the quiet-day radiation, is accounted for in a straightforward manner by the hydrodynamic expansion of the enhanced corona, whose temperature is observed to rise to three or four times 10^6 °K, and higher, at such times. The expansion velocity of a region of the solar corona at 4×10^6 °K reaches about 1150 km/sec, and the density may be $10^2/cm^3$ at the orbit of Earth, which would seem to fill all the requirements for the enhanced corpuscular radiation. We cannot absolutely rule out the possibility of other dynamical mechanisms contributing to the enhanced radiation, of course, but we do not know at the present time what they might be.

It is evident that a sudden increase at the sun in the rate of coronal expansion will lead to an overtaking of the slower solar wind or corpuscular radiation ahead by the more recent and faster wind behind. Consequently the enhanced solar corpuscular radiation from an enhanced coronal region tends to scoop up the quiet day wind ahead like a snowplow and build up a blast wave with a shock transition at its head. The hydrodynamic behaviour of the blast wave can be computed under certain idealized circumstances to show how its velocity and density vary with time and distance from the sun (PARKER, 1961a). The cosmic ray effects of the magnetic fields carried in the blast wave can be shown to resemble the observed Forbush-type decreases (PARKER, 1961a, 1963) the principal effect being a general sweeping back of the cosmic rays by the advancing front.

There is an interesting point of distinction that should be made here between the shock front and blast wave from sudden enhancement of the coronal temperature and the interference of a steady fast stream with a steady slow stream from the different regions of the corona. The loci of the gas in the fast and slow streams are both Archimedes spirals, as discussed in Section 4, even though the motion of each element of gas is in the radial direction. The faster stream is the more open spiral and crowds increasingly into the slower stream ahead with distance from the sun. A shock transition may develop between the two broad streams (PARKER, 1963; SARABHAI, 1963). There are then, two means by which shock transitions may arise in interplanetary space. One is by a rapid time variation in coronal temperature and the other is by the interaction of streams from a variation of temperature with position around the sun. It was emphasized by SNYDER (1963) that the Mariner II observation of a shock front (1547 UT, October 7, 1962) does not distinguish between the two possibilities because of an unfortunate coincidence in the geometry of the position of the space vehicle, the sun, and Earth at the time the shock was observed. Continuing observation of the solar wind from space vehicles should eventually establish the relative frequency of occurrence of the two kinds of shock transitions.

6. Extension of the Solar Wind into Interstellar Space

It is of some interest to consider the distance to which the solar wind may extend out from the sun into interstellar space. Thus far in dealing with the expansion of the corona near the sun we have considered interstellar space to be a perfect vacuum. It is not precisely a vacuum, however, and the interstellar hydrostatic pressure must be considered if an estimate of the extent of the wind into space is to be considered. Very roughly the situation is that the impact pressure of the solar wind of number density $N(r)$ and velocity v is $N(r) Mv^2$. The velocity is approximately constant at large distance from the sun because gravitational effects are essentially negligible. From the condition (2) for conservation of matter it follows that $N(r)$, and hence the impact pressure, decreases outward from the sun like $1/r^2$. If p_i denotes the effective interstellar pressure, the solar wind is slowed down by the interstellar medium at the radial distance r_1 where the impact pressure becomes comparable to p_i. Then using (2) and denoting by r_E the radius of the orbit of Earth, we have

$$r_1 \simeq r_E \left[\frac{N(r_E) Mv^2(r_E)}{p_i} \right]^{\frac{1}{2}}.$$

It is a straight forward matter to show from the hydrodynamic equations that the solar wind goes through a shock transition to subsonic velocities in the vicinity of r_1. The velocity declines through the shock transition to something of the order of one or two hundred km/sec. If some such numbers as $N(r_E)=5/cm^3$ and $v(r_E)=500\,km/sec$ are employed, with an interstellar pressure of 4×10^{-12} dynes/cm^2 corresponding to $B^2/8\pi$ for a magnetic field of 10^{-5} gauss, the result is $r_1=70\,AU$. Other numerical values are possible, of course, and an estimate most anywhere between 10 and 10^3 AU can be obtained. The point we wish to make here is that the effect of the interstellar pressure does not make itself felt until far beyond where the hydrodynamic expansion of the corona, leading to the solar wind, has run its course. In terms of the analogous de Laval nozzle, the gas flow is exhausting into so nearly perfect a vacuum that the back pressure is completely negligible so far as the flow in the nozzle itself is concerned. The shock transition, back down to subsonic velocities of about one fourth the supersonic velocity in the nozzle, takes place only at a very large distance from the exhaust end of the nozzle.

Detailed hydrodynamic calculations of the solar wind and its transition to subsonic velocities have been presented elsewhere (PARKER, 1961b, 1963). AXFORD, *et al.* (1963) have recently made the very interesting point that charge exchange of the tenuous solar wind in the vicinity of r_1, with neutral atoms of the interstellar gas diffusing inward from interstellar space, may possibly alter the simple hydrodynamic picture presented here. It is even conceivable that under certain circumstances a strong shock transition might be avoided altogether. This will have no effect on the solar wind in the inner solar system, of course, but it is an interesting question and may have some consequence for modulation of the galactic cosmic ray intensity.

7. The General Phenomenon of the Stellar Wind

The general hydrodynamic property of the tightly bound solar corona to expand continually into a supersonic solar wind suggests that the coronas of stars other than the sun might exhibit similar expansion properties (PARKER, 1960, 1963). The basic requirement for a perceptible stellar wind is that the star possess a corona of ionized gas with a thermal energy which is not less than about 0.03 the gravitational energy. This corresponds to about 0.5×10^6 °K for the sun, and since the stellar mass divided by the radius varies but little over the main sequence, the same approximate minimum temperature would apply throughout much of the main sequence.

The convective zone beneath the solar photosphere is currently believed to be the energy source for the solar corona (see discussion and references in OSTERBROCK, 1961 and WHITAKER, 1963). Consequently we would infer that perhaps most any star with a convective zone near its photosphere may have an extended corona and a stellar wind. We would also infer that perhaps any star exhibiting vigorous disordered mass motions in its atmosphere might have sufficient atmospheric heating to have an extended corona. Altogether, then, there should be a wide class of stars, of both early and late types, both on and off the main sequence, which one would suspect of having stellar winds. The stellar winds may in many cases be very much more vigorous than the wind from the sun. The question of the existence of stellar winds in stars other than the sun is an intriguing one on which the continuing study of stellar spectra and atmospheres may hopefully shed some light.

References

AXFORD, W. I., DESSLER, A. J., and GOTTLIEB, B.: 1963, *Astrophys. J.* **137**, 1268.
BEHR, A. and SIEDENTOPF, H.: 1952, *Z. Astrophys.* **32**, 19.
BIERMANN, L.: 1951, *Z. Astrophys.* **29**, 274.
BIERMANN, L.: 1952, *Z. Naturforsch.* **7a**, 127.
BIERMANN, L.: 1957, *Observatory* **107**, 109.
BILLINGS, D. E.: 1959, *Astrophys. J.* **130**, 961.
BILLINGS, D. E.: 1963, *The Solar Corona* (ed. by J. W. EVANS), Academic Press, New York, p. 226.
BLACKWELL, D. E.: 1960a, *Endeavour* **19**, 14.
BLACKWELL, D. E.: 1960b, *J. Geophys. Res.* **65**, 2476.
BLACKWELL, D. E.: 1960c, *Scient. Amer.* **203**, no. 1, 54.
BLACKWELL, D. E. and INGHAM, M. F.: 1961, *Mon. Not. Roy. Astron. Soc.* **122**, 113.

BONETTI, A., BRIDGE, H. S., LAZARUS, A. J., LYON, E. F., ROSSI, B., and SCHERB, F.: 1962, Paper Presented at COSPAR Meeting, Washington, D.C.

BRIDGE, H., DILWORTH, C., LAZARUS, A. J., LYON, E. F., ROSSI, B., and SCHERB, F.: 1962, *J. Phys. Soc. Japan* **17**, Suppl. A-II, 553.

CAHILL, L. J. and AMAZEEN, P. G.: 1962, *J. Geophys. Res.* **67**, 3547.

CHAPMAN, S.: 1957, *Smithsonian Contr. Astrophys.* **2**, 1.

CHAPMAN, S. and FERRARO, V. C. A.: 1931, *Terr. Magn. and Atm. Electr.* **36**, 77 and 171.

CHAPMAN, S. and FERRARO, V. C. A.: 1932, *Terr. Magn. and Atm. Electr.* **37**, 147 and 421.

CHAPMAN, S. and FERRARO, V. C. A.: 1933, *Terr. Magn. and Atm. Electr.* **38**, 79.

CHAPMAN, S. and FERRARO, V. C. A.: 1940, *Terr. Magn. and Atm. Elect.* **45**, 245.

CLAUSER, F. H.: 1960, *Nuovo cimento,* Suppl. **22**, No. 1.

COLEMAN, P. J., DAVIS, L., SMITH, E. J., and SONETT, C. P.: 1962, *Science* **138**, 1099.

DOLGINOV, S. and PUŠKIN, N.: 1959, *Pravda,* July 15.

ELSÄSSER, H.: 1954, *Z. Astrophys.* **44**, 274.

FORBUSH, S. E.: 1954, *J. Geophys. Res.* **59**, 525.

GRINGAUZ, K. E., BEZRUVKIH, V. V., OZEROV, V. D., and RYBČINSKIJ, R. E.: 1960, *Soviet Phys. – Doklady* **5**, 361.

HEPPNER, J. P., STOLARIK, J. D., SHAPIRO, I. R., and CAIN, J. C.: 1960, *Space Research* 1 (ed. by H. KALLMANN-BIJL), p. 982.

LÜST, R., MEYER, F., TREFFTZ, E., and BIERMANN, L.: 1963, *The Solar Corona* (ed. by J. W. EVANS), Academic Press, New York, p. 21.

MacDONALD, M. J. and ROBERTS, N. O.: 1960, *J. Geophys. Res.* **65**, 529.

MITROPOLSKAJA, O. N.: 1959, *Astron. Ž.* **36**, 224.

MOGILEVSKY, E. I.: 1963, *The Solar Corona* (ed. by J. W. EVANS), Academic Press, New York, pp. 271–280.

MORRISON, P.: 1956, *Phys. Rev.* **101**, 1397.

MUSTEL', È. R.: 1959, *Astron. Ž.* **36**, 215.

MUSTEL', È. R.: 1963, *The Solar Corona* (ed. by J. W. EVANS), Academic Press, New York, pp. 255–269.

MUSTEL', È. R. and MITROPOLSKAJA, O. N.: 1959, *Observatory* **79**, 15.

NEUGEBAUER, M. and SNYDER, C. W.: 1962, *Science* **138**, 1095.

NOBLE, L. M. and SCARF, F. L.: 1963, *Astrophys. J.* **138**, 1169.

OSTERBROCK, D. A.: 1961, *Astrophys. J.* **134**, 347.

PARKER, E. N.: 1958a, *Astrophys. J.* **128**, 664.

PARKER, E. N.: 1958b, *Phys. Rev.* **109**, 1874.

PARKER, E. N.: 1960, *Astrophys. J.* **132**, 821.

PARKER, E. N.: 1961a, *Astrophys. J.* **133**, 1014.

PARKER, E. N.: 1961b, *Astrophys. J.* **134**, 20.

PARKER, E. N.: 1962, *Space Sci. Rev.* **1**, 62.

PARKER, E. N.: 1963, *Interplanetary Dynamical Processes,* Interscience Publishers, New York.

PARKER, E. N.: 1964, *Astrophys. J.* **139**, 72, 93, 690.

SARABHAI, V.: 1963, *J. Geophys. Res.* **68**, 1555.

ŠKLOVSKIJ, I. S., MOROZ, V. I., and KURT, V. G.: 1960, *Astron. Ž.* **37**, 931.

SCHLÜTER, A.: 1950, *Z. Naturforsch.* **5a**, 72.

SIMPSON, J. A.: 1951, *Phys. Rev.* **81**, 895; **83**, 1175.

SIMPSON, J. A.: 1954, *Phys. Rev.* **94**, 426.

SIMPSON, J. A.: 1960, *Astrophys. J. Suppl.* **4**, 378.

SONETT, C. P., JUDGE, D. L., and KELSO, J. M.: 1959, *J. Geophys. Res.* **64**, 941.

SNYDER, C. W.: 1963, private communication.

STÖRMER, C.: 1955, *The Polar Aurora,* Clarendon Press, Oxford.

STUMPFF, P.: 1956, *Astron. Nachr.* **286**, 87.

VAN DE HULST, H. C.: 1953, *The Sun* (ed. by G. P. KUIPER), University of Chicago Press, Chicago, Ill.

VSEHSVJATSKIJ, S. K.: 1963, *The Solar Corona* (ed. by J. W. EVANS), Academic Press, New York, pp. 271–280.

WHITAKER, W. A.: 1963, *Astrophys. J.* **137**, 914.

WURM, K.: 1943, *Mitt. Sternw. Hamburg-Bergedorf* **8**, No. 51.

Discussion and Questions

L. Tonks: In the hydrodynamic problem the temperature is not a free variable, the temperature is determined by the conditions in the duct etc. Now it seems to me that there is some condition left out. You need another equation here. I was wondering what the physics of this situation is.

E. Parker: The equation which you mention is called the energy equation. There are several possible forms of the energy equation but there is no way of telling just what processes are determining the temperature. So the procedure I have followed is to take what little is known from observations concerning the temperature $T(r)$ and plug in $T(r)$ as an observed function of radial distance. I would give my right arm to know what the energy equation is so we can compute $T(r)$.

L. Tonks: Well, close in to the corona wouldn't it be the heat transfer that would probably dominate? Then outside it would probably be something else.

E. Parker: Well, we have solved the problem analytically for the hypothetical case that heat is generated only at the base of the corona and transported outward only by conduction. Now this is not necessarily the correct solution because who knows how much wave dissipation there may be far out in the corona. But in any case, the solutions are just about what I have illustrated for you here. The temperature declines first as $r^{-4/7}$ and then flattens off to a complicated function. However the point I want to emphasize is that whatever you seem to do with the temperature, whether you assume it is uniform, or like a heat flow equation, or even increasing due to wave dissipation, the velocities you obtain are of the order of the 500 km/sec that I showed on the board and the behavior of the hydrodynamic functions follow qualitatively the sketches I have shown. I think we can say at this point that we simply don't know what the temperature is and what determines it. This means that the problem of heating the corona is still a very fundamental problem and a lot has to be done on it before we can be more definitive about the temperature.

A. Kantrowitz (AVCO): I'd like to draw another phenomena which may have an analogue. Namely, that if you do have a finite back pressure, there will be a normal shock. Perhaps we can make a guess as to the position of such a shock wave out from the sun. Then in that case I may point out that the radial function will not be critically dependent upon the pressure but some small power.

E. Parker: Clauser pointed this out in 1960 and since that time we have made some calculations of models where you fit the shock in between the boundary and the solutions of the hydrodynamic equation. Rather than go into any detail let me just point out how the numbers come out. The flow will go outward into space becoming more and more tenuous until finally it reaches a point where the interstellar pressure is equal to the stagnation pressure. And it is a question simply of estimating how far out that will occur. The velocity of flow of the solar wind is practically constant. The only thing that happens is that the density drops off something like $1/r^2$. If you now suppose that the interstellar pressure is the pressure of a 1 gamma interstellar field, and if you take four particles per cc as the density at the orbit of the earth and 500 km/sec as the velocity, you find that the shock should occur somewhere in the vicinity of 70 astronomical units. I would say the uncertainty in that number is probably a factor of 5 or less. The point is that the shock in the wind is far beyond the orbit of Earth. The shock transition at 70 AU should heat the gas tremendously because of the 500 km/sec velocity and there are a lot of interesting things which can occur.

S. Bowyer: You mentioned that there is no reason to assume the flow is radial. This morning Dr. Snyder gave the indication that he thought it was radial.

E. Parker: It is just a question of emphasis by what we mean radial or not quite radial. It surely must be approximately radial. What I had in mind though was the fact that, if for instance the corona near the equator of the sun is hotter than it is over the poles, and there seems to be some indication that this is the case, then you would expect the equatorial flow tube to spread out like this from the equatorial plane, which would pull the density down faster than simple radial flow. This doesn't affect the velocity very much but it certainly affects the density. I agree that the flow must be approximately radial, exactly how radial I don't know. I expect at some times it may be far enough from radial as to give some trouble in the plasma measurements.

Unidentified: In some earlier work, I believe you were associated with it, Simpson postulated the mixing up of magnetic fields somewhere outside the earth's orbit between major and minor planets. Can you say how the solar wind may or may not be affected by these mixing fields?

E. Parker: I can only give you several possibilities. There are many things in the solar wind suggesting that it may be turbulent and gusty. Observations of Dr. Snyder shown earlier this morning showed that it is certainly gusty. Let me recite some of the possible instabilities. There are the mirror and hose instability which tend to keep the pressure nearly isotropic, but in maintaining the pressure isotropic they do put waves and wiggles in the magnetic field. There is a distinct possibility that the corona is rather filamentary particularly at the sun. Many of the Soviet workers have emphasized this important point. If that is the case the flow along each streamer will be as I have indicated here according to Bernouilli's equation. But if neighboring streams have different velocities, you get other instabilities and all I can say is that this will be a problem that will have to be solved entirely *ad hoc* from observations. You can write down the various theoretical possibilities but you have to wait for observations to tell which are important. You have to know the temperature precisely, the field, the density, etc.

C. Snyder: On this question of the direction of the plasma flow, I just want to comment that our plans for Mariner III which is of course a year away are to have a plasma probe on board which will be able to measure the flow direction. Furthermore we will have a wider angle of acceptance. So I think this problem will be pretty well settled then if someone doesn't do it before then.

Unidentified: May I elaborate a little more on my question. Some of the indications are that Saturn or Jupiter respond to activity on the sun as if particles go directly there or with very little encumbrance, so that there is probably no region where they are slowed down. Yet in the description of solar flares, the Chicago group said they behaved as though it were a resonant cavity, in which the cosmic ray particles emitted by the flare were trapped and behaved as if they were diffusing out through a magnetic field region.

E. Parker: We have to distinguish between the particles which make up the solar wind and what you call the cosmic ray particles. I firmly believe, and as Prof. Biermann showed from some cometary evidence, that the solar wind moves unencumbered to Jupiter carrying with it any and all magnetic fields. So there is a lot of tangled field extending out to Jupiter. That field is moving out at something like 500 km/sec. Now the cosmic ray particles move at somewhere near the velocity of light so it hardly sees this 500 km/sec velocity. It would see a nearly static barrier even though that barrier were moving unencumbered in the solar wind. Do I make myself clear on this?

H. Zirin: I think your point about the structure of the corona outside an eclipse shows that it is very irregular. The densities above active regions are perhaps 10 times what they are outside. In fact when you get down toward minimum there are very few active regions and you could scarcely observe the corona at all. In fact there is even the possibility that if the corona is connected with the chromospheric network that you have a very fine structure of 30000 km in diameter. So it becomes very difficult as you said to disentangle the spherically symmetric corona from the disordered corona which is a creature of solar activity.

E. Parker: I am not sure that I understood your remark about the spherically symmetric corona.

H. Zirin: I meant that there isn't any such thing.

E. Parker: Yes. The slide I had shown demonstrated this. That's why fortunately one can integrate the equations of motion along each particular flow line without much regard for the neighboring flow lines. I think the best point of view at the present time, is to average over several streamers and figure average values of the flow. We have looked at a few idealized models of streamers and, as I mentioned, the results are qualitatively the same as with a homogeneous corona.

SOLAR PROTON EXPERIMENTS

K. W. OGILVIE

Space Sciences Division, Goddard Space Flight Center, Greenbelt, Mld.

Abstract. This paper discusses some of the experiments which have been made upon solar particles since 1960. Present knowledge of the abundance of the solar cosmic rays is treated, but the main emphasis is upon the energy region between 1 and 30 MeV. The significance of the exponential rigidity form for the spectrum of solar protons is discussed, and some new results from rocket observation of the November 1960 events are presented. Observations of increases in the solar proton flux at the time of the sudden commencement are discussed, and observations of cut off changes by Injun I are briefly treated.

Introduction

Since August 1960 a good deal of progress has been made in the study of solar particle events. This has largely been made possible by the increased use of rockets and satellites. It is quite clear, for instance, that the occurrence of such events is quite common; particles are probably accelerated in every major flare, and are able to be observed by an instrument with adequate sensitivity and in the proper position with respect to the sun. This in itself is important, since the acceleration mechanism now has to satisfy an additional boundary condition; namely, that is a usual rather than a rare phenomenon.

In particular I should like to touch on the following points:

(1) Elucidation of the charge spectra of the particles in several events.

(2) Increasing knowledge of the form of the particle spectra.

(3) The sudden commencement increase effect, in which a high intensity of low energy particles suddenly arrives at the time of the sudden commencement of the magnetic storm, following a flare which produces observable particles.

(4) The use of solar protons to study the changes in the cut off rigidities of the earth's field occurring at the time of the main phase of the magnetic storm.

Very little progress known to me has been made with the problem of the acceleration of particles in solar flares, which is intractable theoretically and very difficult experimentally.

1. The Charge Spectrum of Solar Cosmic Rays

One important area in which real progress has been made lately is the determination of the charge spectrum of solar cosmic rays.

Using emulsion techniques, the Minnesota group has measured the P/α ratio for several events (FREIER, 1963). These emulsion measurements by balloon are confined to the rigidity interval 0.8–1.4 BV. In their paper on exponential rigidity spectra

FREIER and WEBBER, 1963) they give a table of

$$\frac{(dJ/dP)_0^p}{(dJ/dP)_0^\alpha}$$

which represents the ratio of solar protons to solar α-particles at constant rigidity if we assume an exponential rigidity spectrum.

TABLE I

PROTON TO α-PARTICLE RATIOS IN THE SAME RIGIDITY INTERVAL
FOR SOME EVENTS

(after FREIER)

Event	P/α
Feb. 23, 1956	1.4 ± 0.5
May 10, 1959	1.0 ± 0.5
July 10, 1959	1.4 ± 0.3, 5 ± 1
July 14, 1959	1.0 ± 0.5
July 16, 1959	1.0 ± 0.5
May 4, 1960	⩾ 50
Sept. 3, 1960	40, 30
Nov. 12, 1960	1.0 ± 0.5
Nov. 15, 1960	1.0 ± 0.5, 2 ± 1
July 18, 1961	6 ± 1, 6 ± 1

It will be seen (Table I), that these ratios are often of the order of unity, but are sometimes anomalously large (September 3, 1960). They find that the value of P_0 in the equation

$$dJ/dP = (dJ/dP)_0 \exp(- P/P_0)$$

is the same for the α-particles and protons in the same event.

An extensive series of measurements using rocket borne emulsion detectors was made by BISWAS et al., 1962 at various times during the November 1960 events. The abundances of hydrogen, helium, carbon, nitrogen, oxygen, neon and larger nuclei were measured, as well as an upper limit for light nuclei. The magnitude of the event allowed good measurements to be made, and the charge resolution for medium nuclei was found to be 0.27 of a charge. Very careful corrections were made for the form of the trajectory, particles approaching the emulsion from below, etc.

The integral and differential energy spectra of solar protons in the kinetic energy interval from 14.5 to a few hundred MeV were obtained. Concentrating first upon the form of the spectrum the authors found it possible to fit a power law in total energy above 35 MeV, but with a change of slope at that energy (Figure 1).

The spectrum of α particles could also be expressed as a power law in total energy, and the flux of α-particles at 100 MeV/nucleon agreed with that obtained on November 13 by NEY and STEIN (1962). The charge spectrum shows interesting features (see Table II).

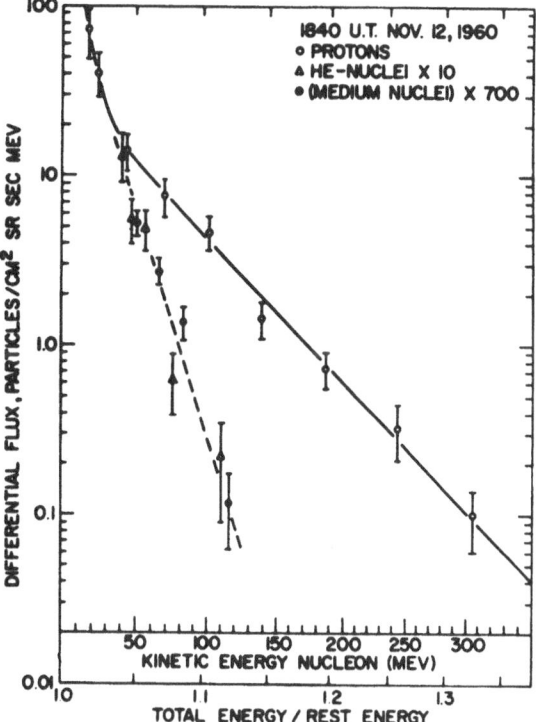

Fig. 1.   Differential energy spectrum of protons, α-particles, and medium nuclei for a rocket flight in the November 12, 1960 events (FICHTEL).

TABLE II

RELATIVE ABUNDANCES OF NUCLEI IN SOLAR COSMIC RAYS, THE SUN, AND GALACTIC COSMIC RAYS WITH A BASE OF 10 FOR OXYGEN

Nuclei	He	Be, B	C	N	O	F	Ne	11NA–18^A	19^K–28^{Ni}
Solar Cosmic Rays*	1100	< 0.2	6	≲ 2	10	< 0.3	1.5	1.3	≲ 0.5
Sun**	?	< 0.01	6	1	10	< 0.01	?	1	0.007
Galactic Cosmic Rays*	360	11	18	≲ 8	10	< 1	2.9	7.3	3.6

* The uncertainty in the values in this line varies from 10 to about 30%.
** The uncertainty in the values in this line is of the order of a factor of 1.5 to 2.

(1)   One of the striking characteristics of the charge spectrum of galactic cosmic rays is the high relative abundance of heavy nuclei. In this study the ratio of particles with $11 \leqslant Z \leqslant 18$ to the medium nuclei in solar cosmic rays was found to be 0.08 ± 0.03 compared to 0.35 ± 0.10 for galactic cosmic rays extrapolated back to their source.

(2)   There was no positive evidence for light nuclei. This is similar to the situation in the sun, but not to galactic cosmic rays.

K. W. OGILVIE

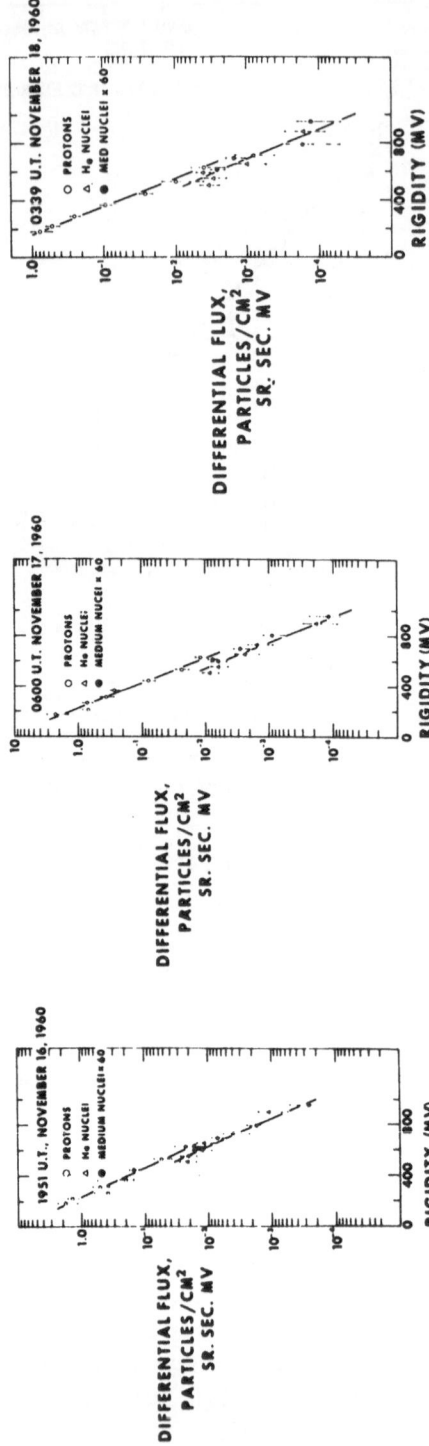

Fig. 2.   Differential exponential rigidity spectrum – November 15 event (FICHTEL).

(3) In the medium group O is more abundant than C contrary to the situation in cosmic rays and similar to spectroscopic measurements on the sun, while N is less abundant than C or O.

(4) The energy spectra of the He nuclei and the medium nuclei had the same slope, much steeper than that of the proton component. The rigidity spectra were similar, showing clearly that rigidity is the preferred parameter here. Since the He and medium nuclei have the same energy/nucleon spectrum, the He/M ratio is not a function of energy and is found to be about 5 times larger than for galactic cosmic rays. The He/P and M/P ratios are functions of energy; rapidly decreasing with increasing energy.

Although the energy/nucleon spectra for protons and multiply changed particles were different at different times, the relative abundance of protons with respect to the others remained very constant.

In the later firings made into the November 15 events the result were generally similar. An exponential rigidity spectrum could be successfully fitted to the results (Figure 2) and the parameter P_0 found to be independent of the atomic weight but in the energy range covered this is not a particularly stringent test.

To summarize this work, the medium and heavy nuclei, when compared in the same velocity interval, show relative abundances which are characteristic of the sun and not of galactic cosmic rays, and this result is obtained at five different trials in two different events. The He nuclei have the same energy/nucleon spectra as the medium nuclei, indicating perhaps that all these groups had the same velocity spectra at the source. The different velocity spectrum for protons supports the idea that the medium and heavy particles were fully ionized at the source.

One must emphasize that two of the differences between the relative abundances of solar and galactic cosmic rays i.e. the M/H ratio and He/M ratio would be enhanced by fragmentation. Thus it appears that "solar" particles cannot be the sole source of galactic cosmic rays.

In these experiments on the change spectra of solar cosmic rays the flux of deuterons and tritons was certainly less than 1%, and there was no positive evidence for any at all.

TILLES, DE FELICI and FIREMAN (1900) have studied the quantity of tritium by analysis of parts of the discoverer satellites and also of rockets flown during the November 1960 events. The amount found is hard to account for by the effects of reactions caused by cosmic rays, trapped particles, or of solar protons or α-particles. The abundance ascribed to the solar stream is very uncertain, but they consider the most likely source to be the He$^4(P, 2P)$H^3 reaction in the acceleration region. If this is true, as the authors point out, this quantity is another source of information about the properties of the acceleration region.

Some evidence has been obtained about the presence of electrons in the solar stream. On September 4, 1960, EARL (1962) flew a cloud chamber containing lead plates for several hours above Minneapolis during a part of the September 3, 1960 event. Apart from measurements of the spectrum, and its changes due to changes in the geomagnetic cutoff, he made a search for electrons. He found no electrons between 450 MeV and 800 MeV, while observing some 7000 protons due to the solar flare with momentum above 450 MeV/C. The abundance of electrons in the solar beam was thus much less than the corresponding figure for galactic cosmic rays ($\sim 1\%$) in similar rigidity intervals. Thus if the sun accelerates electrons to a high energy, much higher than is required merely to give the required radio emission, at least they do not always get away from the sun. Although the September 3 1960 event was not a typical one, this is rather likely, especially at higher energies, since the radiation loss of a relativistic electron is proportional to $E^2 B^2$. MEYER and VOGT (1961) also studied the September 3, 1960 solar particle event with a balloon borne apparatus flown on September 8 and 15, 1960. They conclude that electrons were not stored in interplanetary space in a similar way to the accelerated protons.

2. Solar Particle Spectra

Historically it is largely because of its ease of manipulation that a power law was introduced to describe the energy spectrum of solar protons. For example, ANDERSON et al. (1959) introduced a differential spectrum of the form $(K/E^n)\,dE$ to deduce the proton spectra from the absorption curves obtained during the ascent of balloons in the August 1958 events.

Some theoretical support for this form of spectrum was thought to come from the fact that a Fermi acceleration mechanism leads to a power law in (total energy/ rest energy). At energies above about 100 MeV and over a restricted energy range a constant exponent can usually found for experiments of moderate accuracy. However, it became clear from radio (BAILEY, 1900) and low energy (OGILVIE et al., 1962) measurements that in general a power law spectrum should not be used with a constant exponent n to estimate the flux at a low energy (of the order of a few MeV) from the knowledge of the flux at higher energies.

It is of considerable interest to know the form of the spectrum of solar particles as a function of energy and of charge. If one has this and can extrapolate back to the source spectra, it may enable us to eliminate some of the possible acceleration mechanisms.

FREIER and WEBBER (1963) have recently shown that the spectra of solar particles and protons may often be very well represented over the rigidity range shown in Figure 3 by an exponential rigidity law of the form

$$dJ/dP = (dJ/dP)_0 \exp(- P/P_0),$$

where P is the rigidity, or momentum per unit charge. Let us consider this spectrum.

The differential and integral spectra now have the same functional form and $J = J_0 e^{-P/P_0}$, where $J_0 = P_0\,(dJ/dP)_0$. The quantity rigidity is used because of the observations that α-particles, protons and in one case medium nuclei, have the same relative rigidity spectra, whereas their energy spectra are not the same. The rigidity of a particle, its momentum per unit charge, is proportional to its radius of curvature in a magnetic field, and in a constant field particles with the same rigidity follow the same paths.

To the extent to which this picture is accurate the characteristic rigidity P_0 (t) is a function of time only and its changes reflect the changes in shape of the spectrum. The exponential rigidity spectra are not applicable in general if t, the time between

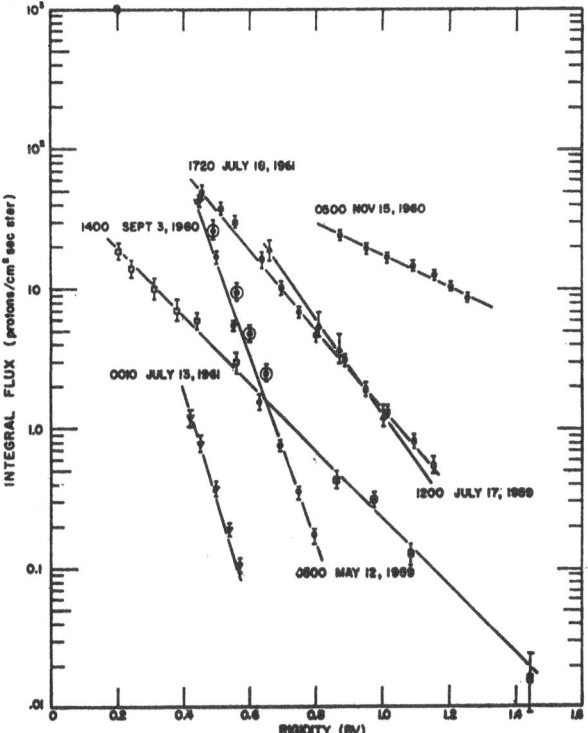

Fig. 3. Integral proton spectrums are shown as exponentials in rigidity at selected times for six different solar flares. Data points taken from counter ascents are shown as solid symbols; those taken with emulsions as open symbols (WEBBER).

observation and the time of the flare, is less than a few times t_D, which is the time taken for particles to reach the earth from the sun at a particular rigidity. Thus one might expect deviations from the exponential form at low rigidities soon after the flare, even if this form fits the observations at higher rigidities.

The exponential form predicts a definite intensity, J_0, of particles above $P=0$, and the question is then raised whether one can relate this quantity to the magnitude of the other geophysical phenomena in the same event, such as the geomagnetic storm.

Figure 3 shows the data from several events plotted as integral exponential rigidity spectra by Webber and Freier. The good agreement will be noted, especially the September 3, 1960 event, for which measurements at low energy were available from rockets (DAVIS *et al.*, 1961). As a check on the assumption of spectral shape, the authors calculate the riometer absorption predicted, and make a comparison, some-what hampered by the presence of cut off changes, with the observed absorption. The spectrum deduced for 0500 November 15, 1960 is remarkable for its flatness, $P_0 = 350$ mV, but nevertheless gives fair agreement when subjected to this test of comparison between calculated and observed riometer readings.

Since an α-particle has twice the mass to charge ratio of a proton, the obser-vation (Figure 4) that the same value of P_0 describes the differential spectra of both particles and protons in at least some events is an indication that this form of spectrum has some fundamental importance. The dotted portions of the curve show the deviations caused by cut off changes.

However one must consider carefully what is meant by J_0. The differential response of a riometer begins to drop off below 20 MeV (195 mV) and falls precipitously below 10 MeV. (As an example consider the values of J_0 given in Figure 4). At energies of the order of 30 MeV, particles move from the sun to the earth along individual orbits. At energies of the order of a few KeV they are characteristic of the magnetic storm particles and move with the body of the plasma. The intervening

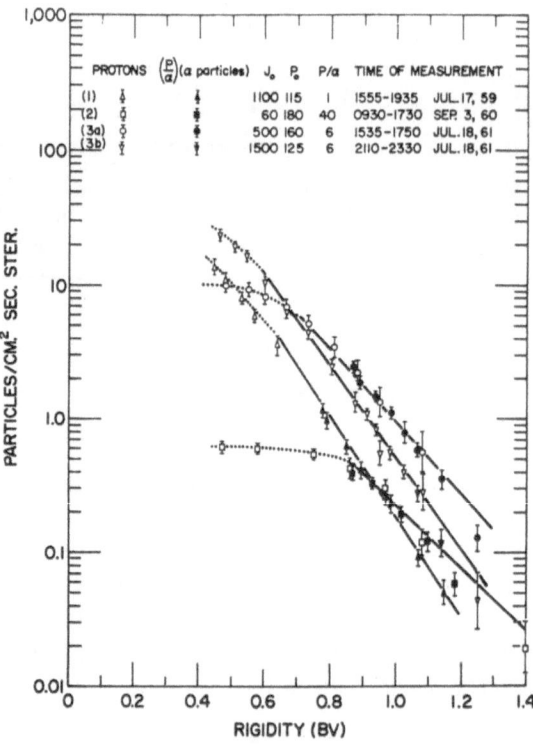

Fig. 4. The integral flux of solar protons and solar α-particles multiplied by the (P/α) ratio, plotted as a function of rigidity for three of the flares shown in Figure 1. The dotted parts show the effects of the geomagnetic cutoff (WEBBER).

region is one of great interest and the value of J_0 lies in this region. The energy density of the particles in this region is such that their motion is controlled by the inter-planetary field, and it is qualitatively correct to say that events with the largest intensity of low energy particles are associated with the largest magnetic storms. Webber and Freier put forward the idea that the solar cosmic ray spectrum forms a high energy tail on the distribution of storm particles.

In this connection D. A. BRYANT and the author have recalculated the results obtained (OGILVIE et al., 1962) by means of the rocket shots made by NASA at the time of the November 12, 1960 event, to see what sort of a fit is given by an exponential rigidity spectrum.

The instrumentation of the rocket flights into the November 1960 events (OGILVIE et al., 1962), consisted of a geiger counter, a scintillator with a Cs I crystal, sensitive to protons in the energy range 2 MeV to 160 MeV, and a ZnS phosphor scintillator sensitive to protons with energies between 200 KeV and 2 MeV. It is mostly the results of the Cs I detector which we shall discuss. A pulse height analysis was per-formed upon the pulses from this counter by altering the load of the phototube using a motor driven switch, followed by an amplifier with a fixed trigger level. This system has been fully described before, but it is necessary to point out that since such a system introduces overlapping energy intervals with none of the end points common to all levels, the appropriate method of analysis is to assume a spectral form and compute the ratios of the rates in each interval for a range of one or more parameters and then compare the computed and observed ratios, Figure 5. Since in this case we have four levels, three ratios are defined, and to get a meaningful result we can put in two parameters only.

For the recalculation of the power law spectrum we have used n, the exponent in a power law, $N = N_0 E^{-n}$, and a cut off energy E_c, and for the rigidity analsyis we have used P_0 the characteristic rigidity and Wp the proportion of α particles to protons in a given rigidity interval. Figure 6 shows the Ottawa neutron monitor counting rate during the November 1960 events, together with the firing times of the first three rockets and other relevant details of the events. The rockets fired into the November 15 event were between 41 and 91 hours after the flare (Table III).

The observations on November 12 all took place during the magnetically active part of this very complicated event. A class 3 flare at 1000 UT on the November 10 which had no Type-IV radio emission and gave rise to no detectable particles pro-duced the sc at 1844 UT on November 12, and this occurred during the arrival at

TABLE III

Rocket	Δt (time after flare)	P_0	J_0
1013	41 hrs 46 m.	75 ± 5 mV	1230 ± 350
1014	51 hrs 53 m.	65 ± 5	1600 ± 500
1026	73 hrs 32 m.	75 ± 5	410 ± 60
1027	97 hrs 32 m.	95 ± 5	55 ± 15

the earth of particles emitted by the flare of 1320 UT on the 12th. Good fits could be found to power law spectra for the results of the three rockets fired into the November 12 event, and the first of these is illustrated in Figure 7.

The November 15 event firings occurred at times of magnetic quiet, between 41

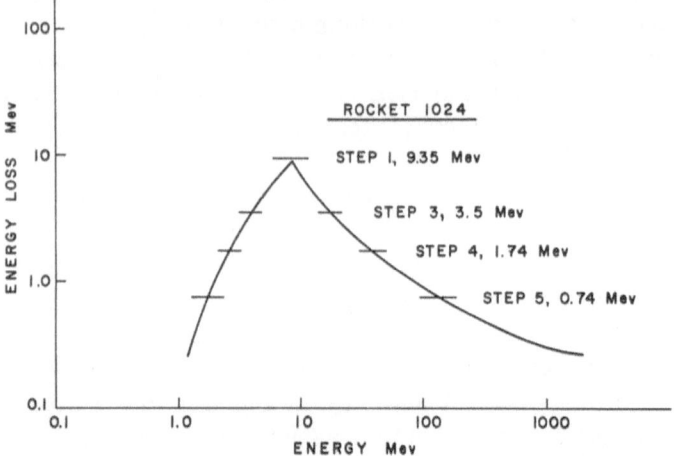

Fig. 5. This figure illustrates the energy intervals which correspond to the pulse height analysis carried out during the rocket flights.

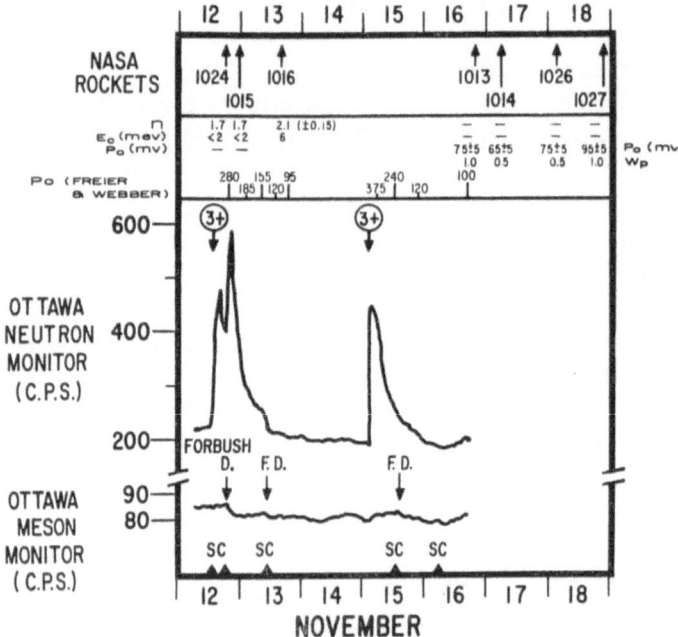

Fig. 6. A diagram showing the times of firing of rockets during the November 1961 event. Also shown are the neutron and meson monitor rates, and parameters determined both by Ogilvie and Bryant and by Freier and Webber.

and 91 hours after the flare. An exponential rigidity spectrum provides a good fit to the observed data, and the details of the parameters used are shown in Table III. In the case of these results Wp, the ratio of α-particles to protons in the same rigidity interval, was found to be unity.

We conclude that the extrapolation of spectra measured above energies of the order of 15 MeV down to zero rigidity defines values of J_0 which, although useful parameters for the description of the spectra at higher rigidities, should not be taken as necessarily representing the intensity of the lowest rigidity flare particles. At some times and in some events the intensity inferred in this way will be much too low to represent the real situation.

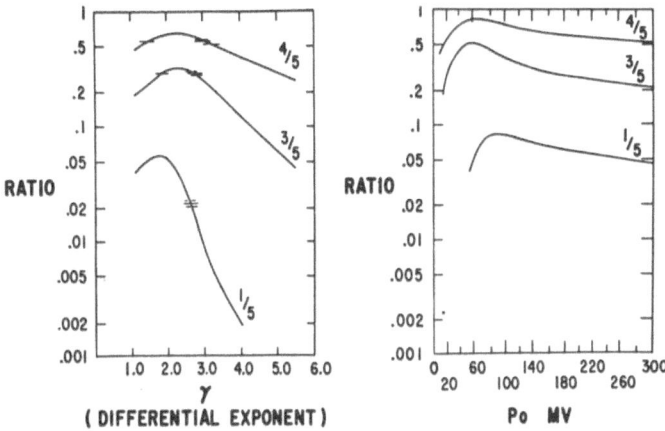

Fig. 7.   This figure shows the calculated ratios as a function of power law exponent and P_0, with the observed values for the rocket 1024 plotted as horizontal lines. These should cut each calculated curve at the same value of γ or P_0 to indicate a good fit. Note the behavior of the ratio 1/5.

3. Experimental Progress with Absorption Intensity at the Time of the Sudden Commencement

A very striking effect was noticed and described by Bryant et al. in analyzing the Explorer XII observations of the solar storm following the flare of September 28, 1961 (BRYANT et al., 1962). In Figure 8 we see the profile of this event in various energy ranges. Of the rise to the maximum and the first part of the decay, we shall have nothing to say here. However the cloud of solar plasma reached the earth late on the 30th, at 2108 UT producing a Forbush decrease and sudden commencement magnetic storm. At 1930 UT, before the sudden commencement, the lower energy solar protons stopped decreasing in intensity and a new rapid increase took place. In the 9–21 MeV energy range the magnitude of this increase was a factor of ten larger than the original increase on the 28th, but no increase was noted in the intensity of protons in the 200–300 MeV energy range. As shown in Figure 9, both the intensity of 9–21 MeV particles and the >5 MeV particles showed a series of well correlated and rapid fluctuations, lasting for about three hours. BRYANT et al. (1962) have shown

K. W. OGILVIE

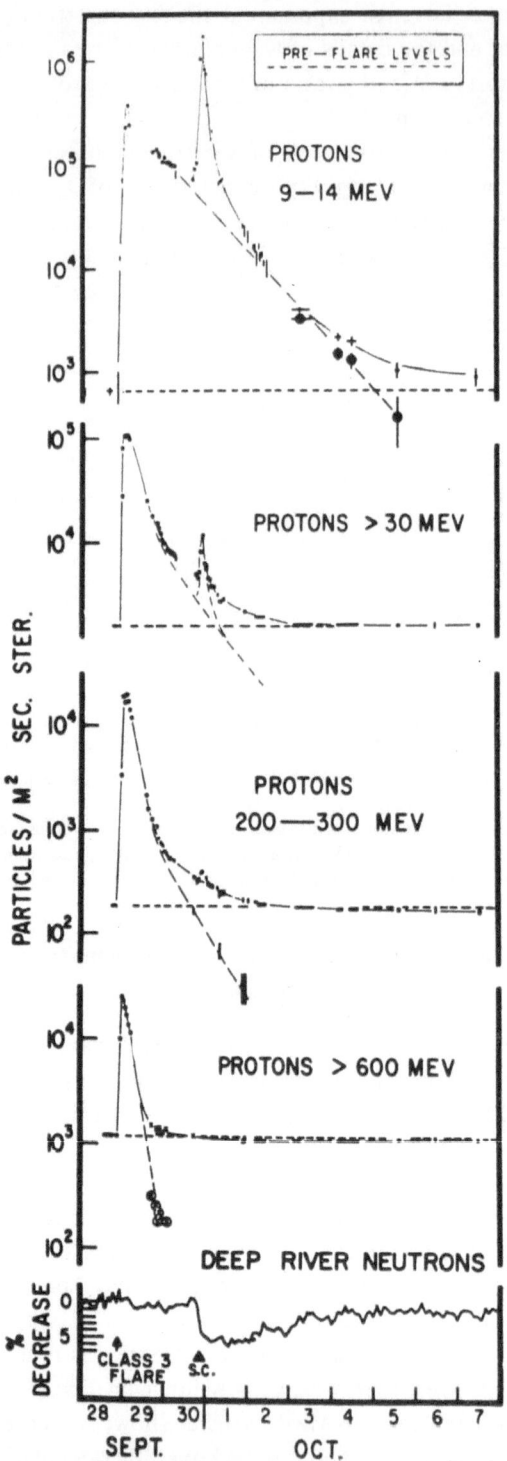

Fig. 8.   The time profile of the September 30, 1961 event in various energy intervals (Bryant *et al.*).

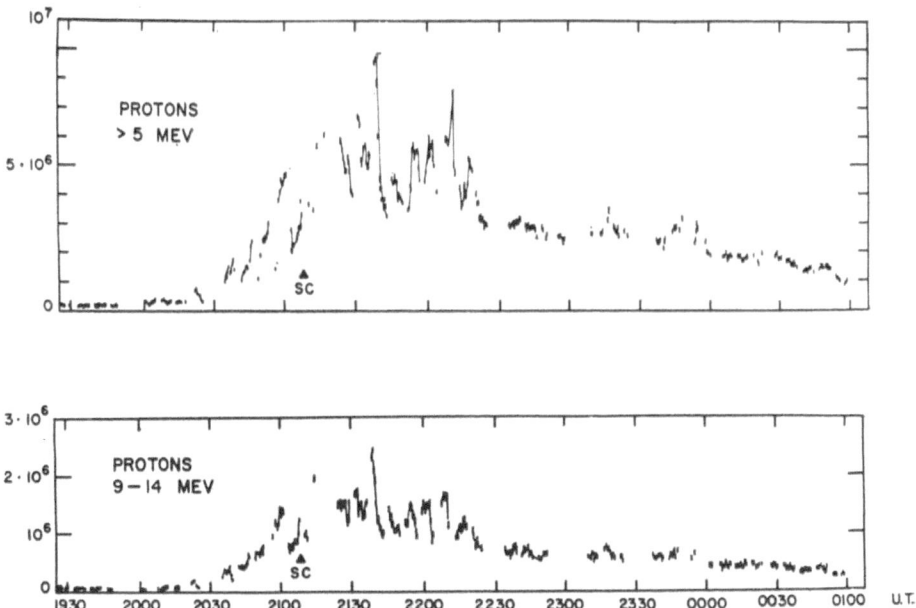

Fig. 9.   Time profile of the sc increase on a larger scale, showing fluctuations
(Bryant *et al.*).

that this effect is partly due to spatial anisotropy, and partly due to flux change.

An increase at about this time has also been reported by Hoffman, Davis and Williamson as a result of measurements carried out with the ion-electron detector on Explorer XII. The maximum flux was about 2×10^5 protons/cm sec ster above 140 KeV.

This behavior is not a usual feature of solar particle events, in fact only two other events February 11, 1958, and May 8, 1960 can be considered as certainly of the same type. These are reported by AXFORD and REID (1963).

At this point it is important to make a distinction between this type of event which Axford and Reid refer to as pre sc enhancement, and ones which show a small increase in riometer absorption at stations located in the auroral zone at the time of the sudden commencement. In a long paper ORTNER *et al.* (1962) have discussed cosmic noise absorption at the time of sudden commencement of magnetic storms. The sudden onset of absorption of about 4 db at stations close to the auroral zone on both the day and night sides of the earth is called SCA. Data from 71 SCA's were studied. No increase was observed on riometers at Resolute Bay or King Salmon during this search, so this increase effect appears to be confined to the stations close to the auroral zone. Included in the period are sixteen instances when polar cap absorption was in progress at the time of the sc.

ORTNER *et al.* (1962) put down this effect to the ionization from electrons which they assume precipitated into the atmosphere of the earth along the boundary between the magnetosphere and the solar plasma, but point out that an alternative explanation, involving dumping from the radiation belt is also possible, but unlikely. The general picture of electron bombardment is supported by the observations by BROWN *et al.* (1961) of an X-ray burst at balloon altitude coincident with the sudden commencement on June 27, 1960.

It must also be pointed out that yet another effect, a decrease in absorption at the time of a sc, has been observed (HULTQVIST *et al.*, 1959). This is thought to be due to 'contraction of the polar cap', essentially a cut off increase, at the time of the initial phase of a geomagnetic storm.

We can state then that the effect we are considering here, an increase in the flux of solar protons at low energies at the time when the boundary of the solar plasma cloud is passing the magnetosphere, is a fairly rare occurrence. It must be emphasized that the satellite Explorer XII was outside the magnetosphere at the time of these observations.

Explanations of this effect have been offered both by BRYANT *et al.* (1962) and by AXFORD and REID (1962, 1963).

The first authors suggest that trapping between supposed regions of enhanced field at the boundary of the plasma cloud connected by regions of weaker field (Figure 10), could cause solar protons to be trapped, and so conveyed from the sun to the earth or even collected on the way, so that the observed high density at low rigidity would occur as the front crossed the point of observation.

Reid and Axford, on the other hand, discussing the storm of February 11, 1958 (Figure 11), at first suggested the following mechanism. They envisage the plasma

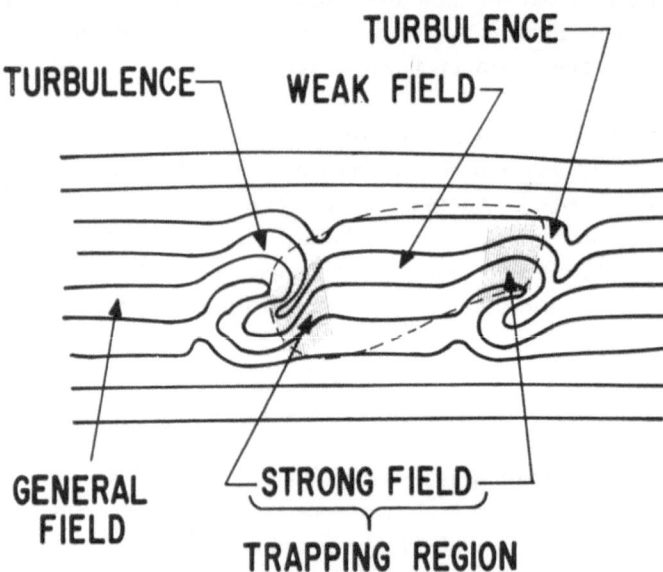

Fig. 10.   This figure shows a mechanism by which particles having energies of the order a few MeV may be conveyed from vicinity of the sun to the earth.

cloud as acting like a piston in interplanetary space. If the interplanetary field is predominately radial the passage of this shock will produce a 'kink' in the field lines. Since the gas B contains no energetic protons, we should expect the absorption to decrease rapidly following the entry of the earth into B. They suggest that the increase at the time of the sc is due to an acceleration of solar particles in the region ahead of the shock wave, the particles being reflected back and forth between the two ends of the field line at the boundary. The probability of such an effect being seen on the

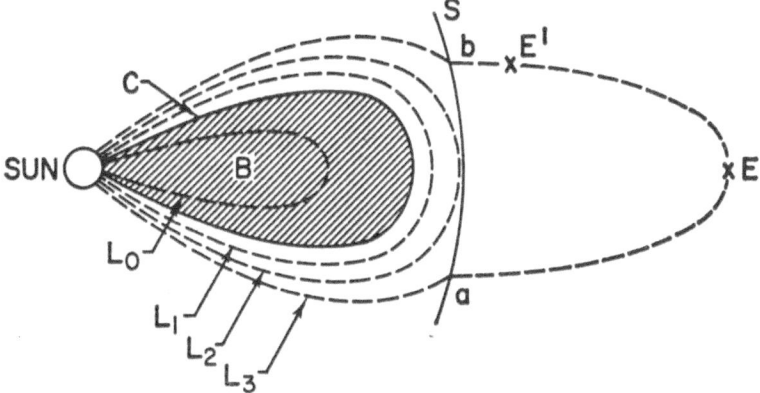

Fig. 11.   Distortion of the solar magnetic field due to the expanding bubble (B). The contact surface (C) separates the bubble from the previously existing interplanetary gas. A shock wave (S), driven by the expansion of the bubble, distorts and compresses the magnetic field in its passage but cannot affect the field ahead. (AXFORD).

earth is a crucial factor here, since any theory must predict such an acceleration in certain cases only. In this connection it must be noted that the magnetic storm of February 11 was a very intense one, whereas that of September 1961 was rather small. The corresponding shock wave on February 11 must have been appreciable for a long time, and the earth probably entered region B while the division between them was still well marked.

However, this mechanism would predict an increase beginning many hours before the sudden commencement and Reid and Axford have modified their theory. They now consider it likely that reflection of particles takes place between the moving shock wave and the standing shock in the solar wind upstream of the magnetosphere. In this event the increase takes place close to the earth and not in a large region of interplanetary space. This also limits the upper energy for which the effect can take place, in agreement with observation. The time before the sudden commencement at which this effect can begin to work is limited by particle drift to about one hour; the particle cannot spend any longer because the rotational component of the inter-planetary field sweeps it out of the accelerating system.*

* Magnetic field experiments on the satellites Explorer XII and IMP-I, by CAHILL (1963) and NESS (1964) have observed this shock as the sudden onset of a disordered region between the interplanetary and geomagnetic fields. (Added in proof.)

This mechanism, however runs into difficulties in trying to explain the absence of a Forbush decrease at these low energies.

It is not possible to make a definite decision between these two theories, since this depends to some extent on their relative improbability. In a small event when the shock wave becomes an acoustic wave before reaching the earth, acceleration on the Reid and Axford model would not be correlated with the sudden commencement, whereas the model by BRYANT *et al.* (1962) requires special field configurations which have not been observed in an unequivocal way. The Reid and Axford model seems to be amenable to a calculation of the profile of the increase as a function of the position of the earth with respect to the advancing front, and perhaps this should be done. It is possible that the event of July 13–14, 1961, is of this type; the event will be discussed later.

4. Cutoff Changes as Revealed by Solar Proton Experiments with the Satellite Injun I

The Satellite Injun I (PIEPER *et al.*, 1962) carried two silicon pn junction detectors sensitive to protons of energy between 1 and 15 MeV, and a geiger counter sensitive to protons with energy above 40 MeV. In the usual situation most of the particles registered by the pn junctions will be at the lower end of the rather large energy range studied.

In July 1961, after the 11th of the month, a period of activity started, which continued until the 28th of the month. There were six distinct storms, and Injun made 40 passes during these times. As the satellite moved toward the pole the earth's field allowed protons with progressively lower energies to reach the 1000 km altitude. The flux which is observed therefore increased to a plateau maximum corresponding to large L values. The satellite was in an orbit with inclination 67° and apogee and perigee 998 km and 881 km respectively, so it reached as far north as a region of space in which $L=12$. For the purpose of argument the authors define the cutoff as that point (L value) where the unidirectional intensity is 10/sec and thereafter rises rapidly as one goes in the direction of L increasing. The range of longitude covered was 120° to 350 °E, and so the geomagnetic latitude scale in the diagrams has a fairly well defined meaning. After the main phase of the storm had begun isotropy was normally observed, in agreement with the results of others. Determination of the spectral shape could not be undertaken.

Let us take as an example the event of July 13. Figure 12 shows the variation of proton flux with L value for various times. The prestorm passes and those on the 14th and 16th of July present exactly the appearance which one would expect; namely, the cutoff is reduced during the main phase of the geomagnetic storm from about $L=6$ to a maximum of about $L=3$, while the intensity of the solar protons then reduces monotonically with time. This is shown in Figure 13, and it is interesting to see that the cut off is not increased during the initial phase of the magnetic storm. We must notice the very rapid drop in intensity which occurred late on the 13th, and the exponential recovery. Returning now to the previous diagram (Figure 15), we see that this corresponds to the double humped curves for 1546 UT and 1734 UT on the 13th.

It is possible (RAY, 1963), that this type of cut off variation with L can be explained in terms of a model of the magnetosphere described by Ray and similar to that which

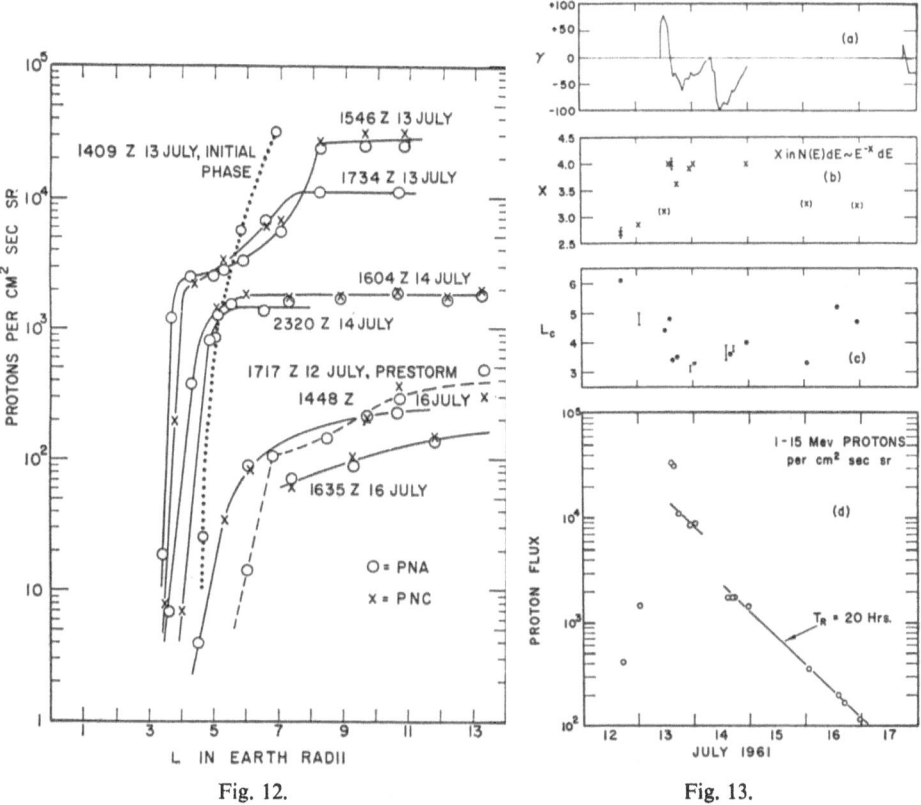

Fig. 12.   Fig. 13.

Fig. 12. *L* (magnetic shell parameter) dependence of 1- to 15-MeV solar protons observed by Injun I in relation to the storm of July 13.

Fig. 13. Analysis of storm of July 13 for 1- to 15-MeV protons. (a) Variations in the horizontal component of magnetic intensity derived from observations at four low-latitude stations, measured in gamma (10^{-5} gauss). (b) Time variation in the exponent in power-law description of solar proton spectrum. (c) Time variation in geomagnetic cut off L_e in earth radii. (d) Time dependence of plateau flux.

has been suggested by Rothwell and by Parker. Assuming the magnetic field at such a point as that occupied by the satellite to be made up of contributions due to the earth and to a ring current, in order to produce such an L value versus intensity curve it is necessary to add the assumption that beyond some distance the field becomes 'rough', that is, not dipole in character at all. This change in the character of the field is ascribed to the loading of plasma into it. However the correspondence

of the observations with this theory is not verified at the present time since the parameters in the theory have not been determined. It is an attractive view however,
since we know from the study of increases at the time of the sc that the solar plasma
sometimes contains energetic particles.

This event is possibly of the sudden commencement increase type, but one cannot
be certain of this because of the ambiguity due to the satellite Injun being inside the
magnetosphere.

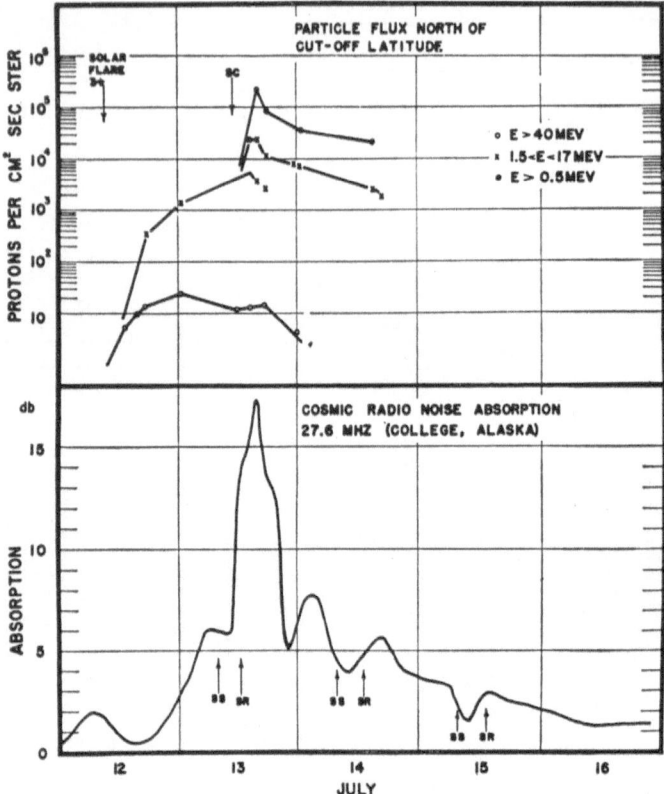

Fig. 14.   Time variation of the proton fluxes and the cosmic radio absorption during the July 13–14
event.

The apparent excess of low energy protons which one can see clearly in these
diagrams (Figure 13, 14) is associated with the solar plasma arriving at the time of the
sudden commencement storm. This again focuses attention on the relationship of
the lowest energy region of the solar protons to the solar plasma.

The storm encountered on July 18 is a peculiar one in that there was an increase
observable on the Deep River neutron monitor, and a flux at Injun I which indicates
a flat spectrum in the region below 40 MeV. Figure 15 shows the different behavior
of the flux as a function of L and time. Such large changes in the cutoff were not
observed as on July 13–14.

Changes in cut off are presumably produced by alterations in strength and position of the ring current, particles arriving from the sun augmenting it and reducing the critical L value; followed by a subsequent decay in its strength, possibly due to charge exchange (DESSLER *et al.*, 1961). In some magnetic storms the cutoff change is large and in others smaller, while in a few storms the solar plasma front contains particles with energies of order 1 MeV, as well as the low energy particles which perhaps

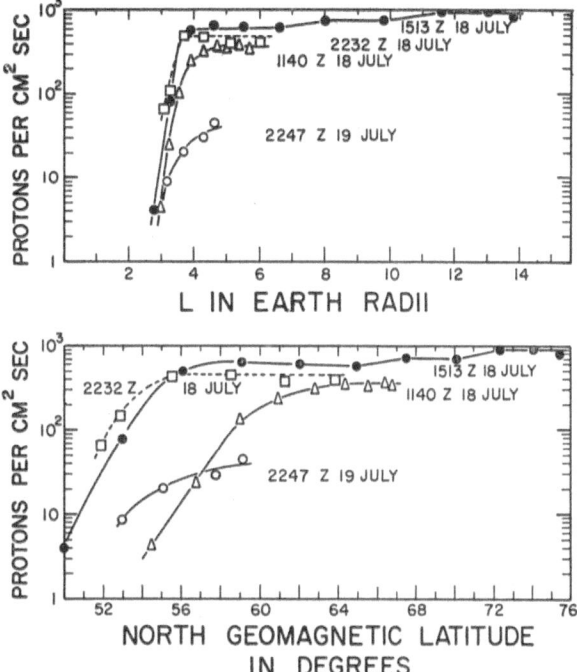

Fig. 15. Latitude and L dependence of solar protons above 40 MeV observed by Injun I in relation to the storm of July 18.

augment the ring current. Changes in cut off do not occur to an appreciable extent before the arrival of the plasma, and the intensity increase of the lower energy (~ 1 MeV) solar particles depends upon the characteristics of the plasma front.

(1) Its energy density, which determines the degree to which it can deform the earth's field, and provide acceleration if one accepts Reid and Axford's view.

(2) Its ability to trap particles (of order 1 MeV energy) within it. The energetic protons in these events exhibited an exponential decay; there is much evidence that they are not trapped within the geomagnetic field.

Note Added in Proof

Since this article was written there has been a considerable increase in our knowledge of energetic particles arriving at the earth from interplanetary space. Much of this concerns those particles having energies of the order of a few MeV which were emphasized above, and has resulted from experiments of F. B. McDonald and his co-workers, using the satellites Explorer XII and XIV.

For instance, solar particle propagation has been classified into four modes, two associated with propagation at the individual particle velocities, and two associated with propagation at the velocity of the enhanced solar plasma. The first of these is the direct way, observed by sea level monitors, but for some events particles with travel times in the range 1 to 100 hours show a strong velocity dependence in their intensity variation. These form the second class and the velocity dependent propagation in the events September 28, 1961, November 1962 and October 23, 1962 were studied in the following way.

The observed differential intensities centered about energies in the range 22 to 230 MeV were obtained as a function of time (BRYANT *et al.*, 1965), for each event. When these were plotted against the distance travelled by the particle, by multiplying the particle velocity by the time from the flare, it was found that a set of normalizing factors existed which reduced the distance variations of the various energy intervals to a common curve for each event. The distance range in question was about 100:1. The normalization factors, one for each energy interval, represent a measure of the spectrum for the corresponding event, which was found to have the form of a power law in kinetic energy. This procedure does not work for all events, but when it does there is no reason to suppose it invalid to extrapolate this spectral form back to a distance much less than 1 AU, in other words to the source of the particles or at least where they leave the influence of the sun. The reader is referred to Bryant's paper for a discussion of fluctuation phenomena observed, but it is clear that the properties of the source region and the propagation medium may sometimes be studied separately by these techniques.

Particles arriving at the earth in a manner determined by the motion of the enhanced solar wind plasma do so about two days after the flare. Such clouds are found to contain particles with energies above 1 MeV, some of which may be obtained by the acceleration of plasma particles themselves, and some derived from the solar active region.

Observations are sometimes made of recurrent events which occur when an active region passes control meridian on successive 27 day solar rotations, leading to the postulation of long lived particle streams in interplanetary space (BRYANT, 1963).

In connection with the observations of the increase in particle intensity at the time of the sudden commencement on September 30, 1961 described above, PARKER (1965) suggests that the particles were of interplanetary origin. Accelerations in the shock transition formed by the moving blast wave is postulated, by analogy with other shock acceleration processes, for example by the impact of the solar wind against the geomagnetic boundary.

These new developments again show the importance of the interesting energy region between a few keV and a few MeV.

References

ANDERSON, K. A. *et al.*: 1959, *J. Geophys. Res.* **64**, 1133–1147.
AXFORD, W. I. and REID, G. C.: 1962, *J. Geophys. Res.* **67**, 1692–1696.
AXFORD, W. I. and REID, G. C.: 1963, *J. Geophys. Res.* **68**, 1793–1802.
BAILEY, D. K.: 1962, *J. Geophys. Res.* **67**, 391.
BISWAS, S., FICHTEL, C. E., and GUSS, D. E.: 1962, *Phys. Rev.* **128**, 2756.
BISWAS, S., FICHTEL, C. E., GUSS, D. E., and WADDINGTON, C. I.: 1963, *J. Geophys. Res.* **68**, 3109.
BROWN, R. R. *et al.*: 1961, *J. Geophys. Res.* **66**, 1035–1041.
BRYANT, D. A., CLINE, T. L., DESAI, U. D., and McDONALD, F. B.: 1962, *J. Geophys. Res.* **67**, 4983.
BRYANT, D. A., CLINE, T. L., DESAI, U. D., and McDONALD, F. B.: 1963, *Phys. Rev. Letters* **11**, 144.
BRYANT, D. A., CLINE, T. L., DESAI, U. D., and McDONALD, F. B.: 1965, to appear in *Astrophys. J.*
CAHILL, L. J. and AMAZEEN, P. G.: 1963, *J. Geophys. Res.* **68**, 1835.
DAVIS, L. R., FICHTEL, C. E., GUSS, D. E., and OGILVIE, K. W.: 1961, *Phys. Rev. Letters* **6**, 492–494.
DESSLER, A. J., HANSON, W. B., and PARKER, E. N.: 1961, *J. Geophys. Res.* **66**, 3631.
EARL, J. A.: 1962, *J. Geophys. Res.* **67**, 2107–2117.
FREIER, P.: 1963, *J. Geophys. Res.* **68**, 1805–1810.
FREIER, P. S. and WEBBER, W. R.: 1963, *J. Geophys. Res.* **68**, 1605–1929.
HULTQVIST, B. J. *et al.*: 1959, *Tellus* **11**, 319–331.
MEYER, P. and VOGT, R.: 1961, *J. Geophys. Res.* **66**, 3950.
NESS, N. F., SCEARCE, C. S., and SEEK, J. B.: 1964, *J. Geophys. Res.* **69**, 3531.
NEY, E. P. and STEIN, W. A.: 1962, *J. Geophys. Res.* **67**, 2087–2105.
OGILVIE, K. W., BRYANT, D. A., and DAVIS, L. R.: 1962, *J. Geophys. Res.* **67**, 929–937.
ORTNER, J. *et al.*: 1962, *J. Geophys. Res.* **67**, 4169–4186.
PARKER, E. N.: 1965, *Phys. Rev. Letters* **14**, 55.
PIEPER, G. F. *et al.*: 1962, *J. Geophys. Res.* **67**, 4959–4981.
RAY, E. C.: 1963, Private Communication.
TILLES, D., DeFELICI, J., and FIREMAN, E. L.: 1961, *Phys. Rev.* **126**, 1935.
ZMUDA, A. J., PIEPER, G. F., and BOSTRÖM, C. O.: 1961, *J. Geophys. Res.* **66**, 3631.

ROUND TABLE DISCUSSION

Participants. J. W. Townsend (Chairman), H. Alfvén, W. E. Behring, L. Biermann, J. Dungey, F. Hayden, C. C. Lin, S. I. Pai, E. Parker, C. Snyder.

J. W. Townsend: To start this off, I think we had the beginnings of an interesting discussion between Dungey and Alfvén about whether the null points and the aurora are in any way connected. We have a bit of experimental evidence, for a change, and a bit of theoretical input. Would you care to say a few words?

J. Dungey: Well, I think I have to confess to the prejudice that I think observations in space are more useful to plasma theoreticians than experiments in the laboratory. But I think Prof. Alfvén gets a special dispensation because his experiments are so beautiful. Dr. Cladis also performed some experiments and got results that agreed with my viewpoint, so he gets a special dispensation too. In the first order approxi-

mation one could consider that particles at the neutral point move straight down into the auroral zone and thus cause aurorae. But I think that anybody who knows anything about the auroral observations knows that what is observed is much more complicated. Many of the particles will be mirrored with motions which can be gotten from the adiabatic theory, they have drift due to non-uniform magnetic field, they also have drifts due to electric fields, and for the sort of energies we have in an auroral primary these seem comparable. The general conclusions I expect to get, although I haven't calculated anything yet, is that you can get aurorae at lower latitudes than you would have by just following a line down from the neutral point. I don't think you can get it to higher latitudes.

H. Alfvén: I should like to show a slide (Figure 3 on page 154) that is very similar to the picture Dr. Dungey has calculated. (Prof. Alfvén showed a slide of an experiment by Lindberg in which a magnetized plasma was fired at a terrella. The plasma magnetic field and the terrella field were antiparallel.)

On the original photograph you see something which is extremely similar to his case. On the other hand we also seem to have precipitation of aurorae on the terrella which has nothing to do with these lines which go directly to the null point. You can have precipitation here without a null point. This is obvious from this slide, and from the slide (Figure 4 on page 154) I showed you earlier with the parallel magnetic field. I should also like to show you another slide (page 136). This is an earlier experiment in which the terrella is just sitting between two plates which give an electric field, and there is no external magnetic field. The plasma impinges on the terrella and we have here zones of precipitation; in this case there is no null point at all because there is no interplanetary field. So all these cases indicate that you can have precipitation on the terrella without any null point. In this case the plasma density is extremely low. In the other case it is much larger than the value one would want in order to reproduce the magnetospheric conditions.

J. Townsend: Prof. Alfvén, does the latitude of the precipitation seem to vary between the two types? Is the direct precipitation at more northerly latitudes than the null point precipitation?

H. Alfvén: The latitude depends on the magnetic field of the terrella and on the electric field. In the last experiments that I have shown, the electric field is produced by the two plates, and, in the experiments that I have shown earlier, by a plasma which is moving and giving a $V \times B$ field. The latitude also depends on the temperature of the plasma.

J. Heppner: Did you state what the ratio of the plasma energy density was to the energy density of the magnetic field in the case where you were shooting magnetized plasma at the terrella?

H. Alfvén: The experiment has been running only a short time and we are not sure about the properties of the plasma, but I have some data here. Far away from the dipole the magnetic pressure is of the same order as the gas pressure.

J. Heppner: The ratio is essentially unity then?

H. Alfvén: Possibly not very far from it.

D. Matthews (Defense Research Board, Canada): I have a question specially directed to Dr. Dungey and Prof. Alfvén. What are your views on the role, if any, that is being played by the shock wave postulated by Spreiter and Jones (*J. Geophys. Res.* **68** (1963) 355)?

J. Dungey: Well, if this is connected with particles, the question is what particles are observed at the boundary? At least in one case the cadmium sulfide detectors saw a large intensity of particles beyond the boundary.

Unidentified: I think there is no evidence that Explorer XII, which I think you are talking about, got out beyond the shock wave.

J. Dungey: That's not the point. Many of the trapped particles cut off sharply at the boundary. I think that the critical question is what particles are between the boundary and the shock, assuming there is a shock. An important mechanism of acceleration is the Fermi mechanism, and when we have turbulence we expect to get the Fermi mechanism. Beyond the boundary we get large amplitude waves and we might expect Fermi acceleration, but do we observe particles there?

J. Townsend: How about a suggestion or question from the floor?

C. Chang: I think Prof. Alfvén's results are greatly different from Dr. Dungey's results or Dr. Chapman's auroral conditions. They are different in a number of respects; one is that the dynamic pressure outside the geomagnetic cavity is comparable to the magnetic field pressure. However the ratio in the other cases observed is very large, about 70 or so, is it not?

N. Ness: Yes, very large in any case.

C. Chang: Therefore I think the phenomena are quite different. Secondly I am wondering about the time element. Is your's a transient or a steady state case? You see Dr. Chapman's theory is concerned with the steady state case rather than the time when the plasma impinges on the geomagnetic cavity. Tomorrow I will develop this more completely and I hope that we can then argue about this more since that is one of the reasons we are here.

J. Cladis (Lockheed Missiles and Space Company): I would like to say just one thing more about the experiments we are doing. Our plasma pulse is approximately 15 microseconds long. When the plasma hits the magnetic field the luminous front remains in that region for a long time after the plasma pulse has decayed, as though all the particles are trapped within the magnetic field. The rate at which the luminosity front moves against the magnetic field is comparable to the Alfvén speed, assuming that nearly all the plasma is trapped on the lines of force.

L. Biermann: I would like to raise a question which is connected with Prof. Snyder's talk and also with what I discussed myself. From the cometary observations we see that during the time of a storm, the direction of the solar wind may change by a number of degrees, perhaps six or so. Therefore I would like to ask Dr. Snyder whether it isn't possible that during such events the measured flux might be too low? In connection with this I have another question. You mentioned that the velocity of the vehicle gets up to 40 km/sec although most of the time it is 35 km/sec or so; i.e., roughly one tenth of the solar wind particle velocity when the solar activity is low. So this angle that you mentioned is of the order of 6 degrees. Now a report in *Science* says that there was some small angle between the axis of the vehicle and the direction toward the sun. If one combines all these factors there might be a considerable depression of the value of the flux.

C. Snyder: This is an important observation of yours, that the direction of the plasma changes during a storm. The question is what way does it change? A deflection in one direction could take the particles all the way out of our look angle, but a deflection in the other direction would partially compensate for the aberration angle and increase our mesasured current. Can you say which way it goes?

L. Biermann: The normal direction of the tail is the one given by the radial direction of the sun combined with an angle which may be considered in the first order as an aberration, and against this reference you observe fluctuations of 5 degrees or more.

C. Snyder: Does this simply mean that the velocity of the wind becomes so high that the tail becomes nearly radial?

L. Biermann: No, as the velocity becomes higher it deviates more strongly from the usual direction.

C. Snyder: I see. Well then, I can only say that it is perfectly clear that, if the plasma we are seeing is not radial, we are getting the wrong value for the flux. There is no question about that. But let me point out that we do not feel that we know the magnitude of this flux to within better than a factor of 2 or 3 in any case. From the theoretical curves I have shown on the third slide you have to get off about 4 degrees or so before the detection efficiency is down by a factor of 2. The azimuthial velocity of the probe when it started out was somewhat smaller than the azimuthial velocity of the earth, since it was shot backwards from the earth. It started below 27 km/sec and it rose gradually up to 40 km/sec. It is not necessarily true that the error we make on account of this aberration angle makes the flux smaller, for the following reason. When we attempt to calculate the flux we correct for the aberration angle using the angular acceptance function of the probe. This function is not known with great accuracy, and it may be that we are overcorrecting. All I can say is that we feel that this is an order of magntiude experiment. I don't think that the aberration angle corrections are making an order of magnitude difference, but I certainly would say they might make a difference of a factor of 2 or 3.

D. Malaise (Astrophysical Institute, Belgium): In connection with Dr. Biermann's question, I measured the angle of several comet tails over a number of years, and for comet 1959k, the variation of the angle with time had a magnitude of 13 degrees, and was observed for a period of 13 days. More recently I observed another comet for a few months, and when the comet was 1 AU away from the sun, the variation was 2 or 3 degrees but when very near, that is 0.65 AU, the variation was 20 degrees. In the case of the first comet, on one day the variation was ahead of the motion of the nucleus by 8 degrees, indicating that the direction of variation is directly affected by the direction of the solar wind.

L. Biermann: May I comment on this? The angle I am referring to is the angle that Hoffmeister refers to as the primary tail. One has to be careful that it is this part of the comet and not the further extension, which may be greatly affected by turbulence. May I take it that your measurements of angles refer to what Hoffmeister calls primary tails?

D. Malaise: Yes, it was a very long tail of 3 million km and was the principal tail of the comet. I have some photographs here.

C. Snyder: Dr. Biermann, do I understand it that you haven't seen any deflections as large as the 20 degrees that this gentleman sees?

L. Biermann: The figures I was quoting were from some of Hoffmeister's work. I do not have all the figures in mind, but I think I recall angles up to 10 degrees. I do not remember angles as large as 20 degrees.

D. Malaise: Yes, the measurements of Hoffmeister showed 11 degrees (*Z. Astrophys.* **22** (1943) 265; **23** (1944) 1).

C. Snyder: This seems to be a very serious problem but I don't know how we can handle it at this time. As far as our data is concerned I think we can say that the effect is not serious, but we should keep this in mind for future work. Let me make another comment. Another part of your question, Dr. Biermann, which I forgot to answer was the question about the direction in which the plasma probe was pointed. We were of course aware that we would have some troubles with the aberration angle, and if we were able to do so we certainly would have pointed the probe in a direction so as to correct for this. Unfortunately in this particular mission we had a difficulty which will probably not be repeated again. Halfway through the mission the space craft rolled over, so that if we had pointed it 6 degrees ahead in the first part it would have 6 degrees behind during the later part, and we would have seen nothing. So we were sort of stuck. In fact the plasma probe was pointed toward the sun, and its deviation from the center of the sun never exceeded a tenth of a degree. So we do not have any correction for the direction of the space craft, but we do have a correction for the direction of the solar wind, which I'm afraid I do not know how to make.

J. Warwick: I have a comment and also a question for Prof. Biermann. My comment is that it seems to be extremely useful to have these interpretations from the comets inasmuch as they give us another source of information beside the geophysical observations. I would like to point out that there is another possible detector for the solar plasma and that is the planet Jupiter, which is known to possess a strong magnetic field and trapped radiation belts that are in many respects analogous to those of the earth. My question is, has Dr. Biermann made any comparison between comets and their activation during solar events, and the radiation from Jupiter, particularly in the decimetric range?

L. Biermann: I would like to point out that my very first work on comets was on recurrence phenomena in comet tails and geomagnetic indices. That has been continued to be explored by Rhea Lüst, so I won't comment on that further (*Z. Astrophys.* **57** (1963) 192). As far as the planet Jupiter is concerned we haven't done any work. I agree that it would be quite interesting because Jupiter is more distant from the sun than most comets. Of course Jupiter is always in the ecliptic plane, so in that respect we get more information from the comets.

R. Frost (General Electric): We heard today of plasma density measurements being made at 0.7 AU and Prof. Biermann, I believe it was, mentioned observations made to 0.3 AU from point radio sources. I would like to ask Prof. Biermann how these are going to join up, if at all? It seems to me that 0.3 to 0.7 isn't too great a jump.

L. Biermann: From the radio point source data one can reproduce a value for the density itself. One can see if these values are consistent with the values which one gets by extrapolating the data from the corona. But there is a gap from about 10 solar radii until 0.7 AU or so where relatively little observation can be made. Radio observations as such cannot be used to bridge this gap.

F. Sylvestri (USAF): Could you comment on the advantages to be gained with telescopes mounted on balloons, rockets and satellites?

Unidentified: Well, there are three reasons why you would want to put a telescope in a satellite. First, the earth's atmosphere may be absorbing the radiation of interest. Second, the light in the night sky places a limit on the observations that can be made, particularly in the observation of faint sources. Third, the so-called poor seeing caused by turbulence in the earth's atmosphere. Now if you go in a balloon you can get above poor seeing to a large degree. Very recent work on the planets and some beautiful photographs of the granulation of the sun were taken from a balloon, that could not possibly have been taken from the ground. Now as you go higher you overcome the problem of the night sky and the problem of absorptions. The reason one wants to look at the stars in the UV, X-ray and even gamma ray portion of the spectrum is because of the interesting physical processes that we suspect should be radiating in this portion of the spectrum. In short we are looking through a narrow window on the earth and missing very large portions of the spectrum.

F. Sylvestri: But from the earth we can see a lot of the sun, and I wonder if this has been exploited to the point where we know enough about the frequencies which do penetrate the earth's atmosphere.

Unidentified: What you are asking is – are enough ground-based solar observations being made, and the answer is certainly not. As our instrumentation and knowledge increase we can always do more. So yes I would have to agree with you. The whole symposium indicates considerable interest in the practical overtones of the mechanism of solar flares and geomagnetic effects. These can be studied from the ground, and should be. I think the satellite provides you with an opportunity to look at different parts of the phenomena that you cannot see from the earth. I do not think, for example, that you could have detected the Van Allen belts from the earth. You might have deduced their presence, and in the case of a high altitude explosion you might have deduced from synchrotron radiation that trapped particles were present. We had a very stimulating discussion about where the edge of the magnetosphere is, and whether there is a gap between that and the turbulent shock region. We do not know the answers to these questions and the only way I know to get them is with satellites.

J. Townsend: I'd like to put a question to Dr. Biermann. Last fall when I gave a talk at GSFC, a member of the audience made a point that he felt that the solar wind as we were seeing it with velocities of 400 to 800 km/sec was quite incapable of explaining the comet tail data because the reaction cross-sections were too low at these velocities. He felt that there must be, in addition to what we see, another solar wind that has, as I recall, 200 eV. Would you comment on this?

L. Biermann: I don't quite know what is being referred to. What I discussed was the acceleration mechanism, i.e. the transfer of momentum from the solar wind to the comet tails. Perhaps ten years ago I proposed that ionization was largely due to an exchange of charge, and since it happens that CO^+ is visible and also N_2^+, it seemed at the time that the presence of these two in the tails might have some significance. Since that time we have found out from experiments that the ionization of CO^+ and N_2^+ cannot be brought about by exchange of charge between neutral molecules and particles from the sun. Other than this I can't make any comment.

H. Alfvén: As I mentioned after your talk, Dr. Biermann, we have some experiments which show that the interaction between an ionized plasma and a neutral gas gives much more ionization than one would expect.

L. Biermann: Yes, but that would only mean that in addition to the sources of ionization which we have already considered we might have some new ones. But I don't see what this has to do with this case.

J. Townsend: If I remember the discussion, he felt that the flux that they observed was insufficient to steer the tail. He felt there must be a higher flux and he presumed that it was at a lower energy.

L. Biermann: If that is the point then one must compare the momentum flux involved. That I have done a number of times. Your measurement of the solar wind indicates a flux of about 10^8 protons/cm^2/sec at the lowest level. With the comet data you should take into account the fact that magnetic fields are present that can cause transfer of momentum between adjacent particles in the whole volume of the tail. Then the flux coming from the sun should be enough to accelerate the tail if there are not too many invisible ions. That would be consistent with the picture which I have indicated this morning.

PART III

MAGNETOSPHERE, MAGNETOPAUSE AND
TRAPPED RADIATION

ON THE PENETRATION OF INTERPLANETARY PLASMA
INTO THE MAGNETOSPHERE

H. ALFVÉN, L. DANIELSSON, C.-G. FÄLTHAMMAR and L. LINDBERG

Royal Institute of Technology, Stockholm

Abstract. Theoretical and experimental evidence is presented to indicate that the usual magneto-spheric models may need essential modification. Theoretically it is concluded that the low-density magnetospheric plasma, in contrast to an ideal magnetofluid, may support strong electric fields even along the magnetic field lines. This can profoundly influence the penetration of interplanetary plasma. Laboratory experiments show that a plasma can penetrate much closer to a terrella than what would correspond to the magnetosphere boundary in closed magnetosphere models.

1. Introduction

One of the fundamental problems in connection with aurorae and magnetic storms is the penetration of the interplanetary plasma into the earth's magnetosphere. This penetration seems difficult to account for on the basis of hydromagnetic models, where the plasma is pictured as a perfectly conducting fluid with a "frozen-in" magnetic field. It is the purpose of the present paper to point out that *the low density plasma in the magnetosphere need not necessarily behave like a magnetofluid.*

Some properties of a low-density plasma in an inhomogeneous magnetic field are discussed in Section 2 of the paper and it is concluded that the magnetic field lines in the magnetosphere need not be equipotential lines. This is most important, because electric fields across the magnetosphere can permit part of the interplanetary plasma to penetrate deep into the magnetosphere (ALFVÉN and FÄLTHAMMAR, 1963). The depth of penetration is determined by the electric and magnetic fields and the temperature in the incoming plasma and not by the dynamic pressure as in hydromagnetic models.

An exact mathematical analysis of the motions in the magnetosphere is too complicated to be attempted at this stage. However, idealized models with simple geometry can be valuable for shedding light on certain aspects of the phenomena. This approach is used in Section 3. Model experiments represent another approach which we believe is most important as a complement to the theoretical considerations. In Section 4 some recent experiments are discussed which may be of relevance for understanding some of the problems involved.

As the behaviour of the plasma depends on whether the plasma magnetic field has a component parallel or antiparallel to the dipole field, there is a possibility that two categories of magnetic storms exist, which is considered in Section 5. This should be checked by observations.

Future observations by space-craft-carried instruments may provide evidence to decide between the model discussed here and the hydromagnetic models. Possible experiments are suggested in Section 6.

2. Electric Fields in Low-density Magnetic Plasmas

By *low-density magnetic plasma* we mean a plasma where the mean free path λ is much larger than the characteristic dimension l_c, which, in our application is the separation of the magnetic mirrors confining the plasma.

In an electron-proton plasma the effective mean free path of an electron measured in cm is given to the order of magnitude by

$$\lambda \approx 10^4 \frac{T_e^2}{n_e} \tag{1}$$

where T_e and n_e are the temperature (°K) and number density of electrons (cm^{-3}).

Let us consider the magnetosphere of the earth, in particular a magnetic field line which intersects the equatorial plane at, say five times the earth's radius i.e. at 3×10^9 cm. The density in the equatorial plane is probably of the order 100 cm^{-3} or less. If we put $l_c = 3 \times 10^9$ we find that $\lambda \approx l_c$ if the temperature is about 10^4 degrees. If the temperature is higher, the magnetosphere should be counted as a low-density region. No reliable temperature measurements seem to exist and it can perhaps not be excluded that during undisturbed conditions the temperature may be low. However, during magnetic storms the magnetosphere is invaded by a presumably hot plasma ($10^5 - 10^6$ degrees) ejected from the sun, and this plasma is heated still more by magnetic compression when it penetrates into the magnetosphere.

If we put the cross-section of the magnetosphere equal to 10^{20} cm^2, an interplanetary plasma with a flux density of 10^8 protons cm^{-2} sec^{-1} each with an energy of 10^3 eV carries an energy of 10^{31} eV/sec to the magnetosphere. If the outer region of the magnetosphere has a volume of 10^{30} cm^3 and contains 100 particles/cm^3, the energy needed to heat it to 10^4 degrees (≈ 1 eV) equals 10^{32} eV. Hence the interplanetary wind carries energy enough to make the outer magnetosphere a "low density region" in 10 seconds. Therefore, *at least during magnetic storms the outer magnetosphere is likely to be a low-density region.*

As discussed in a different context elsewhere (ALFVÉN and FÄLTHAMMAR, 1961) there is in a low-density plasma no relation of the simple form $i = \sigma E$ between the current density and the electric field. Suppose that there is only an electric force parallel to **B**. Then an individual electron obeys the equation of motion

$$m_e \frac{dv_{\parallel}}{dt} = e_e E_{\parallel} \tag{2}$$

and the corresponding current density $i_{\parallel}$ is given by

$$\frac{di_{\parallel}}{dt} = \frac{n_e e_e^2 E_{\parallel}}{m_e} . \tag{3}$$

This shows that in the low-density case the conductivity $\sigma_{\parallel}$ defined by $i_{\parallel}/E_{\parallel}$ has no meaning. Even if $E_{\parallel} = 0$ we may have $i_{\parallel} \neq 0$, and for $E_{\parallel} \neq 0$ the current at a given instant may be zero or even antiparallel to $E_{\parallel}$.

It is interesting to study the case when the magnetic field is inhomogeneous in such a way that the guiding centres of the particles oscillate along a field line between two mirror points at a distance l_c. Suppose that the plasma consists of electrons and one kind of positive ions, and that their magnetic moments are μ_e and μ_i. In the presence of an electric field $E_\parallel$ parallel to $\mathbf{B}$ they are acted upon by the average forces

$$f_{e_\parallel} = -\mu_e \frac{dB}{ds} + e_e E_\parallel \tag{4}$$

and

$$f_{i_\parallel} = -\mu_i \frac{dB}{ds} + e_i E_\parallel \tag{5}$$

parallel to the magnetic field.

If under the influence of this force a particle oscillates with the velocity $v_{k_\parallel}$ between the points s_1 and s_2, it spends the time

$$dt = ds/v_{k_\parallel} \tag{6}$$

on the line element ds. Its half-period is

$$\tau_k = \int_{s_1}^{s_2} ds/v_{k_\parallel} . \tag{7}$$

Hence it gives rise to an average space charge

$$dq_k = \frac{e_k\, ds}{\tau_k\, v_{k_\parallel}} \tag{8}$$

on the line element ds. (The subscript k is used to distinguish between different kinds of particles.) If N_i positive ions and N_e electrons oscillate simultaneously between the same mirror points, the space charge at each line element is $N_i dq_i + N_e dq_e$ and if N is so large that we have a plasma (which must be quasi-neutral), we must have $N_i dq_i + N_e dq_e = 0$ and also $N_i e_i + N_e e_e = 0$. According to (8) this means that $\tau_i v_i = \tau_e v_e$, or, if we introduce $\alpha = \tau_i/\tau_e$,

$$v_{e_\parallel} = \alpha v_{i_\parallel} . \tag{9}$$

Here α is a constant which can have any value (because electrons and ions need not have equal energies).

Equation (9) and the equations of motion

$$m_e \frac{dv_{e_\parallel}}{dt} = f_{e_\parallel}$$

$$m_i \frac{dv_{i_\parallel}}{dt} = f_{i_\parallel}$$

give

$$f_{e_\parallel} = m_e v_{e_\parallel} \frac{dv_{e_\parallel}}{ds} = \frac{\alpha^2 m_e}{m_i} m_i v_{i_\parallel} \frac{dv_{i_\parallel}}{ds} = \frac{\alpha^2 m_e}{m_i} f_{i_\parallel}. \tag{10}$$

If we assume $e_i = -e_e = |e|$, we obtain by introducing (4) and (5) into (10)

$$- |e| E_\parallel - \mu_e \frac{dB}{ds} = \frac{\alpha^2 m_e}{m_i} \left(|e| E_\parallel - \mu_i \frac{dB}{ds} \right) \tag{11}$$

or

$$E_\parallel = - K \frac{dB}{ds} \tag{12}$$

with

$$K = \frac{1}{|e|} \frac{\mu_e/m_e - \alpha^2 \mu_i/m_i}{1/m_e + \alpha^2/m_i}. \tag{13}$$

Introducing (9) and $\mu = W_\perp/B$ we find the following expression for the *invariant K*

$$K = \frac{W_{i_\parallel} W_{e_\perp} - W_{e_\parallel} W_{i_\perp}}{|e| B (W_{i_\parallel} + W_{e_\parallel})}. \tag{14}$$

We have used the notations $W_\parallel = m v_\parallel^2/2$ and $W_\perp = m v_\perp^2/2$ with the indices e and i referring to electrons and positive ions.

Equation (12) shows that in a low density plasma the electric field parallel to the magnetic field is zero only when $(dB/ds = 0)$ (e.g. homogeneous field) or when the relation

$$\frac{W_{i_\parallel}}{W_{e_\parallel}} = \frac{W_{i_\perp}}{W_{e_\perp}} \tag{15}$$

is satisfied. This relation means that the helices of the ions and of the electrons have the same pitch angle. If (15) is satisfied, so that $K = 0$, both kinds of particles oscillate with the same amplitude in the absence of an electric field. However, if $K \neq 0$, so that $W_{i_\parallel}/W_{i_\perp} \neq W_{e_\parallel}/W_{e_\perp}$, electrons and ions would oscillate with different amplitude, if there were no electric field. As the quasi-neutrality of the plasma requires that $N_i dq_i + N_e dq_e \approx 0$ everywhere, particles of both kinds must oscillate with equal amplitude. This can only be achieved by setting up an electric field.

We integrate (12) between two points A and C where the magnetic field strengths are B_A and B_C. The voltage difference between C and A is

$$V = V_C - V_A = K (B_C - B_A). \tag{16}$$

We introduce

$$\gamma = B_C/B_A. \tag{17}$$

Then (16) becomes

$$|e| V = (\gamma - 1) \left[\frac{W_{i_\parallel} W_{e_\perp} - W_{e_\parallel} W_{i_\perp}}{W_{i_\parallel} + W_{e_\parallel}} \right]_A \tag{18}$$

where the expression within the brackets refers to the point A.

EVALUATION OF THE VOLTAGE IN A SIMPLE CASE

The resultant voltage depends on how the particle population is injected, because this influences the values of $W_{i_\parallel}$ etc. in (18). As an illustration we consider the following simple model. (The actual injection mechanism in the magnetosphere is not yet known.)

A line of force of the geomagnetic dipole field intersects the ionosphere at C (Figure 1). Outside the ionosphere there is a low-density plasma, which is produced by injection of hot plasma in the equatorial plane (at A in Figure 1) and evaporation of low-energy particles from the ionosphere. For the sake of simplicity we assume that in the equatorial plane only particles of one sign, say electrons, are injected. At C, where the field strength is B_2 ($B_2 > B_1$), there is a source emitting ions with negligible energy. The emission takes place as soon as there is a voltage difference between the plasma at C and the source. In this way the voltage of the plasma at C is fixed by the voltage of the source. The injection of electrons at A lowers the voltage at this point, and as a consequence positive particles are emitted from the source at C, until Equation (18) is satisfied. Then we can calculate the voltage difference from (18).

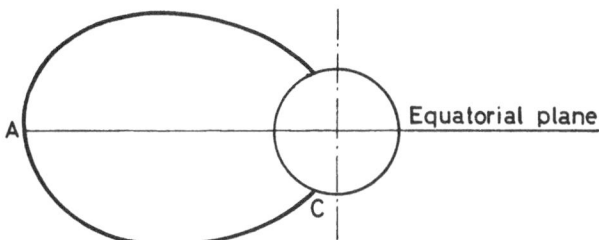

Fig. 1.   The earth and a geomagnetic field line.

As the positive ions have been accelerated by the voltage V before they reach A, we have $W_{i_\parallel} = |e| V$, and $W_{i_\perp} = 0$. Hence we obtain from (18)

$$|e| V = (\gamma - 1) \frac{|e| V W_{e_\perp}}{|e| V + W_{e_\parallel}} \tag{19}$$

which besides the solution $V = 0$ also gives

$$|e| V = (\gamma - 1) W_{e_\perp} - W_{e_\parallel}. \tag{20}$$

If in a medium-density plasma (by which we mean a plasma where $\lambda \ll l_c$ but still $\lambda \gg \varrho_L$, ϱ_L being the Larmor radius) we have two similar sources on the same field line, we obtain a conduction current carrying electrons from the source A to C and ions from C to A. This current, which tends to annihilate any voltage difference between A and C, is proportional to V. As in cosmical medium-density plasmas the conductivity usually is large, the current will often prevent any large voltage difference between A and C. Contrary to this, in a low density plasma the voltage difference V given by (20) is produced before any appreciable current can flow. If we inject electrons at A, they cannot reach C unless their energy is increased by V to such an extent as to displace the mirror point to C. Only if the voltage difference exceeds the value given by (20) can electrons reach C so that a current is produced. Further, as the ions emitted from C oscillate along the field line without collision, they cannot carry any average current when a stationary state is reached.

In our model a magnetic field line, which intersects the equatorial plane (field strength B_A) at 5 or 10 earth radii reaches the ionosphere at a point where the field strength (B_c) is so much larger that γ is 100 or 1000. This means that under our simplifying assumptions there may be a voltage difference between the equatorial plane and the ionosphere which exceeds the equivalent voltage of particles injected at A by a very large factor.

In case the collision frequency in the plasma is finite, the oscillating particles emitted from C will loose their energy and accumulate at A, whereas the particles emitted from A will have a chance of being scattered into the "loss cone" so that they move along the field lines to C. In this way a current is caused so that the voltage V given by Equation (20) is eliminated. For finite collision frequencies the voltage V is large only under the condition that the "life time" of the plasma is shorter than the collision time. By "life time" we mean the time before the plasma has drifted away from the region we consider, and new plasma has been injected.

From the above considerations we conclude that in the magnetosphere there may exist a voltage difference between the point where a field line intersects the equatorial plane and the point where it intersects the ionosphere. This voltage difference is not necessarily equal for different magnetic field lines, which means that there may well exist electric fields in the outer parts of the magnetosphere even if the ionosphere is considered as an equipotential surface.

RIGOROUS ANALYSIS OF THE ONE-DIMENSIONAL CASE

The case of a low-density plasma situated in a straight narrow tube of force with magnetic mirrors at both ends has been studied in detail by PERSSON (1963). In contrast to the simple considerations given above Persson's analysis takes into account the velocity distribution of the particles. From this analysis it is found that the electric field vanishes only if the pitch-angle distributions of electrons and ions are identical. When they differ appreciably electric fields of the order of magnitude given by (12) prevail.

3. Theoretical Conclusions from Simple Models

For an analysis of the outer regions of the magnetosphere, the strength and direction of the interplanetary magnetic field is essential. We shall confine our discussion to two simple cases.

(A) The interplanetary field immediately outside the magnetosphere is parallel to the magnetosphere field in the equatorial plane.

(B) The fields are antiparallel.

CASE (A)

Without any plasma the magnetic field consists of a dipole field superposed on a homogeneous field. The field strength in the equatorial plane is given by

$$B = B_0 + a/R^3. \tag{21}$$

The motion of a charged particle in the equatorial plane can be treated exactly. If there is a homogeneous electric field E in the equatorial plane, the closest approach to the dipole, of a particle with energy W is essentially

$$L = \left(\frac{\mu a}{|e| E}\right)^{1/4} \tag{22}$$

where a is the magnetic moment of the dipole field, e is the charge of the particle, and μ is the adiabatic invariant of the particle. For a typical electron in a plasma with electron temperature T_e the adiabatic invariant is

$$\mu_e = \frac{kT_e}{B}. \tag{23}$$

Notice that in the present case *the plasma is stopped by diamagnetic repulsion, not by the magnetic pressure.* For the actually observed values of magnetic fields and velocity and for reasonable values of the interplanetary electron temperature (so far observationally unknown) this means that the plasma can penetrate to depths where the auroral-zone magnetic field lines cross the equatorial plane.

The motion of the plasma will produce, by charge separation, additional electric fields modifying the externally applied field. Using again a two-dimensional model KARLSON (1962) has analysed this effect and concluded that the general character of the motion remains essentially unchanged.

The magnetic field given by (21) is modified considerably by currents produced by the plasma. For example the front side of the magnetosphere is compressed considerably by the dynamic pressure of the plasma.

Also the topology of the field may be changed. In the undisturbed case there is a sphere with the radius $(2a/B_0)^{1/3}$ on the surface of which the field is everywhere parallel to the spherical surface. This surface goes through the singular points $B=0$, and can be considered as separating the magnetosphere from the interplanetary field.

The two-dimensional case treated by Karlson may give a first-order description of the motion of the interplanetary plasma near the equatorial plane. It has sometimes been argued that in the actual three-dimensional case there could not exist an electric field across the magnetosphere because it should be short-circuited along the magnetic field lines. That need not be true. As was found above in Section 2, the low-density plasma may support a large potential difference along the strongly convergent field lines that end in the ionosphere. On the other hand, near the singular field line separating the magnetosphere from interplanetary space, there is no strong field-strength gradient parallel to **B**. According to (12) this means that no appreciable electric fields should exist there. However, this exceptional region is small, and the incoming plasma may change the magnetic structure near the neutral point of the singular field line in such a way that at least part of the plasma can penetrate.

CASE (B)

DUNGEY (1961) has directed the attention to the importance of the antiparallel case.

In the undisturbed case the magnetic field in the equatorial plane is $B = a/r^3 - B_0$. It is zero in the equatorial plane along the circle

$$r_0 = (a/B_0)^{1/3}.$$

If similar to case A we treat the motion of charged particles in the equatorial plane, we find that no penetration through this circle is possible. On the other hand field lines from the polar caps go to infinity, and as pointed out by Dungey, an interplanetary plasma may penetrate along these field lines to the ionosphere.

Plasma moving parallel to the equatorial plane but a few earth's radii to the north or to the south of it will experience an increasing magnetic field. The phenomena produced may be somewhat similar to the conditions in the equatorial plane in the parallel case.

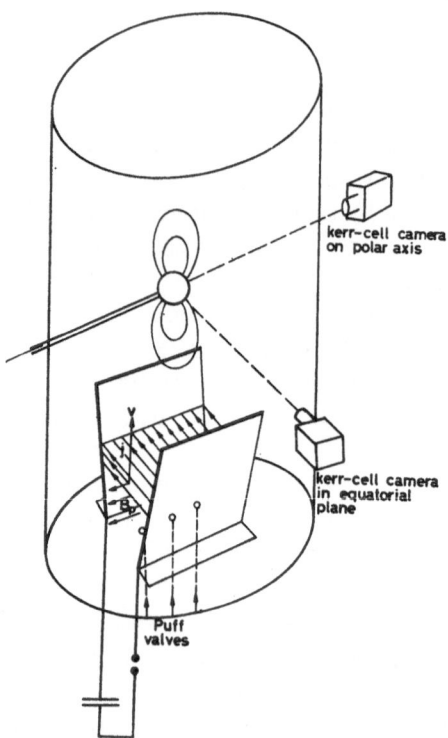

Fig. 2.   Schematic sketch of the apparatus in the Lindberg and Danielsson model experiment.

4. Model Experiments

In order to study what happens when a magnetized plasma is shot towards a magnetic dipole field a laboratory experiment has been made by Lindberg and Danielsson. It can be considered as a development of the Birkeland-Malmfors-Block experiments.

APPARATUS

The main parts of the device are shown in Figure 2. A terrella of radius $R=1.2$ cm is situated in a volume filled with hydrogen at low pressure $(0-10\,\mu)$. The terrella gives a field of 10000 gauss at its equator. The muzzle of a plasma gun with plane electrodes of height 30 cm and width 20 cm is placed 10 cm below the terrella. At the bottom of the gun there are three puff-valves, which are able to let in a cloud of hydrogen gas very fast. Some time after opening the valves, when the cloud has expanded a few centimeters upwards between the electrodes, a discharge is fired through the cloud. The high discharge current, 10–50 kA, ionizes the cloud into a plasma. The current

TABLE I

APPROXIMATIVE DATA

Plasma Quantities (undisturbed plasma)

Magnetic field	$B_p = 300$–700 gauss
Velocity of the plasma	$v_p = 2$–5×10^6 cm/sec
Temperature	$T = 10000$–50000 °K
Number density	$n = 10^{15}$–10^{16} cm^{-3}
Sound velocity	$v_s = 10^6$ cm/sec $\left[\left(\dfrac{\gamma \cdot kT}{m}\right)^{\frac{1}{2}}\right]$
Magnetohydrodynamic velocity	$v_{MH} = 10^6$ cm/sec $\left[\dfrac{B_p}{(4\pi \cdot nm)^{\frac{1}{2}}}\right]$
Gyroradius of the ions	$\varrho_i = 0.4$ cm
Gyroradius of the electrons	$\varrho_e = 0.01$ cm
Gyrofrequency of the ions	$\omega_i = 5 \times 10^6$ sec^{-1}
Gyrofrequency of the electrons	$\omega_e = 10^{10}$ sec^{-1}
Magnetic pressure	$p_B = 10^4$ dyn/cm^2 $[B_p{}^2/8\pi]$
Dynamic pressure	$p_d = 4 \times 10^4$ dyn/cm^2 $[nm \cdot v_p{}^2]$
Static pressure	$p_s = 10^4$ dyn/cm^2 $[nkT]$

Terrella Quantities

Terrella radius	$R = 1.2$ cm
Magnetic field at the surface in the equatorial plane	$B_{eq} = 10^4$ gauss
Radius of the forbidden region	$L \sim 1.2R$
Radius of magnetosphere	$R_s = 2$–$3R$

Transformation Factors

Length	10^{-9}		Time	10^{-10}
Dipole field	3×10^4		Plasma field	10^7
Potential	10^{-3}			

gives rise to a magnetic field in the plasma and the $j \times B$ force accelerates it upwards. In the vicinity of the terrella the velocity of the plasma is 2–5 cm/μ sec, the magnetic field in the plasma 300–700 gauss (originating from the discharge current), the depth and width of the plasma 10×20 cm (the current density about 100 A/cm^2). See further Table 1. The number density and temperature (and from these derived quantities) are only estimated.

Two different cases are studied:

(A) the plasma field is parallel to the dipole field in the equatorial plane,

(B) the fields antiparallel.

The diagnostic methods used are kerrcell-photography and magnetic probe measurements.

PARALLEL FIELDS

In the case when the equatorial field of the terrella is parallel to the field in the plasma a dark region like Figure 3 is observed between 11 and 14 μsec after firing the gun.

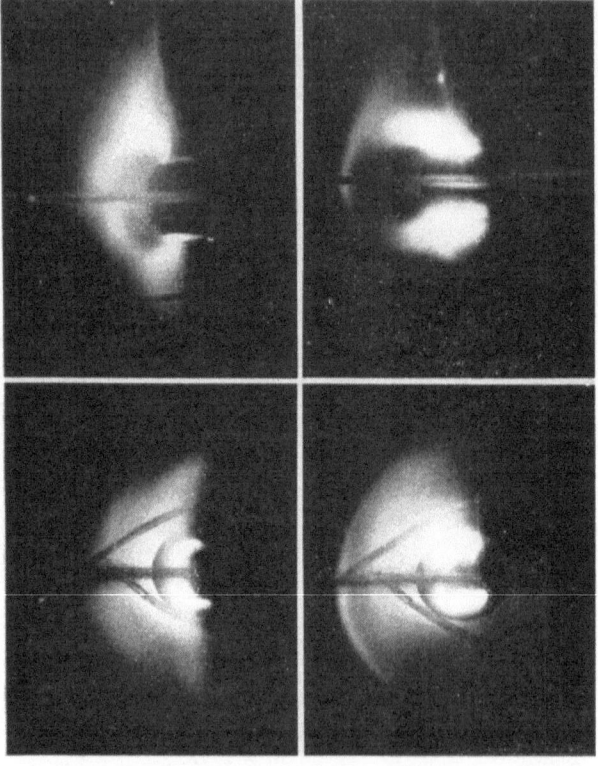

Fig. 3.   Fig. 4.

Fig. 3.   Simultaneous kerrcell-photographs, from two perpendicular directions, of the penetration of a plasma with parallel magnetization.
Above: seen from evening side. Below: seen from the above north pole.

Fig. 4.   The same as Figure 3, but with antiparallel magnetization.

Compared with a dipole line its border is considerably deformed as if "compressed" by the plasma. When it first appears, it extends to about 1.6 R in the equatorial plane, but expands to 1.8 R before it disappears. (All distances are measured from the centre of the terrella) On the backward or "night" side traces of a similar dark region are sometimes seen. The lower picture in Figure 3 is taken simultaneously with the upper one, but from the pole axis. It shows a luminous ring around the pole at a certain latitude possibly corresponding to an "auroral zone".

Magnetic probes measuring the field components parallel to the terrella axis are located in the equatorial plane at 1.7 R in the directions 6^h 12^h 18^h and 24^h counted with the plasma gun (representing the sun) in the 12^h direction. – When the plasma approaches the terrella they all show an increase, followed by a smooth decrease, which can be interpreted as a temporary compression of the terrella field. On the dayside a probe, located at $3R$, shows a more steep front, Figure 5a. The magnitude of the field change is usually 150–200% of the undisturbed plasma field.

ANTIPARALLEL FIELDS

When the equatorial field of the terrella is antiparallel to the field in the plasma the dark zone gets quite a different shape, Figure 4. The border is rather like the expected shape of the singular field lines, although somewhat compressed by the plasma. Also in this case the zone is visible between 11 and 14 μsec after firing the gun, and its radius in the equatorial plane increases during this interval from $2.6R$ to $3.3R$.

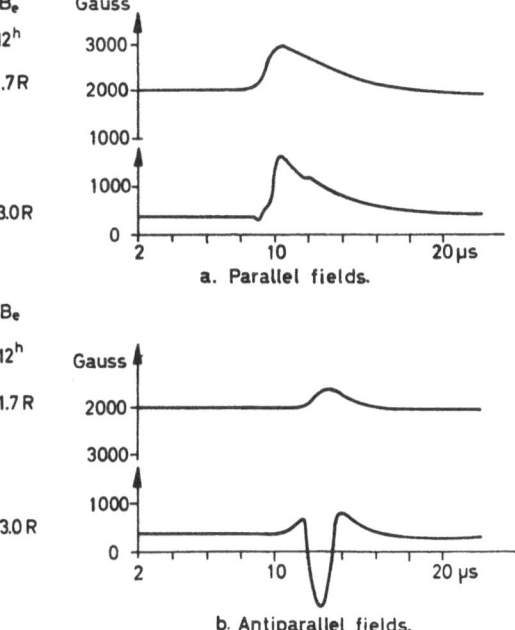

Fig. 5.   Oscillograms showing the magnetic field variation at two points in the direction of the plasma source. (a) Parallel case. (b) Antiparallel case.

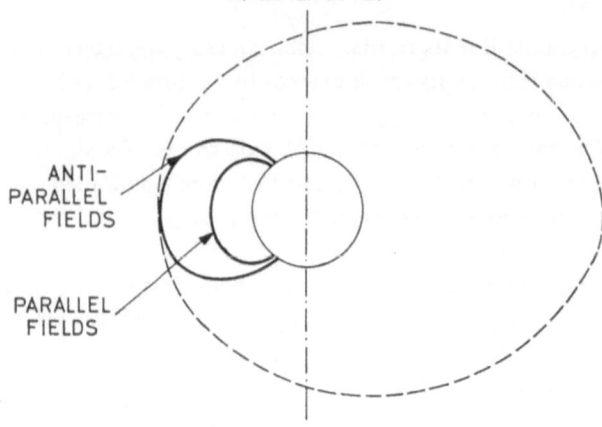

Fig. 6. Limits of the penetration of plasma in the two experimental cases (full lines). In these experiments the „magnetosphere border" would approximately correspond to the dashed line.

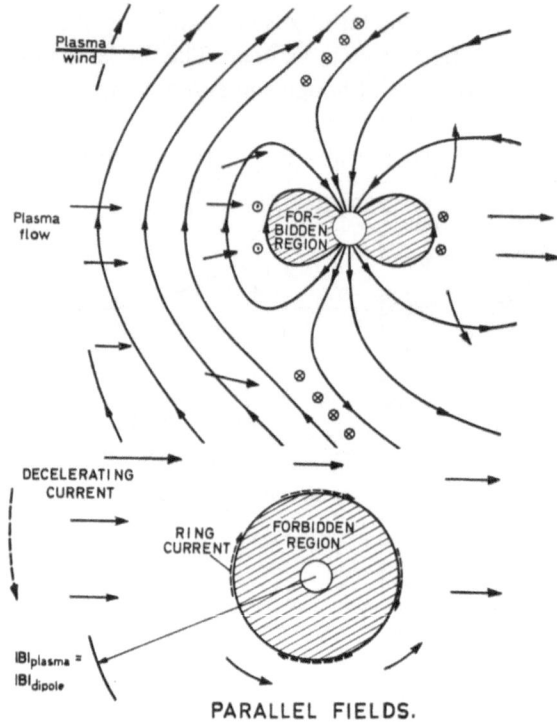

Fig. 7. Possible configuration of magnetosphere field.

Magnetic probes in the equatorial plane show a compression followed by a decay and this occurs nearly simultaneously in all points. Of special interest are the two probes on the day side, at $1.7R$ and $3.0R$, Figure 5b. The inner one shows just the compression mentioned, while for the outer one the compression is interrupted and followed by a peak of reversed field, indicating that the plasma and its field surround the probe for about 1 μsec, before being reflected. The amplitude of this peak varies very much from shot to shot. The rapid reversal of the field shows that a strong current flows in a thin sheath separating the terrella field from the plasma field. We have here a phenomenon which is similar to the Cahill discontinuity (CAHILL and AMAZEEN, 1963).

The fact that zones with sharp boundaries are observed for as long as 3 μsec may indicate that diffusion is not too serious during this time, which corresponds to a movement of the plasma about 15 cm ($=12R$).

Neither in the parallel nor in the antiparallel case is the plasma stopped at a limit which can be identified with the "border of the magnetosphere" in the closed magnetosphere models, cf. Figure 6.

5. Storms and "Antistorms"

It is of special interest to note that both with parallel and antiparallel fields the plasma hits the terrella at high latitudes. It is possible that in the two cases the auroral phenomena and polar disturbances do not differ very much. On the other hand we

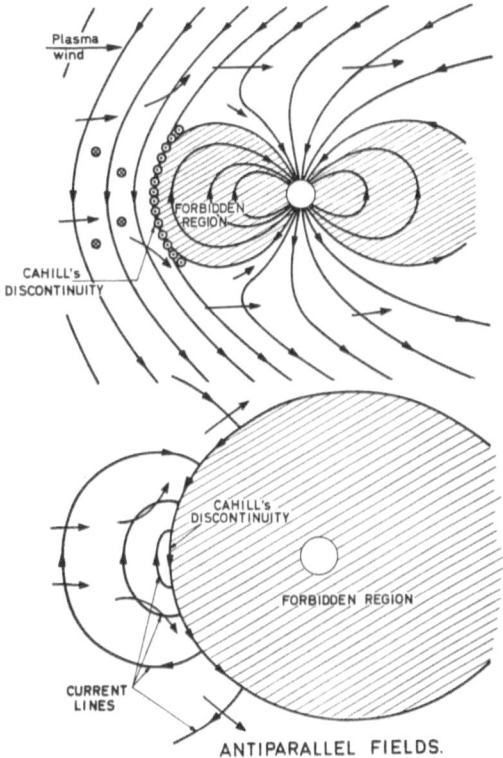

Fig. 8. Possible configuration of magnetosphere field.

should expect that some features of the phenomena are different in the parallel and antiparallel cases.

Hence one may tentatively suggest that there are two different types of storms, which we may call *parallel storms* and *antiparallel storms*.

A parallel storm is produced when a plasma with parallel magnetization hits the magnetosphere.

An antiparallel storm is produced by a plasma with antiparallel magnetization.

The polar phenomena (including the aurora) may not be very different, but the parallel storms should have a larger equatorial disturbance than antiparallel storms.

Possible magnetic field and current distributions in the two cases is shown in Figures 7 and 8.

6. Observational Tests

A decision between the frozen-in model and the electric field model of the magnetosphere could be made in the following ways:

(1) Determination of number density n and temperature T, and calculating the mean free path λ. The frozen-in model requires $\lambda \ll l_c$, the electric field model requires $\lambda \gg l_c$ (for each kind of particle).

(2) Determination of the pitch angle distribution of both electrons and ions. Only if they are identical can the electric field be neglected.

(3) Direct measurement of the electric field.

(4) Measurement of the plasma drift $v_\perp$ at two points on the same line of force in order to see whether the field lines are frozen-in.

References

ALFVÉN, H. and FÄLTHAMMAR, C-G.: 1963, *Cosmical Electrodynamics, Fundamental Principles,* Clarendon Press, Oxford.
BLOCK, L.: 1958, *International Astronomical Union Symposium* 6.
BOSTICK, W. H.: 1962, *Phys. Fluids* 5, 1305.
CAHILL, L. J. and AMAZEEN, P. G.: 1963, *J. Geophys. Res.* 68, 1835.
DUNGEY, J. W.: 1961, *Phys. Rev. Letters* 6, 47.
HELMER, J. C.: 1963, *Phys. Fluids* 6, 723.
KARLSON, E.: 1962, *Phys. Fluids* 5, 476.
KARLSON, E.: 1963, *Phys. Fluids* 6, 708.
KAWASHIMA, N. and ISHIZUKA, H.: 1963, *J. Phys. Soc. Japan* 18, 763 (1963).
PERSSON, H.: 1963, *Phys. Fluids* 6, 1756.

Discussion and Questions

E. Hones (IDA): I wonder did I interpret your earlier remarks correctly that the coupling between the magnetosphere and the ionosphere is very poor even at very short radial distances above the ionosphere?

H. Alfvén: It depends upon the relation between the mean free path and the characteristic length. If the mean free path is much smaller than the length of the line of force then we have frozen-in conditions, (if the conductivity of the ionosphere is high enough). But so far out in the magnetosphere that the mean free path is long compared to the linear dimensions then the theoretical analysis which I have made is valid. I should like to mention that the condition for this is that the temperature should be higher than about ten thousand degrees.

D. Beard (University of California): How do you evaluate your stream pressure for your streams that hit the dipole?

H. Alfvén: It is made from the experimental data. The velocity is very easy to measure, the density is a little more difficult, but it can be measured. We can also check this through the compression of the magnetic field in the antiparallel case by watching what we believe is the Cahill discontinuity as it goes in and out. Under these assumptions we can calculate a pressure. But I should not like to draw very far-reaching conclusions now because these experiments have been running only a couple of months and it is necessary to have much more experimental data before any definite decision can be made.

H. Petschek (AVCO): In the case of parallel fields where the plasma went in beyond the position of pressure balance, suggests a low conductivity plasma. Now the way that plasma guns operate, there are many reasons why the conductivity should be low, such as low temperature which would not be the same as the magnetosphere case. Do you have any idea of what the conductivity is?

H. Alfvén: I think you can judge this by the difference between the parallel and antiparallel case. The plasma stops very nicely in the antiparallel case but not in the parallel case. This shows that the penetration cannot be due to the conductivity. We have other measurements which indicate that the plasma takes 15 μsec to decay whereas these phenomena occur in 1 μsec.

J. Wolfe (Ames Research Center): How did you overcome the problem of properly scaling the quantities involved in the experiment?

H. Alfvén: It isn't properly scaled. It is just an attempt to reproduce the same type of phenomena as should be important in the magnetosphere. The properties of the plasma are such that the theories of the magnetosphere should be applicable to this case too as there is nothing in these theories that should exclude this case. However, to overcome the problem of scaling we are presently letting the parameters vary over wide ranges. We have as a matter of fact made earlier experiments with a much lower plasma density and found similar phenomena.

E. Parker: I would like to be sure that I clearly understand the conclusions you have drawn from the experiment. Are they (1) that the plasma penetrates sometimes farther than the pressure balance indicates and (2) that it should all the time precipitate over the poles?

H. Alfvén: Yes, in the case of parallel magnetic fields the plasma seems to penetrate farther toward the terrella. It seems that in both cases we have precipitation over the poles.

NULL POINTS IN SPACE PLASMA

J. W. DUNGEY

Department of Physics, Imperial College, London

Abstract. A special phenomenon occurs at a hyperbolic null point in a plasma and leads to the formation of a thin sheet of intense current. This may be the essential feature of solar flares and a current sheet of this sort appears at the magnetospheric boundary. A first approach to the theory is made by hydromagnetics and some orders of magnitude can be obtained. An interesting feature is the „reconnection" of lines of force. When the current density has grown, the effective resistivity may much exceed the normal value and various kinds of instability may occur. Individual particles gain energy in the current sheet and their trajectories will be discussed.

The suggestion that there is something special about a magnetic null point in a plasma was first made by Giovanelli, as a result of watching hundreds of flares on a spectro-heliograph. Giovanelli noticed that flares are more frequent in complex spot groups, containing spots of different polarity jumbled together, and formed the impression that the positions of flares were those where a null point might be expected in the atmosphere above the spot group. He investigated the conditions required for electrons to run away, owing to the decrease of their collision cross-section with increasing energy, and suggested that the observed features of flares are caused by an electric discharge. The electric field would not differ essentially from a uniform field and the particles would gain energy in the simplest possible way by moving through a potential drop. Such conditions are the easiest to set up in the laboratory, but it will be seen that in space they are only possible in the region of a null or of a topological generalisation of this to be mentioned later.

The first simple consideration to notice is that the occurrence of a discharge implies a very large current density compared to the estimates one obtains for curl **B** from the dimensions of sunspots. In a simple-minded picture the electrons are all rushing one way and the protons the other. The problem of the motion of the particles near a null is an awkward one, as will be seen, and the simple-minded picture can barely be improved on as yet. Supposing, then, that the difference between the mean velocities of electrons and protons is of the order of the velocity of light, the current density j must be nec where n is the electron density (very nearly the same as the proton density in a plasma). Now **j** determines curl **B** and we represent |curl **B**| by B/b, where B is an order of magnitude and b is a length to be explained shortly. Then Maxwell's equations give

$$b = B/4\pi ne = \left(\frac{B}{600 \text{ gauss}}\right)\left(\frac{10^9 \text{ cm}^{-3}}{n}\right) \text{metres} . \tag{1}$$

This is sufficient to show that b is very small on the solar scale, and we must now consider more carefully what the significance of b is. Since curl B involves the spatial variation of B, b is a measure of the scale of this variation. Because b is small the

Chang & Huang (eds.), Proc. Plasma Space Sci. Symp. All rights reserved.

discharge must be confined to a region which is small in one dimension. A thin sheet, not necessarily plane, satisfies this requirement and turns out to be the required configuration. Then the small-scale b is the thickness of the sheet, the field must change substantially between opposite sides of the sheet, and this change of course involves a component parallel to the sheet since curl B is to be large and div B always vanishes.

The problem now is how to set up a thin current sheet in an astrophysical plasma, which behaves hydromagnetically. The pictorial method which represents hydromagnetics by treating lines of force as elastic strings is very useful for this purpose. It is well known that the freezing of the field into the material is equivalent to

$$\mathbf{E} + \mathbf{v} \wedge \mathbf{B}/c = 0 \tag{2}$$

and this will be valid, if a large current density is *not* achieved. The other well-known aspect is the tension $B^2/4\pi$ in the lines of force derived from the equation

$$\rho \frac{d\mathbf{v}}{dt} = (\mathbf{B} \cdot \nabla)\mathbf{B}/4\pi - \nabla(p + B^2/8\pi). \tag{3}$$

It may be noted here that Equations (2) and (3) represent the laws of electromagnetic induction. On asking "Can we set up a high current density using induction?" one may remember Lenz's law, which answers positively not, since it says that, if current starts to grow, induction will oppose that growth. Lenz knew nothing of hydromagnetics, however, and his law is normally used for rigid conductors and hence needs checking for fluids. The elastic string model is well suited for such a check. If a current sheet flows across the lines of force, a field such as that shown in Figure 1 results, the tension gives a catapult force and Lenz is vindicated. A current

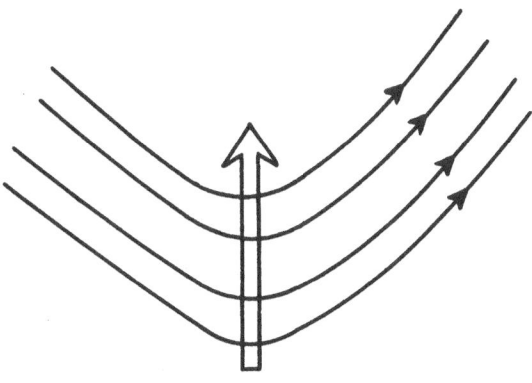

Fig. 1.

flowing parallel to B produces no force, but the field it produces is such that the lines for the total field are twisted. If such a current flows along only a stretch of a line of force and then splays out across the field, the twisting is strongest in that stretch of field line and the tension clearly acts to oppose the twisting and again supports Lenz. Nevertheless, it will now be shown that Lenz's law breaks down at a hyperbolic null.

The field in the neighbourhood of a null may be represented by a Taylor expansion and in a small enough region only lowest order terms involving the first derivatives $\partial B_i/\partial x_j$ need be included. Since curl $\mathbf{B}$ is defined by the antisymmetrical part of the tensor $\partial B_i/\partial x_j$, the current density corresponding to the lowest order terms is uniform. The directions of limiting lines of force are defined by the eigenvectors ∂x_i satisfying

$$\frac{\partial B_i}{\partial x_j}\,\delta x_j = \lambda \delta x_i \tag{4}$$

for some eigenvalue λ. At a hyperbolic null there are three directions and not all of them can have the same sign for λ because div $B=0$. Figure 2 shows the lines of force in a plane containing two of these directions with values of λ of opposite sign. The lines of force are hyperbolae in the small region considered. To test Lenz's law the component of current perpendicular to the paper must be considered.

If the current is zero, the tensor $\partial B_i/\partial x_j$ is symmetrical and then the limiting lines of force are perpendicular. Thus the sign of the component of current perpendicular to the paper depends on which angle between the limiting lines shown is acute. In Figure 2 the current is flowing into the paper and this leads to forces in the directions shown by the thick arrows. The effect of induction is therefore to reduce the acute angle between the limiting lines of force and the picture suggests that the current density increases in intensity. This is not quite rigorous, but can be confirmed as follows.

In the region of the null Equations (2) and (3) can be Taylor-expanded to give the partial time derivatives of space derivatives at the null. This procedure is of course always possible, but does not normally lead to a closed set of equations. In this case a closed set of equations is obtained in $\partial B_i/\partial x_j$, $\partial v_i/\partial x_j$ and ϱ, if the null point is initially a stagnation point and if ∇p is omitted entirely. The effect of ∇p will be discussed later. Here we are still concerned with checking Lenz's law, and, since pressure is nothing to do with electromagnetic induction, Lenz's law should still be valid even for a perfectly compressible plasma. Now, having got the closed set of equations, it turns out merely from examining the signs of the terms that all the dependent variables increase in magnitude indefinitely (DUNGEY, 1953). In particular curl $\mathbf{B}$ increases, and so Lenz is nailed.

The forces in Figure 2 may be regarded as the attraction of parallel currents and may be compared with a pinch discharge. The difference in a pinch tube is that the null is elliptical and the magnetic force can be balanced by pressure. The forces in Figure 2 represent compression in the horizontal direction, but stretching in the vertical direction, and are therefore harder to balance by pressure. An equilibrium can be sought in the following formulation. Consider a two-dimensional model in which the field has a limiting line in the form of a figure eight as shown in Figure 3. The static-equation corresponding to (3) is

$$(\mathbf{B}\cdot\nabla)\mathbf{B} = \nabla\left(4\pi p + \tfrac{1}{2}B^2\right). \tag{5}$$

Now the tension in the lines outside the figure eight squeezes the two loops of the eight together with the result that the parts of the inner lines near the hyperbolic null X are flattened. To be precise, if one consideres an inner line, the curvature at the point P furthest from X is greater than the curvature at the point Q nearest to X. The effect of the tension then makes $4\pi p + \frac{1}{2}B^2$ larger at Q than at P. But p must be constant along a line of force and hence B must be larger at Q than at P. Consider the behaviour of B then as Q approaches X. Since P does not approach a null, B must remain finite, even though the direction of $\mathbf{B}$ reverses, when one crosses X. Consequently, $\mathbf{B}$ must change discontinuously at X, which implies that curl $\mathbf{B}$ has a singularity of the sort that appears to be building up in Figure 2. The conclusion is that plasma pressure is unable to *prevent* the build-up of a discharge, though it is important in the dynamical problem, as will be further discussed. It should at once be remarked that curl $\mathbf{B}$ cannot really have a singularity, but this is prevented by the breakdown of (2) at high current density, rather than by plasma pressure.

Knowing now that the plasma tends to flow in the direction of the $\mathbf{j} \wedge \mathbf{B}$ forces shown in Figure 2, the electric field should be described before discussion of the dynamical problem. The electric field given by (2) is into the paper everywhere around

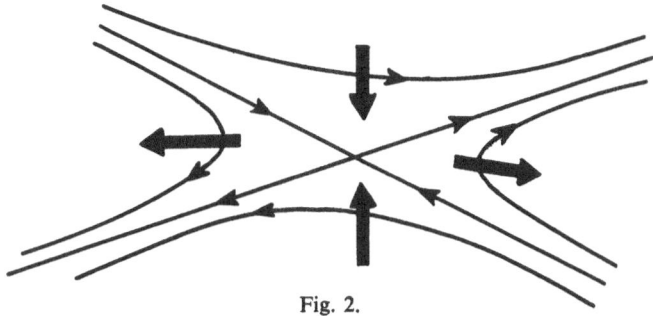

Fig. 2.

the null but of course vanishes at the null. Now some electric field is required to drive the current and this is also into the paper; it may be written $\mathbf{j}/\sigma$, but when j is large the usual theory of conduction may not be valid. Thus $\mathbf{E}$ is everywhere directed into the paper and a steady flow in this region is possible such that $\mathbf{E}$ is uniform and hence $\partial\mathbf{B}/\partial t = -c$ curl $\mathbf{E} = 0$. It is possible to maintain the picture of moving field lines. Far from the null the field lines move with velocity $\mathbf{v}$, but when $\mathbf{j}/\sigma$ is appreciable $\mathbf{E}$ is larger than the value given by (2) so that the field lines must be taken to move faster than $\mathbf{v}$. The effective velocity of the field lines cE/B has a singularity at the null and one must picture the lines as being broken when they reach the null in the obtuse angle and as being reconnected to form lines in the acute angle. This reconnection of field lines is an important feature of the discharge at a null. Field lines can be created or destroyed at an elliptical null, but their reconnection is very slow except at a hyperbolic null or at a topological generalization of this, demonstrated by SWEET (1958). In this generalisation magnetic surfaces have a cross-section containing a region similar to Figure 2, but there can be a component of B perpendicular to the paper and there need be no null.

A simple form of the dynamical problem has been solved exactly by CHAPMAN and KENDALL (1963). Taking the plasma as incompressible they find an exponential growth, with a growth rate of the order of the spatial gradient of the Alfvén velocity. The limitation of their model is that there is no outer boundary and the field diverges at infinity.

Orders of magnitude for the dynamical problem in hydromagnetic form have been discussed recently by PARKER (1963) and his discussion will be outlined but disagreed with in the following. The previous discussion, using Figures 2 and 3, shows that plasma flows towards the current sheet in the vertical direction in Figure 2, and

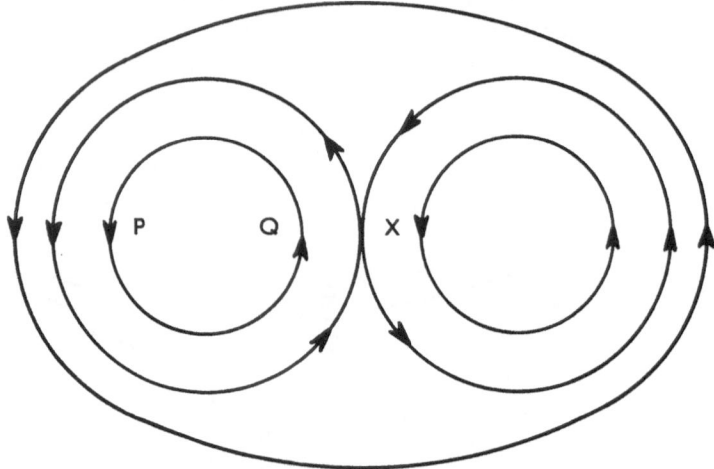

Fig. 3.

flows out in the horizontal direction, clearly in narrow jets. The plasma in the region of the current sheet is compressed magnetically and its gas pressure, which has been neglected previously, may have an important effect on the flow. SWEET (1956) studied this aspect of the flow by considering two flat plates being pressed together and squeezing the plasma between so that it comes out in narrow jets. Parker uses this formulation to obtain numerical estimates, but I believe that the plates should be curved and this makes a big difference to the numbers; I hope to settle the question by two-dimensional computation. The size of the plates is characterized by a large length L and the distance between them by the small thickness l. The inflow has a small speed v and the outflow a larger speed V, which is taken to be the speed of sound in the thin layer, and this is comparable to the Alfvén speed outside. Parker discusses the compressible case, but this is not essential, and for the incompressible case the condition of continuity gives him $Lv=lV$. I will differ from him in taking the plates to have radius of curvature L, which means the velocity is a maximum at distance $\sim(lL)^{\frac{1}{2}}$ from the centre and then continuity gives

$$\frac{v}{V}=\left(\frac{l}{L}\right)^{\frac{1}{2}}. \tag{6}$$

Previous discussion of the electric field showed that j/σ in the discharge should equal vB/c outside and this leads to the other equation used by Parker

$$v = c^2/l\sigma \tag{7}$$

the right-hand side being the velocity of diffusion of a field of scale 1.

From (6) and (7)

$$v^3 = V^2c^2/L\sigma \tag{8}$$

and

$$l^3 = Lc^4/v^2\sigma^2 \tag{9}$$

and these will be used to obtain estimates for solar flares, with the reservation that the effective value of σ may be reduced from Spitzer's value by runaway effects and double stream instabilities, reduction of σ increasing both v and l. Reasonable guesses are $L = 10^4$ km and $V = 100$ km/sec. Giovanelli has suggested that the discharge is above the chromosphere, so perhaps the coronal temperature of 10^6 ° should be used, giving $\sigma/c^2 = 10^{-14}T^{3/2} = 10^{-5}$ sec/cm^2, though this may be an overestimate. These numbers give $l = 50$ cm and comparison with (1) suggests that the velocity difference between the electrons and protons is large. They also give $v = 20$ m/sec and hence $E = 2B$ volts/km. (In previous publications I have used V to calculate E, but I accept Parker's criticism on this point.) With fields of hundreds of gauss, and mean free paths of many kilometres, there is no doubt that the electric field is strong enough to give runaways. The total voltage should be of order EL and 10^7 volts is easily obtained, but particles are sometimes observed with energies as high as 10^{10} volts. A transient increase lasting a relatively short time seems most likely, but this is bound up with the theory of the conductivity and the possibility of instabilities. It seems best to try to compute a laminar flow model and then to study instabilities; a variety of instabilities are possible and a great deal remains to be done. The motion of individual particles in the discharge also needs to be discussed, but first let us finish the discussion of flares with some mainly qualitative remarks.

One feature of flares is the rapid onset, and Parker finds the onset of the discharge at a null too slow. During the build-up, however, the electric field at the null may be much less than that outside and (7) is quite invalid. The growth of current may be very slow at first, but at some stage the forces of Figure 2 must dominate and then the time of growth should be $\sim L/V \sim 100$ sec. It is only this final phase of growth that is observed in the emissions, but observations of the magnetic fields by Severny seem to accord with this view.

With regard to flares in general, one remarkable thing is the great number of features which require at least a qualitative explanation, and a discharge which reconnects lines of force shows promise of accounting for them all. The reconnection of lines of force must upset the hydromagnetic equilibrium of the spot group field, and, since this occurs rapidly, a sudden hydromagnetic disturbance is expected to propagate outward from the flare region. In the following table the various features of flares are listed in two groups, one of which is likely to be caused by energetic particles and the other by the hydromagnetic disturbance.

TABLE I

Features due to energetic particles	Features due to hydromagnetic disturbances
"Solar protons"	Magnetic storms
Hα	Moreton waves
Ultraviolet	Activation of prominences
X-rays	Multiple flares
γ-rays	Some radio noises
Synchrotron radio noise	

Finally, some important recent observations by Severny have shown Doppler shifts which he interprets in terms of the same sort of flow pattern as that we have been discussing, including jets moving out in opposite directions.

I will now turn to the motion of individual particles in the neighbourhood of a hyperbolic null. This is important not only for investigating their acceleration to high energies, but also as a first step to understanding the overall behaviour when collisions do not dominate. The neighbourhood of the null is exceptional in this case in that the adiabatic approximation breaks down. Far enough away the adiabatic theory gives particle drifts as follows. The electric field gives a drift which is just given by (2). The

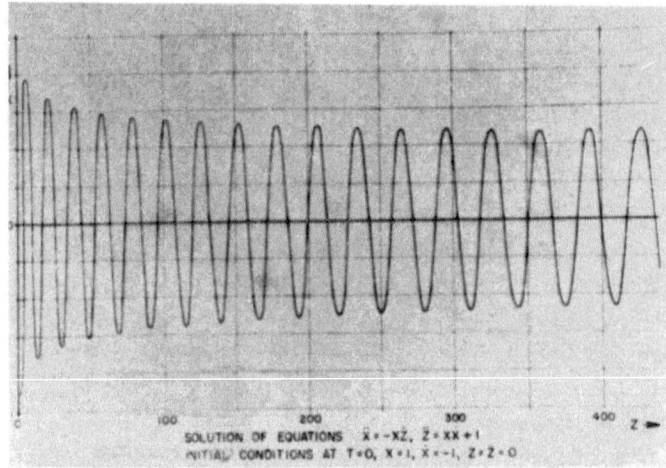

Fig. 4.

two components of drift due to the gradient of magnetic field strength and to curvature of the field lines are both perpendicular to the paper in Figure 2. They are such that the particle gains energy from the electric field when it is near enough to the plane of the discharge (nearer than the limiting lines of force), but loses energy otherwise.

To study the motion of particles in the non-adiabatic region, we may start from a simple case and proceed to successively more complicated ones. The simplest case has no electric field, $B_x = B_z = 0$ and $B_y \alpha x$. The equations can be integrated to give $\dot{x}$ as a function of x^2, and it is seen that the solutions are oscillatory in x. Some trajectories cross the plane $x = 0$ and are easily understood from Figure 4, the curvature of the trajectory changing sign on crossing the plane $x = 0$. The next complication is to add a uniform electric field in the Z-direction. An analytic solution is not then possible, but the equations can be scaled and may be written

$$\ddot{x} = -x\dot{z} \tag{10}$$

$$\ddot{z} = x\dot{x} + 1. \tag{11}$$

Reversing the direction of time reverses the magnetic field but not the electric field. Figure 4 shows a solution computed by T. Speiser of Pennsylvania State University for a particle starting at $x = 1$, with $\dot{x} = -1$, $\dot{z} = 0$, corresponding to Equation (2). The oscillations in x die out very slowly. The particle gains energy, and in this model remains in the current sheath for ever. To be realistic it is necessary to add other weak components to the magnetic field, e.g. $B_y = 10^{-2} z$, which will eventually bring the particle away from the null. A few computations of this sort have been run on an analogue computer by D. Fairfield of Pennsylvania State University. Far enough away from the null adiabatic behaviour is re-established and, if the field

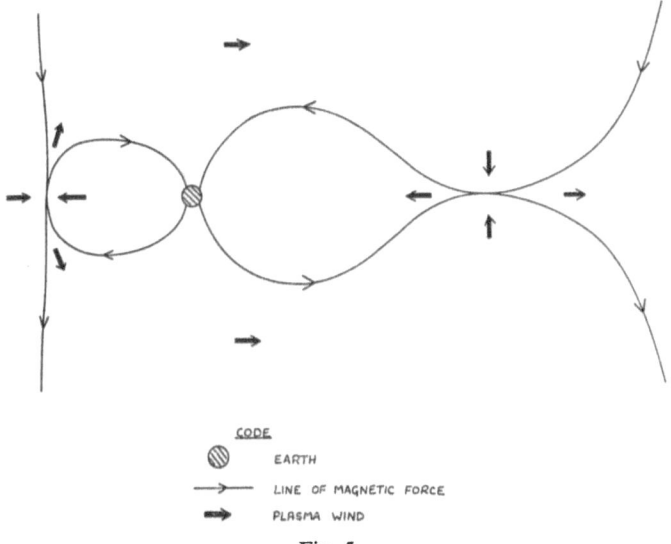

CODE

⊘   EARTH

——→   LINE OF MAGNETIC FORCE

⇒   PLASMA WIND

Fig. 5.

increases with distance, the particle is likely to be reflected by the mirror effect. In the computations, of course, it was necessary to stop when the particle had gone far from the null. It was noticeable that the stopping points clustered near a certain plane. Remembering the definition of the limiting lines of force by Equation (4) this plane was the plane defined by the two limiting lines with the same polarity, i.e. corresponding to values of λ of the same sign. Thus particles accelerated in the discharge tend to come out in a thin fan, but the consequences of the mirror effect further out need further investigation. I should also remind you that instability may lead to rapid time variations in the field, whereas I have discussed only a steady field model.

The role of null points in auroral theory has been discussed previously (DUNGEY, 1962, 1963), the model being shown in Figure 5. Here I will content myself with drawing your attention to the sudden large change in direction found at the magnetospheric boundary by Dr. Cahill from his magnetometer system on Explorer XII. On the nightside Explorer X found a southward tendency in the distant field, but I hope that much more data is going to be presented from Explorer XIV. The magnetospheric boundary offers much the best prospects for observational study of current sheaths and should be ample justification for further theoretical work.

References

CHAPMAN, S. and KENDALL, P. C.: 1963, *Proc. Roy. Soc. (London)* **271**, 435.
DUNGEY, J. W.: 1953, *Phil. Mag.* **44**, 725.
DUNGEY, J. W.: 1962, *J. Phys. Soc. Japan*, Supp. AII, 15.
DUNGEY, J. W.: 1963, *Geophysics* (Les Houches Summer School), Gordon and Breach.
PARKER, E. N.: 1963, *Astrophys. J. Supp.*, in press.
SWEET, P. A.: 1956, *Electromagnetic Phenomena in Cosmical Physics* (ed. by Lehnert), Cambridge University Press, p. 123.
SWEET, P. A.: 1958, *Nuovo cimento*, Supp. Ser X, **8**, 188.

Note added in proof. Since this was written Petschek has made an important advance on the topic. See PETSCHEK, H. E.: 1964, *AAS-NASA Symposium on Solar Flares* (ed. by W. N. HESS, published by NASA), p. 425.

Discussion and Questions

H. Alfvén: In the terrella experiments which seem very similar, an aurora has nothing to do with the null point. The particles seen in an aurora of course are accelerated along the magnetic lines of force but they have nothing to do with the null point and whether the real aurora has anything to do with the null point is of course an open question, but I doubt it very much.

J. Dungey: May I ask, when you say it has nothing to do with the null point, do you agree that the null point is connected by lines of force to the terrella?

H. Alfvén: Yes.

J. Dungey: And there are certain points on the terrella such that if you follow the lines of force from that point you would arrive at a null point.

H. Alfvén: Yes, but you definitely have precipitation along lines of force which never reach the null point.

J. Dungey: I see, but are you saying there is no precipitation along lines of force which do connect to the null point?

H. Alfvén: I am not quite definite but I don't believe it.

J. Cladis (Lockheed): We have made an experiment at Lockheed, Palo Alto, that is very similar to the experiment that was described by Prof. Alfvén. Fire a hydrogen plasma at a magnetic field and take pictures. Our plasma however is not magnetized and we do get the point, rather the luminous belt when the magnetic field pressure equals the plasma pressure. But we did also make extensive magnetic field measurements in the boundary layer and by differentiating these we could arrive at a current distribution. We find that while the plasma does enter the magnetosphere through the neutral point, the current flows in a certain direction around in the equatorial plane so that the current sheath circles the region at which the plasma enters along the lines of force at the surface of the terrella.

THE ARTIFICIAL RADIATION BELT MADE ON JULY 9, 1962*

W. N. HESS

Theoretical Division, Goddard Space Flight Center, Greenbelt, Mld.

Abstract. The available information on the artificial radiation belt formed by the July 9, 1962, high-altitude nuclear explosion is reviewed. Data from Injun (1961o_2). Telstar 1 (1962$\alpha\varepsilon_1$), Traac (1961$\alpha\eta_2$), and Ariel 1 (1962o_1) are combined to form one picture of the artificial belt. The data are consistent to about a factor of 3. The flux map obtained in this way is used to calculate the flux encountered by several satellites. These show reasonable agreement with data on solar cell damage. Crude calculations of shielding are made to indicate the doses received inside various vehicles.

Introduction

On July 9, 1962, at 09.00.09 UT a nuclear explosion of about 1.4 megatons, code named Starfish, was carried out at 400 kilometers above Johnston Island. This explosion produced, as was expected, an artificial radiation belt. Three days afterwards the US-UK satellite Ariel stopped transmitting useful data; on August 2 Transit 4-B stopped transmitting; and Traac stopped on August 14. Instruments on Ariel, Injun, and Traac showed large particle fluxes shortly after the explosion. It took about a month after the explosion to start getting some feeling about the characteristics of the new radiation belt. This is a status report on the new belt using data for a few weeks after Starfish.

Available Data

The information that is available to form a picture of the new radiation belt comes mostly from particle detectors on the Ariel, Injun, Telstar, and Traac satellites. Besides these data, we can use the observed satellite solar cell damage as an integral measurement of the trapped electron flux. Also, some data are available from dosimetry measurements. These data are used to form a picture of the artificial belt one week after Starfish.

Some of the first data about the enhanced trapped particle fluxes after the July 9 explosion came from the X-ray detector (BOYD, WILLMORE, and QUENBY, 1962) on the Ariel satellite. This instrument was not designed to count charged particles, and its efficiency is uncertain. The data from it are quite useful in studying time decay of the trapped flux and in locating contours of constant flux in B-L space.

Data received by the shielded 213 GM counter on Injun have been analysed (O'BRIEN *et al.*, 1962) to give the first picture of the new radiation belt. This counter is the background channel of the magnetic spectrometer, SpB. It has $3\frac{1}{2}$ g/cm^2 of

---

* This paper was already published in *Journal of Geophysical Research* **68** (1963) 667–683.

Chang & Huang (eds.), Proc. Plasma Space Sci. Symp. All rights reserved.

lead shielding and about 1 g/cm^2 of wall and miscellaneous shielding. It was supposed to give the penetrating background to be subtracted from the other channels of the spectrometer. This detector is now called on to provide quantitative information, and it has been calibrated after the fact. This detector is nearly omnidirectional. Fluxes are obtained from the count rates by dividing by the $G_0 = 0.11$ cm^2. Other detectors on Injun will also give useful data sometimes, but often they will be saturated and not usable. So far, few data have been analysed from any Injun detectors except SpB.

Telstar has on it a solid state p-n junction detector (BROWN, 1962) with pulse height analysis that selects electrons in different energy ranges from 0.2- to 1-MeV ranges. A lot of data have been reduced from Telstar for two channels of the electron detector. This detector has given all the available data at large altitudes up to September. It is directional with an aperture half-angle of about 10°. The flux measured is spin-averaged by the vehicle motion. The fluxes are made omnidirectional by multiplying by the appropriate solid angle factor and then using a correction factor between 1 and 2 to correct roughly for the non-isotropic angular distribution.

Traac has on it a 302 GM counter (PIEPER and FRANK, 1962) shielded by 0.265 g/cm^2 of magnesium and 0.40 g/cm^2 of stainless steel, which will count electrons of $E > 1.5$ MeV. This is essentially omnidirectional. Fluxes are obtained by dividing by $G_0 = 0.75$ and correcting for saturation for high count rates.

An analysis of Ariel data (DURNEY, ELLIOT, HYNDS, and QUENBY, 1962) shows that the radiation belt was enhanced out as far as 5 or 6 earth radii.

Analysis of Data

The data from the satellites must be combined to form one over-all picture of the artificial radiation belt. To do this, we will assume that the energy spectrum of the electrons being counted is a fission spectrum. This is certainly the best guess. We will compare the data on this basis and see if, in the regions where direct comparison is possible, there is agreement. This fission energy spectrum $N(E)$ is shown in Figure 1, curve A. Using this electron energy spectrum we can calculate what fraction f of the electrons will be counted by the different detectors. A calibration of the Telstar detectors at Los Alamos in a fission electron beam gives, for the 240- to 340-keV channel, $f = 1/2.8$, and, for the 440- to 680-keV channel, $f = 1/6.0$.

For Injun we have the experimentally determined factor $1/f$ of several thousand determined by comparing detectors on the satellite. This detector has been calibrated at Los Alamos with a fission electron spectrum (PETSCHEK, MOTZ, and TASCHEK, 1962); the factor f determined this way is $1/4000$. We will use this factor in our analysis. The Los Alamos tests show that the detector counts bremsstrahlung from several MeV electrons rather than direct penetrating electrons. (If the shield had been carbon rather than lead, the counter would have counted direct penetrating electrons.)

For Traac, we determine f by considering the penetration of electrons through the detector shield of 265 mg/cm^2 of magnesium and wall of 400 mg/cm^2 of stainless steel.

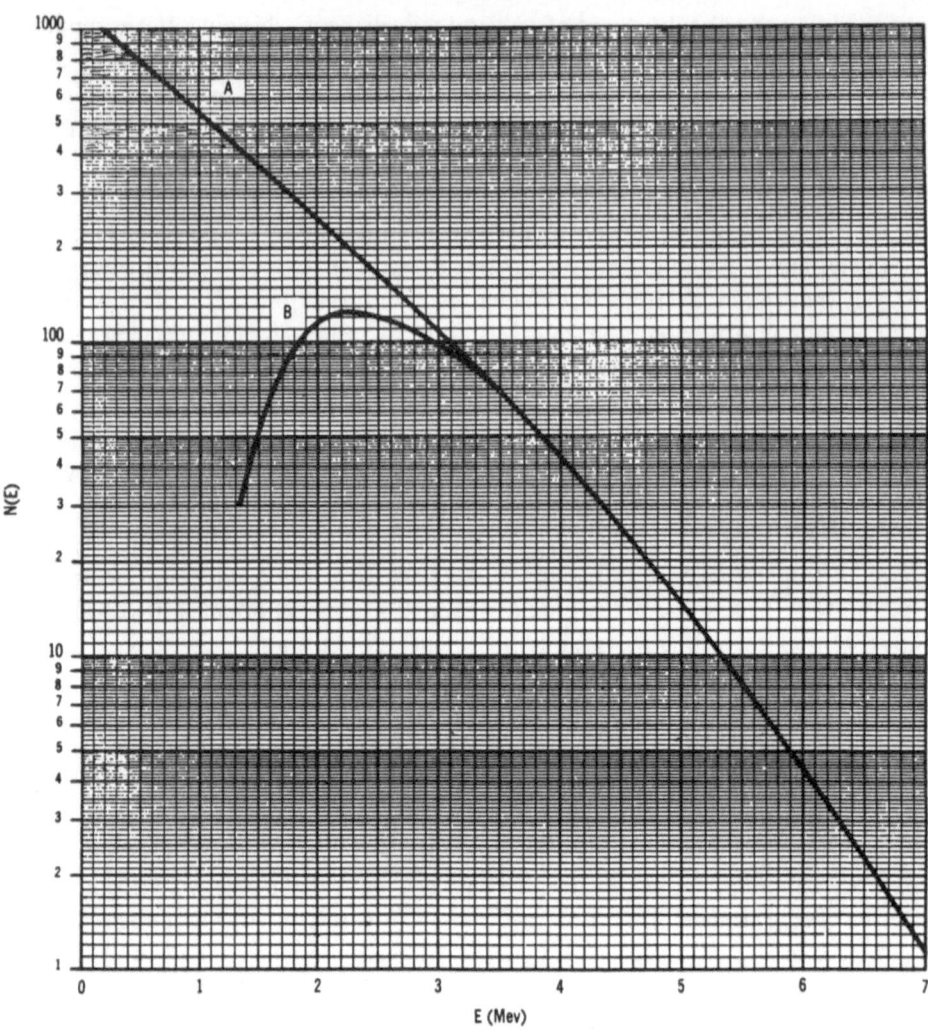

Fig. 1. Curve *A* is the fission energy spectrum and curve *B* the transmission energy spectrum for the Traac GM counter (0.66 g/cm² wall).

Using the range straggling data (MARSHALL and WARD, 1937) for aluminium, we can get the fraction of electrons that penetrate a certain thickness shield as shown in Figure 2. The expression for the extrapolated range *R* is

$$R = 0.526E - 0.094 .$$

This gives the absorber thickness that gives 10 per cent transmission for electrons of energy *E*. For 50 per cent transmission we multiply the energy by 1.38, and for 80 per cent transmission we increase *E* by 1.92. In this way we get the electron transmission spectrum curve B in Figure 1. The integral under this curve gives $f = 1/5.5$ for the Traac counter. The energies of the transmitted electrons are different from curve B, but the number transmitted is given correctly.

Using these factors for the several detectors, we can calculate the total flux of fission electrons. To compare the different detectors, we have plotted the total flux

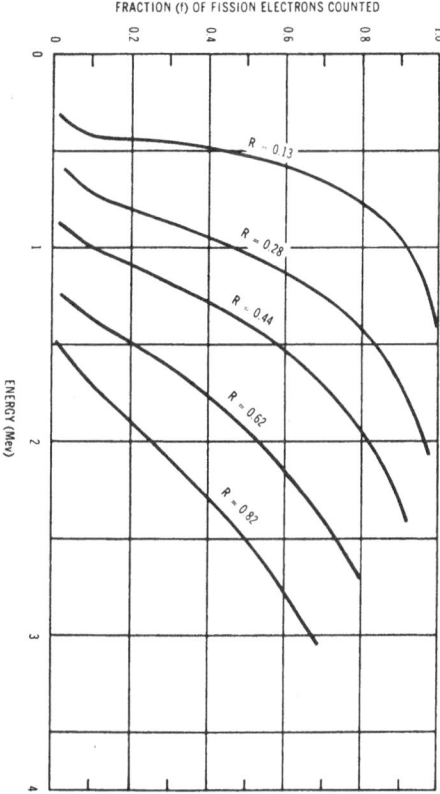

Fig. 2.   The fraction of electrons of different energies that penetrate different shield thicknesses of aluminum.

along several field lines (actually narrow ranges of L) for different values of B. These plots in Figure 3 show that the different detectors agree fairly well in flux values. Avoiding the first day after the nuclear explosion (labelled by the number 0 inside the symbols on the graphs) we can see quite smooth trends in the data. The flux from Telstar may be as much as two times higher than Injun fluxes. Traac and Injun agree quite well where comparisons are possible. In general, the data show agreement to a factor of times 2.

This agreement of the data shows two things: First, because the several detectors give internally consistent results, it seems likely that the detectors are all giving accurate information. Second, it seems that our assumption that the electrons have

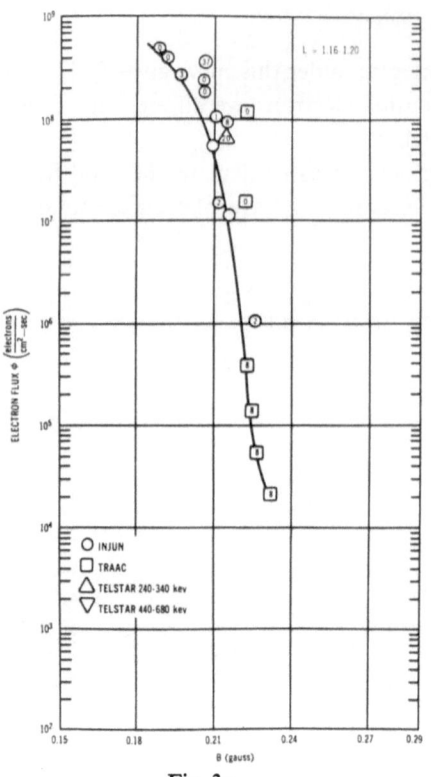

Fig. 3a.

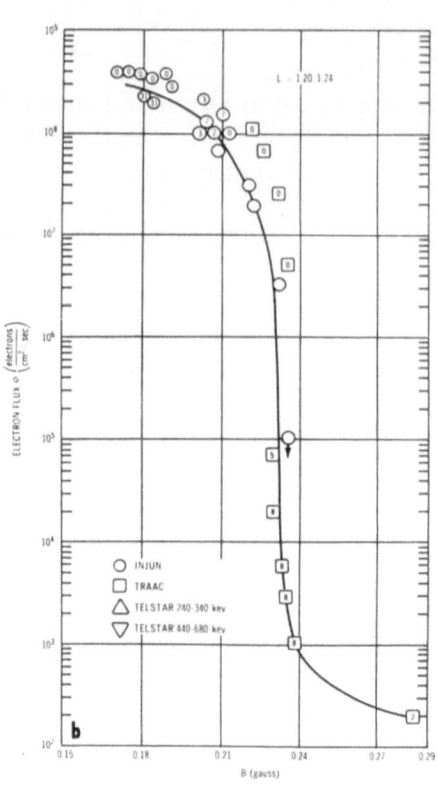

Fig. 3b.

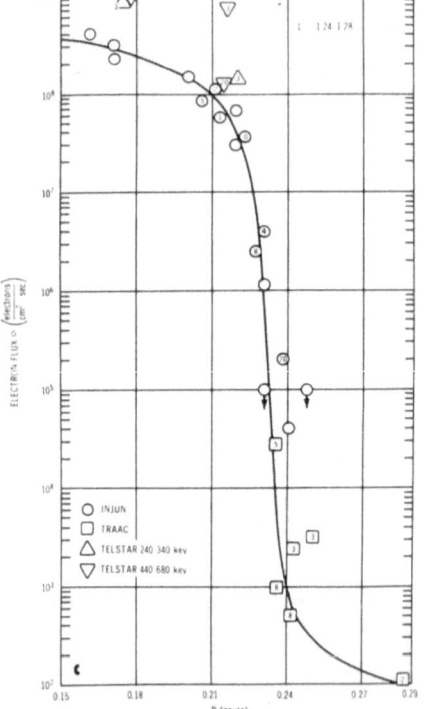

Fig. 3c.

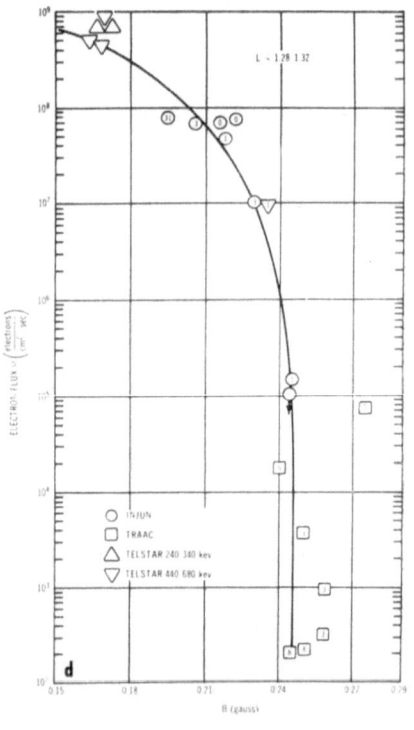

Fig. 3d.

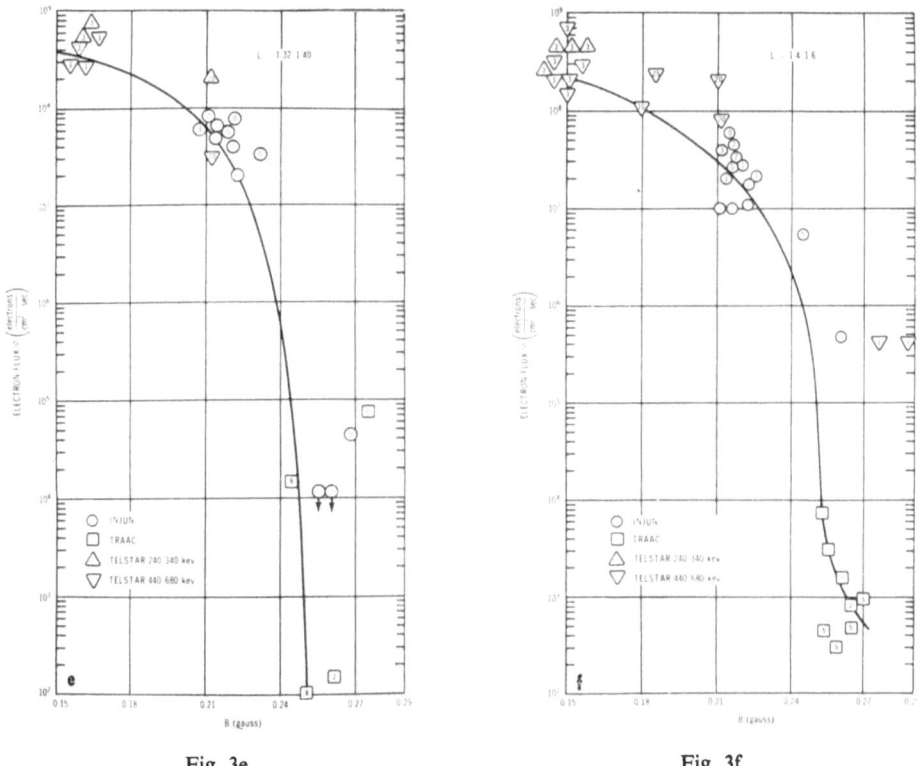

Fig. 3e.   Fig. 3f.

Fig. 3.   The electron flux distributions along different field lines.

a fission energy spectrum is correct, at least in the region of data overlap. It is, of course, possible that the energy spectrum is not a fission spectrum and also that the detectors are not in agreement, but it would have to be a peculiar combination of two such effects that would give the agreement shown here. A comparison of the four channels of the Telstar electron detector also indicates that the energy spectrum is similar to a fission spectrum up to 1 MeV.

Flux Plots

Now we can use all the counter data to construct a composite flux map in B-L space. As McIlwain (1961) has shown these magnetic coordinates are the best way of organizing data about trapped particles. L is constant along a field line in space and for a dipole is the distance from the center of the earth to the equatorial crossing of the line in units of the earth radius (see Figure 4). Values of L are calculated from the real values of the earth's field.

In constructing the flux map for $B > 0.15$ and $L > 2.0$, we use the graphs in Figure 3 to locate the flux contours (Figure 5, region A). Outside of this B-L region the experimental data are essentially all from Telstar. There are several weeks' data from Telstar and considerable redundancy. This flux map in Figure 5 is drawn for about one week after the explosion.* There was considerably more flux at low altitudes at early times. More data are available now than were used to construct Figure 5, and the outer contours are known to be different (BROWN, 1962). Very likely not all of the electrons seen by Telstar are artificial electrons, although enhancements of the

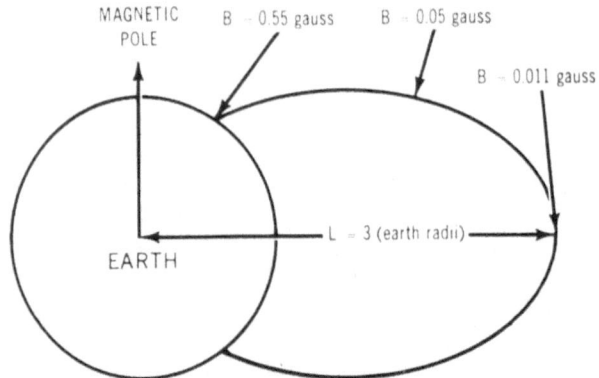

Fig. 4. The *B-L* magnetic coordinate system.

ARTIFICIAL ELECTRON BELT　　(B,L)

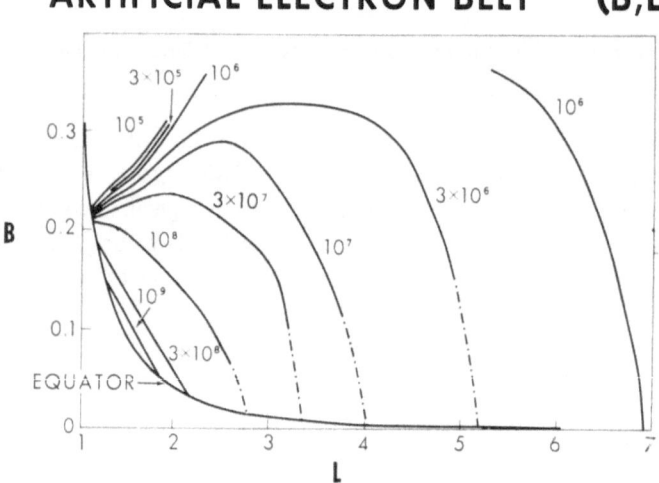

Fig. 5. The *B-L* map of electron fluxes. In region *A* data from the several satellites overlap. Outside this region only Telstar gives data. The dashed curves are extrapolations where no data exist.

* More up-to-date data are given by BROWN and GABBE (1963), including data based on more complete instrument calibration. Their *B-L* map is better than Figure 5 here in that it shows more structure, but the regions of high flux are quite similar. The total number of electrons in Brown's *B-L* plot is smaller by a factor of 3 than here.

radiation belt were seen by Ariel out to 5 or 6 earth radii (DURNEY *et al.*, 1962). The energy spectrum of the electrons in the $\Phi = 10^9$ contour at $L = 1.9$ seems softer than a fission spectrum. We are dealing at this point with an unknown mixture of fission electrons, bomb neutron-decay electrons, and natural electrons. Only Telstar observed these electrons for more than two months after July 9. Subsequent observations with other satellites are complicated by decay and related changes in spectrum.

If we replot Figure 5 in R-λ coordinates where

$$B = (M/R^3)\sqrt{4 - 3R/L} \qquad R = L\cos^2 \lambda$$

we get an equivalent dipole representation of the earth's field shown in Figure 6. The maximum electron flux is about 2×10^9 electrons/cm^2 sec. Integrating to get the total number of electrons stored in the field

$$\int \Phi \, dV = 2 \times 10^{26} \text{ electrons}.$$

About 40 per cent of these electrons lie inside the 10^9 contour, and about 60 per cent of these electrons lie inside the 3×10^8 contour. It is not certain what fraction of these electrons are fission electrons.

ARTIFICIAL ELECTRON BELT

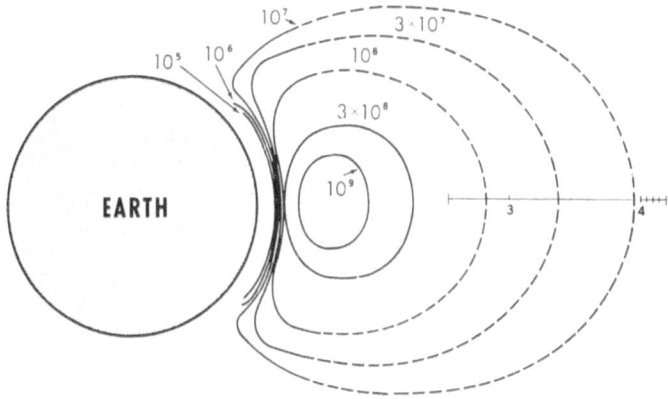

Fig. 6.   The R-λ map of electron fluxes (an ideal dipole representation of the earth's field).

The *B-L* flux map, when replotted in terms of geographic coordinates, gives the flux contours at different altitudes shown in Figure 7. Essentially all the electrons in Figure 7 up to 1000 km are known to be fission electrons.

Vehicle-encountered fluxes

A machine code has been developed that calculates the total number of electrons per cm^2 in the artificial radiation belt that run into a vehicle in space. This is done by calculating a point on the vehicle trajectory, then transforming to *B-L* coordinates,

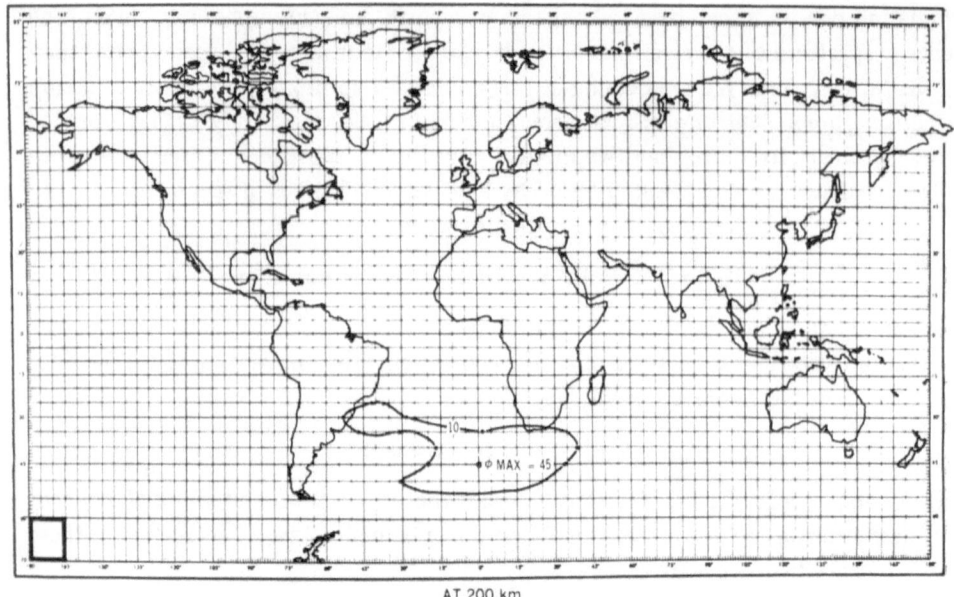

AT 200 km

Fig. 7a.

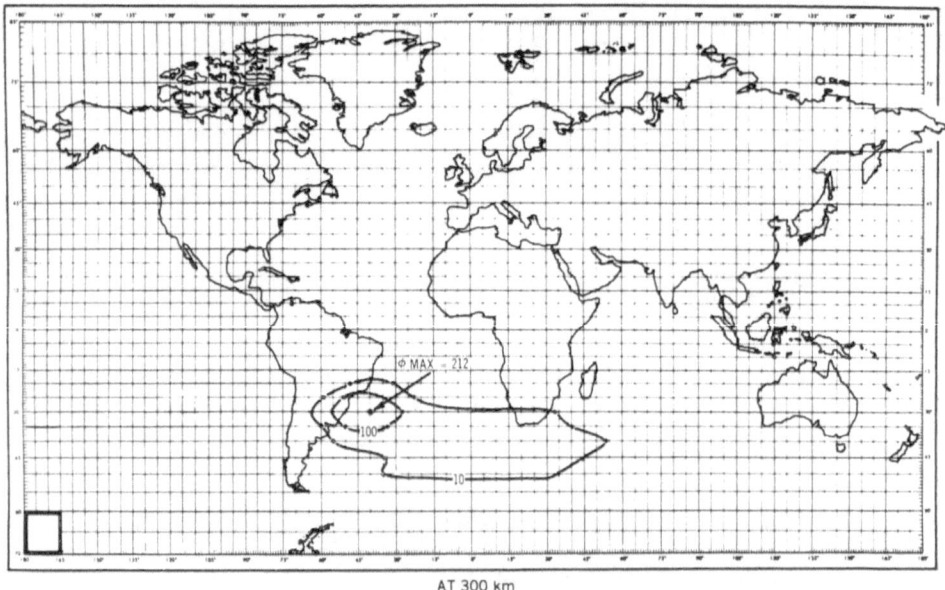

AT 300 km

Fig. 7b.

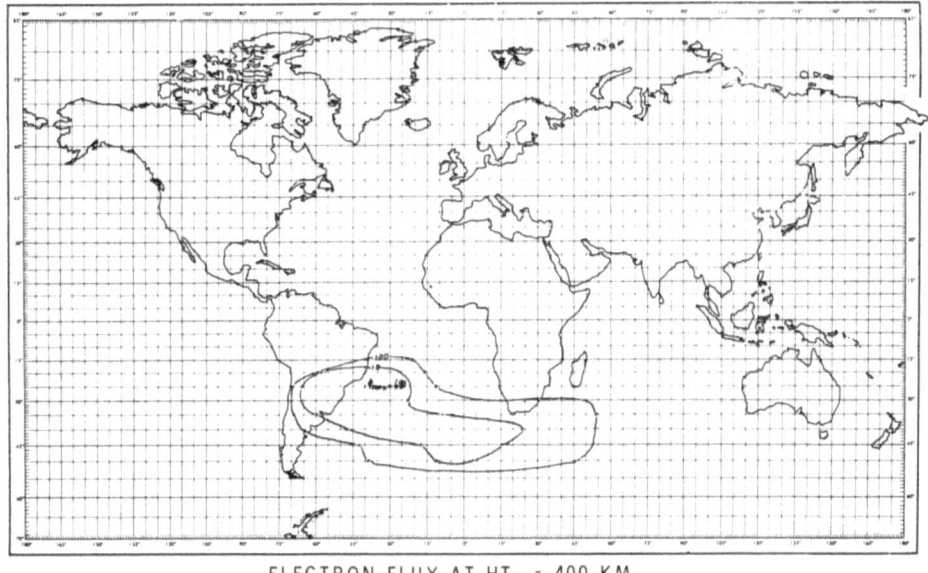

ELECTRON FLUX AT HT. = 400 KM.
electrons/cm.^{2} -sec. x 10^{-5}

Fig. 7c.

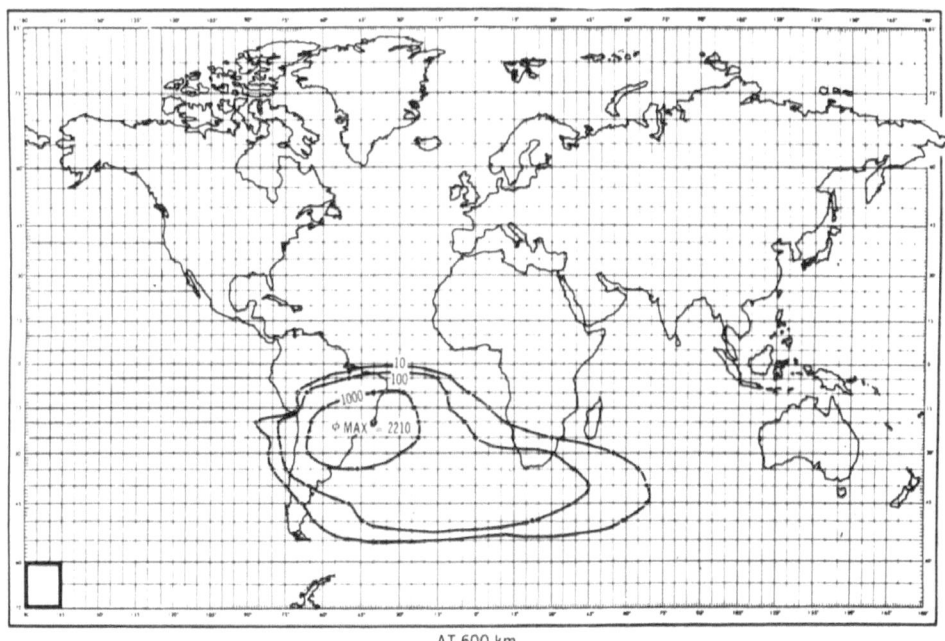

AT 600 km

Fig. 7d.

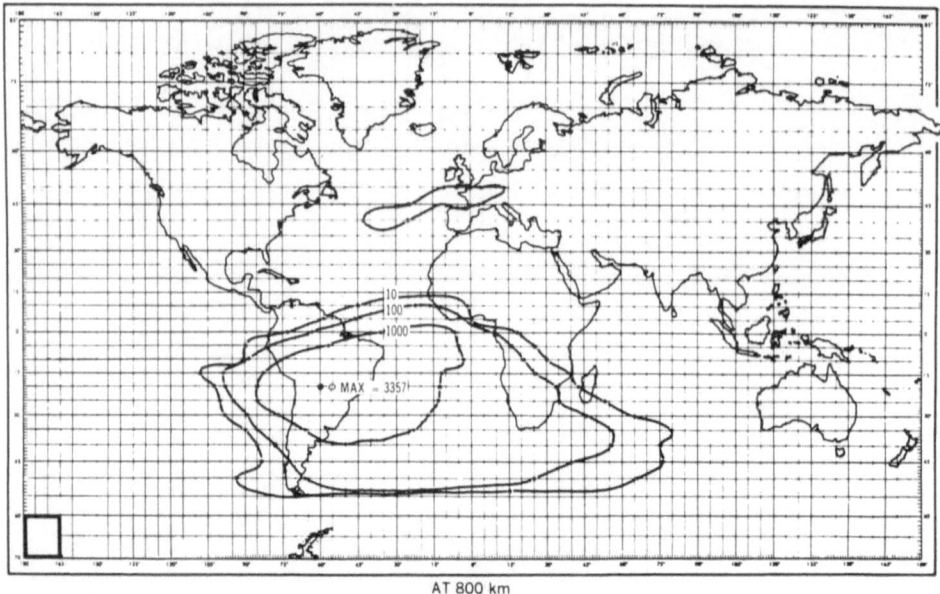

AT 800 km

Fig. 7e.

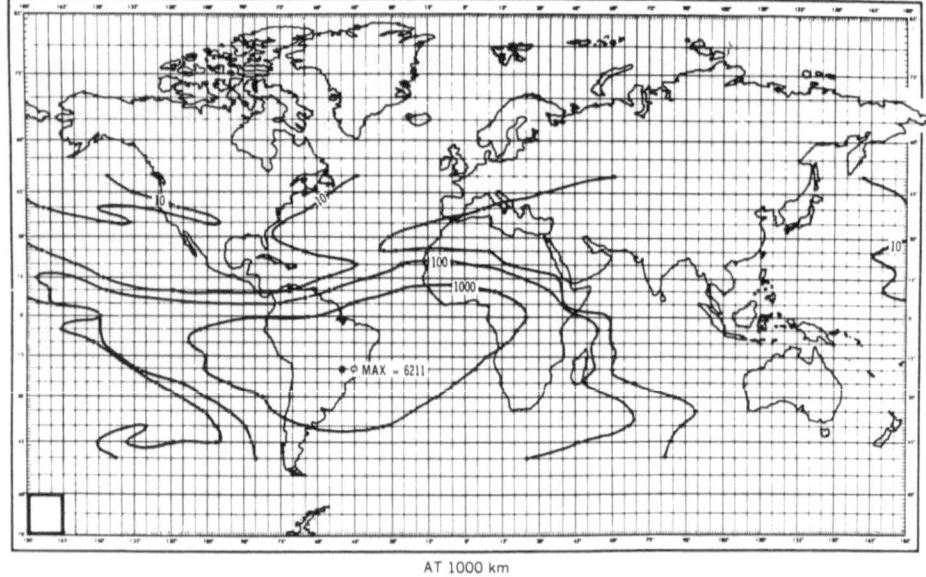

AT 1000 km

Fig. 7f.

Fig. 7.   Electron flux maps at different altitudes above the earth's surface.
Flux is in units of 10^5 electrons/cm² sec.

looking up the electron flux, and integrating along the vehicle orbit. This has been done for all the vehicles listed in Table I and the encountered fluxes determined. These fluxes have been transformed into R/day by using 3×10^7 electrons/cm^2 = 1 R. The orbital elements of these vehicle trajectories are given in Table II.

From the encountered fluxes in Table I, we can learn several things. Let us first consider solar cell damage. The Bell Telephone Laboratories staff (ROSENZWEIG et al., 1962) have studied this problem in considerable detail and have prepared Figure 8, which shows how different type cells are damaged by 1-MeV electron irradiation.

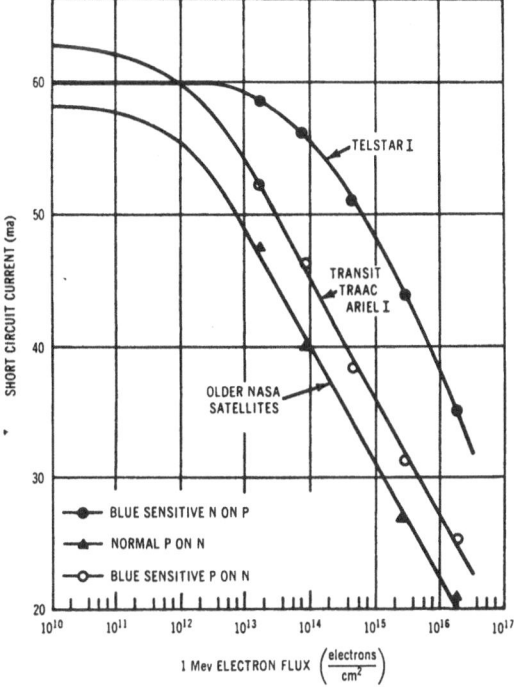

Fig. 8.   Solar cell damage curves.

Above about 0.5 MeV the electron damage is essentially independent of energy. Some care is necessary in using this chart because of the variation in characteristics of solar cells.

About 20 per cent degradation was needed by the blue sensitive p on n type cells on Ariel to produce the observed power supply damage (FRANTA, 1962). This would be caused by about 10^{13} electrons/cm^2 according to Figure 8. In about seven days after the nuclear explosion this flux would be achieved (Table I gives 2.8×10^{12} electrons/cm^2/day for Ariel, of which half would hit the face of the cells). The Ariel power supply started malfunctioning in three and one-half days. This is quite good agreement.

Traac and Transit also had blue sensitive p on n solar cells, but it would take 4×10^{14} electrons/cm^2 to cause malfunction here (FISCHELL, 1962) because the cells

W. N. HESS

were lower efficiency cells to begin with. Table 1 gives 4.5×10^{12} electrons/cm^2/day encountered. Half of these electrons will hit the face of the cells. Traac stopped transmitting in 36 days and Transit 4B in 24 days. Using 30 days as the average, we get a total encountered flux of 0.7×10^{14} electrons/cm^2, only in fair agreement with that required to produce damage.

Telstar used the much more damage-resistant n on p cells because it was to fly routinely through the inner belt protons. Even in the new artificial radiation belt, its power supply lifetime is expected to be considerably longer than one year.

The Telstar solar cells are degrading at a rate that would be produced by 6×10^{12} electrons/cm^2/day of 1 Mev hitting the bare cells (BROWN, 1962). This corresponds to about 1.8×10^{13} electrons/cm^2/day incident on the outside of the 30-mil sapphire covers. Our calculations give $\frac{1}{2} \times 2 \times 10^{13} = 1 \times 10^{13}$ electrons/cm^2/day hitting the cells.

TABLE I

CALCULATIONS ON FLUXES ENCOUNTERED BY

	Ariel	Traac and Transit 4B	Telstar	Tiros 5	Oso 1
Perigee, km	390	960	952	590	552
Apogee, km	1210	1106	5660	971	594
Inclination	54°	32°	45°	58°	33°
Altitude, km					
30 °S latitude	1067	1000	5138	963	594
30 °W longitude			1758		
Calculated R/day					
outside vehicle	110000	180000	800000	46000	27000
Length of machine run					
in satellite days	4	4	4	4	4
Elec/cm²/day	2.8×10^{12}	4.5×10^{12}	2.0×10^{13}	1.15×10^{12}	6.8×10^{11}

* Similar to projects designated internally as Nimbus, S-27, S-48, S-66.
** Six orbits only.

TABLE II

ORBITAL ELEMENTS OF

Vehicle	Ariel	Traac and Transit 4B	Telstar 1	Tiros 5	Oso 1
Epoch; days, hours, min, sec	190, 9, 0, 0	190, 4, 3, 46,506	191, 8, 51, 0	190, 9, 0, 0	190, 9, 0, 0
Semimajor axis, earth radii	1.1254	1.1618	1.5182	1.1224	1.0900
Eccentricity	0.05714	0.009922	0.2430	0.02663	0.003012
Inclination, deg	53.866	32.423	44.803	58.102	32.855
Right ascension of ascending					
node, deg	− 24.881	96.434	− 156.222	− 75.536	154.502
Argument of perigee, deg	− 9.2537	− 51.6890	164.811	118.014	139.136
Mean anomaly, deg	− 86.8833	0.0001	1.1684	− 194.11968	− 164.5453

The observed solar cell degradation on Telstar should be somewhat more than that calculated from the artificial electron belt because slow proton damage probably contributes some to the degradation (BROWN, 1962).

We have neglected the enhanced early time effects here on all the exposed satellites. An appreciable part of the encountered flux may have come in the first few days when the flux was higher than that given in Figures 5 and 6 for +1 week.

Injun and Tiros and other satellites continue to function. Injun has a low duty cycle and Tiros shows some solar cell degradation.

Film badge dosimeter measurements have been made on several space flights. Roughly $10R$/day were measured inside 1.5 g/cm^2 of shielding. To compare this radiation dose with the predictions in Table 1, we must correct for the shielding by using a calculation like that for the Traac GM counter to get f, the fraction of fission

SATELLITES MOVING THROUGH THE ARTIFICIAL RADIATION BELT

Relay	1000-km Polar*	800-km Polar	OAO	Serb	MA-7**	Pogo
1343	1000	800	802	278	160	257
5555	1000	800	817	16668	264	931
50°	90°	90°	31°	17°	33°	90°
4371	1000	755	810	...	261	804
1516						
1.1×10^6	80000	27000	...	...	80	22000
4	4	4	4	4	9.5 hours	4
2.7×10^{13}	2×10^{12}	9×10^{11}	2×10^{12}	1.2×10^{13}	0.24×10^{10}	5.6×10^{11}

VARIOUS SPACE VEHICLES (see Table I)

Relay	1000-km Polar	800-km Polar	OAO	Serb	MA-7	Pogo
305,0,	190,9,	190,9,	153,0,	303,10,	268,14	82,3,55,
0,0	0,0	0,0	0,0	0,0	0,0	32.101
1.5407	1.1568	1.1254	1.1270	2.3284	1.0331	1.0931
0.2143	0.1490×10^{-7}	4.4703×10^{-7}	0.001074	0.5518	0.008552	0.04830
50.0003	90.000	90	30.982	17.0	32.546	90.001
163.708	− 158.175	− 158.175	38.6643	17.2908	75.069	− 73.806
− 167.526	− 180.000	− 180.000	− 68.8243	134.6735	78.188	19.408
7.8219	0.0000	0	36.7572	0	7.6908	2.1956

W. N. HESS

electrons that penetrate the wall. We have calculated f for different thicknesses of shield using the relationship $R = 0.526E - 0.094$, and also using the associated rough-straggling transmission curves in Figure 3. Figure 9 shows a plot of $1/f$ as a function of shield thickness. This is really only good for aluminium, but for lack of better information we will use it for other materials too. For 1.5 g/cm² we get $f = 1/50$ for normal incidence particles. To correct for a distribution of incidence angles we will

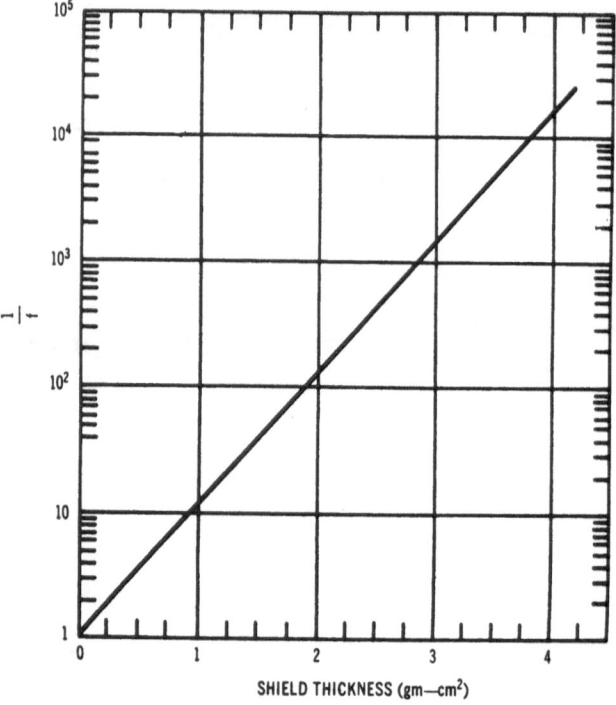

Fig. 9.   The fraction of fission electrons that penetrate different shield thicknesses.

say roughly that about half as many get through. Also, since 2π ster are covered by a much thicker shield, the total factor $f = 1/200$. This says that $10R/\text{day} \times 200 = 2000R/\text{day}$ were incident on the outside of the vehicle. This agrees to within a factor of 2 with the calculated vehicle-encountered flux.

Manned Flight

For a Mercury orbit with an apogee of 264 km, the total encountered flux in six orbits is 2.4×10^9 electrons/cm² outside the vehicle. The breakdown by orbits is shown in Table III. If the apogee is lowered by 30 km to 234 km, the total flux for 6 orbits is reduced to 0.17×10^{10} electrons/cm². If the apogee is raised by 30 km to 294 km, the total flux for 6 orbits is increased to 0.45×10^{10}.

TABLE III

FLUX ENCOUNTERED BY MERCURY

Orbit	Flux per Orbit, electrons/cm²
1	5.0×10^6
2	2.1×10^7
3	4.8×10^7
4	2.9×10^8
5	6.4×10^8
6	1.4×10^9

Appendix 1

SHIELDING AND RADIATION DOSES

Included here for the sake of completeness are some rough calculations on shieldings and dosages.

One consideration that is important in some shielding calculations is bremsstrahlung. The doses delivered by the X rays made by bremsstrahlung will be larger than the direct electron doses for large shield thicknesses.

We can calculate the fraction of the energy of an electron that goes into bremsstrahlung (FERMI, 1950) by

$$E_{\text{brem}}/E_{\text{ion}} = ZE^2/1600$$

For the fission energy spectrum the average energy is about 1 MeV. This gives

$\dfrac{E_{\text{brem}}}{E_{\text{ion}}}$	C	Al	Fe	Pb
	0.004	0.008	0.015	0.050

The energy spectrum of the X rays will be something like Figure 10. There will be a very few X rays up to 8 MeV but not many over 2 or 3 MeV. The low-energy X rays (below about 100 keV) will be absorbed in the shielding. The resultant transmitted energy spectrums will have a peak at about $\frac{1}{2}$ MeV, as in Figure 10. The X rays transmitted through the shield will be quite penetrating. Their mean free path will be roughly 20 g/cm². This means two things. First, they will be hard to absorb, and therefore it will take a lot more shielding to remove them. Second, because they are hard to absorb, they will not be counted efficiently in a particle counter and also will result in less radiation dose. We can now calculate crudely the counting efficiency of the Injun 213 GM counter. From Figure 10 we see that it would only count about 1/20000 of the fission electrons directly. But from bremsstrahlung we get 0.05 of the energy converted to bremsstrahlung, of which half is absorbed in the shield. The mean energy of these X rays will be about $\frac{1}{2}$ MeV. A normal GM counter will detect these X rays with about 1 per cent efficiency. This gives $(0.05) (0.01) = 1/2000$ for the fraction of the electrons counted via bremsstrahlung. This calculation is not very accurate, but it does show that the Injun counter should count electrons via bremsstrahlung with about the observed efficiency.

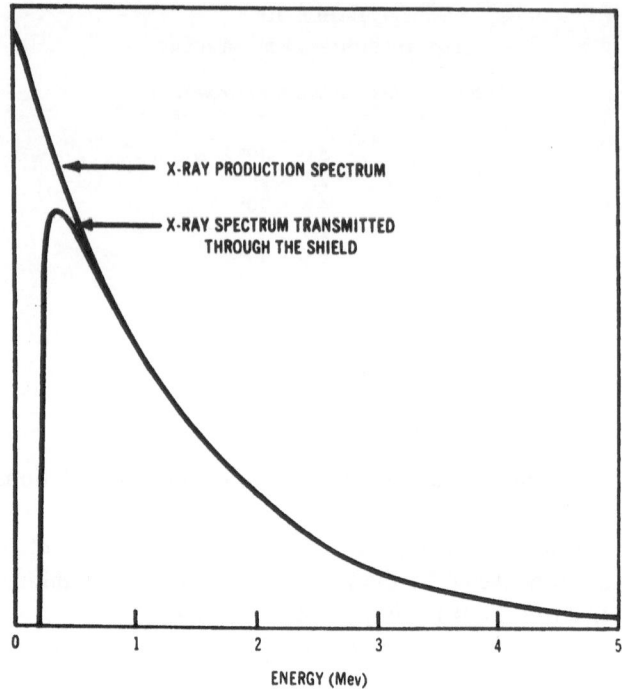

Fig. 10. A crude bremsstrahlung X-ray energy spectrum.

Manned Flight

We must consider the effects of the new radiation belt on manned flights. For Mercury, the total encountered flux for a 6-orbit mission with the MA-7 orbit is 0.24×10^{10} electrons/cm^2 outside the vehicle, or $80R$ outside the vehicle (using 3×10^7 electrons/ cm$^2 = 1R$). The shielding of the vehicle is such that about 1 per cent of this dose is delivered to the astronaut, or about $1R$.

The Mercury dose is almost all received in the South Atlantic hot spot (see Figure 8) and occurs mainly on orbits 4, 5, and 6. The breakdown of the $1R$ dose inside the capsule by orbits is

Orbit	Dose/Orbit
1	$0.003R$
2	0.01
3	0.03
4	0.1
5	0.3
6	0.6

References

BOYD, R. F., WILLMORE, A. P., and QUENBY, J.: 1962, Private communication.

BROWN, W. L.: 1962, Private communication.

BROWN, W. L. and GABBE, J. D.: 1963, 'The Electron Distribution in the Earth's Radiation Belts During July, 1962 as Measured by Telstar'. *J. Geophys. Res.* **68**, 607–618.

DURNEY, A. C., ELLIOT, H., HYNDS, R. J., and QUENBY, J. J.: 1962, Nature **195**, 1245–1248.

FERMI, E.: 1950, *Nuclear Physics,* notes from a course given by Enrico Fermi at the University of Chicago, compiled by J. Orear, A. H. Rosenfeld, and R. Schluter, University of Chicago Press, Chicago.

FISCHELL, A.: 1962, Private communication.

FRANTE, A.: 1962, Private communication.

MARSHALL, J. S. and WARD, A. G.: 1937, 'Absorption Curves and Ranges for Homogeneous β-rays', *Canadian J. Res.* **5**, 15A, 39–41.

McILWAIN, C. J.: 1961, 'Coordinates for Mapping the Distribution of Magnetically Trapped Particles', *J. Geophys. Res.* **66**, 3681–3691.

O'BRIEN, B. J., LAUGHLIN, C. D., and VAN ALLEN, J. A.: 1962, 'Preliminary Study of the Geomagnetically Trapped Radiation Produced by a High-Altitude Nuclear Explosion on July 9, 1962', *Nature* **195**, 939–943.

PETSCHEK, A., MOTZ, H., and TASCHEK, R.: 1962, Private communication.

ROSENZWEIG, W., GUMMEL, H. K., and SMITS, F. M.: 1962, Private communication, and 'Solar Cell Degradation under 1-MeV electron bombardment', submitted to *J. Appl. Phys.*

WELCH, J. A., KAUFMANN, and HESS, W. N.: 1963, Trapped Electron Time Histories for $L = 1.18$ to $L = 1.30$', *J. Geophys. Res.* **68**, 685–699.

Discussions and Questions

H. Alfvén: You gave some figures which seem to indicate that 10^9 is the maximum flux which was observed. Does this mean this is the maximum that can be contained in the field?

W. Hess: 10^9 was the actual observed flux after the Starfish explosion. This is rather smaller than what could actually be contained.

E. Hones: I wonder if you could say anything about the region of the spectrum in which measurements are available and where do we know that a good fission spectrum exists and particularly how far out toward this magic region of 1.7 where all of a sudden the lifetimes change. How far out toward that region do we know there is a good fission spectrum?

W. Hess: Well, I think I'd like to refer his question to Walter Brown. He is making the measurements.

W. Brown: Somewhere in the region about $L=1.5$, the spectrum departs very appreciably from the fission spectrum.

W. Hess: There are experiments coming out from Livermore which tend to say that somewhat inside 1.5 it is not of fission origin.

F. Sylvestri: In that drift I assume they are the most at the points where the magnetic field changes the most, than when they are over the equator.

W. Hess: The drift in longitude?

F. Sylvestri: Yes.

W. Hess: Well, that is rather insensitive to the amplitude of the bouncing motion. It changes only 20 or 30 percent as you change from a particle which is really staying at the equator to one which has a rather large amplitude. It is not a simple thing to explain but it is just not sensitive to that.

F. Sylvestri: I was wondering about an electric field perpendicular to the face of the earth. Would that have any effect?

W. Hess: Well, Dr. Alfvén?

H. Alfvén: It obviously depends on the energy of the electrons. How high is the energy of the electrons?

W. Hess: Typically 1 MeV.

H. Alfvén: I shouldn't think that it would cause very much. If there is a transverse electric field, it would change the circular orbits into elliptical orbits if the energy of the particles are high enough. If the energy is low it will change the orbits more drastically. May I at the same time ask a question? You show that immediately after the explosion the intensity was very large near the equator but it dropped remarkably quickly toward higher values of latitude. Is this in agreement with the assumption that at the point of the explosion that electrons were injected isotropically? I get the impression from the figures this might not be so.

W. Hess: From the calculations it doesn't make a whole lot of difference whether it is isotropic or not. You can't however introduce them all at one pitch angle and try to produce a situation which makes sense. But from the observations and the calculations, you could have a distribution of flux along the lines which show peaks of factors of times 2 or times 3 at any place and that wouldn't change the decay character dramatically. I think it is not especially sensitive to the initial distribution.

H. Alfvén: No, but what I mean is that if you introduce particles of all pitch angles, I should expect more of them to be able to reach high latitudes. But your intensity at high latitudes was down by a factor of something like a hundred or so.

W. Hess: Yes, but the earliest information which I showed experimentally was something like one day after the explosion and by that time you had some very serious changes in the distribution of the various particles. In fact in the first drift in longitude around the earth there was a very substantial loss in particles which I think probably quite clearly obscures what the initial distribution was. As the particles moved from the Pacific across South America into the Atlantic, there is a very substantial loss of the particles at low altitudes as they come into the anomaly. And that I think pretty well wipes out the particles which would have the high latitude mirror points. So that the first time you get a good look at the distribution it has changed so much that you can't see that feature very clearly.

STUDIES OF TRAPPED RADIATION
BY THE TELSTAR I AND EXPLORER XV SATELLITES

W. L. BROWN

Bell Telephone Laboratories, Inc., Murray Hill, N.J.

1. Introduction

This paper on particles in the trapped radiation belts might seem to be out of place at a conference on plasma space science. It does not deal at all directly with the structure of the solar wind or the boundary of the magnetosphere or with the aurora or the ionosphere. On the other hand, plasmas interact extremely broadly with particles and fields and there are important connections between the natural plasmas in space and the particles trapped within the magnetosphere. This paper will point to a few specific cases of this kind in which present understanding is extremely primitive if it exists at all. In addition this paper will deal with the consequences of the special plasmas created by nuclear explosions in space. These sources of new particles, controversial as they are in many respects, have provided some extremely interesting geophysical information on trapping in the radiation belts.

All of the observations of the Telstar I and Explorer XV satellites with which this paper is concerned have been made well within the magnetosphere at maximum radial distances of about 4 earth radii and at latitudes of less than 50°. This is a region in which the motion of charged particles is controlled by the magnetic field of the earth. In order to organize data on these trapped particles MCILWAIN (1961) devised a coordinate system which in effect maps the stable but irregular magnetic field of the earth on a dipole field with the same dipole moment, using the adiabatic invariants of the particle motion. His B,L-coordinate system is illustrated in Figure 1. The magnetic field lines are labelled with L, a coordinate specifying the radial distance of the crossing of the field line through the equatorial plane, and B, the magnetic field intensity on a given L line, a minimum at the equator and increasing in magnitude toward either pole. Position in the equivalent dipole space can be specified by B and L or alternatively by R and λ, the radial distance and dipole latitude as shown in the figure. Results in an R-λ representation are more easily visualized because of the pseudogeographic character of these coordinates. In the R-λ space the surface of the earth is extremely irregular since the magnetic field now serves as the frame of reference.

Charged particles move in spiral paths around the magnetic field lines and bounce between mirror points at magnetic field intensity B_m. In addition, the particles drift in longitude as a result of the radial gradient of the magnetic field and the curvature of the magnetic field lines. In this drift, particles stay on lines of constant L and fill

Chang & Huang (eds.), Proc. Plasma Space Sci. Symp. All rights reserved.

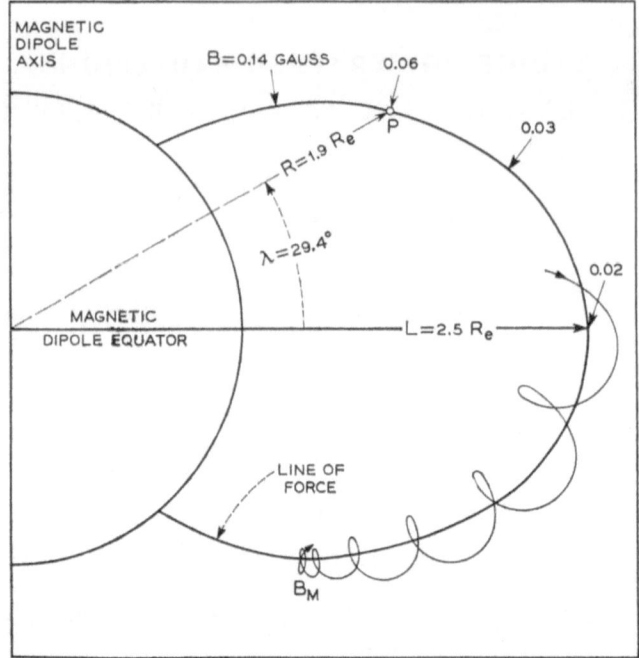

Fig. 1. The *B*, *L* and *R*, λ coordinate system of McIlwain.

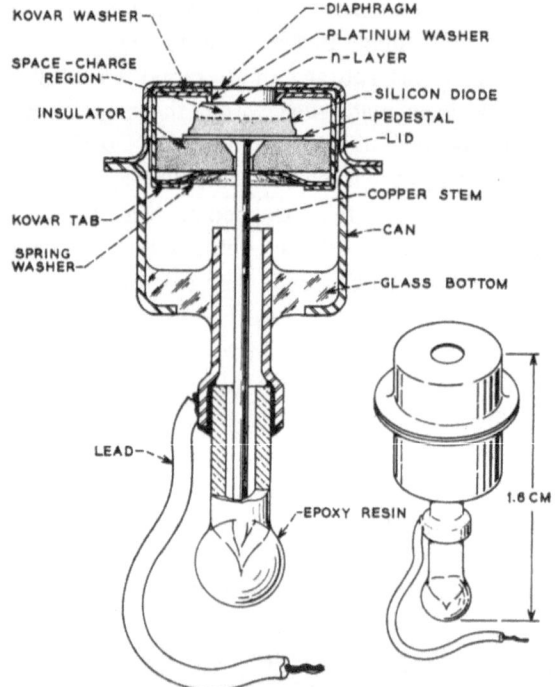

Fig. 2. The cross-section of a silicon *p-n* junction particle detector as used in experiments on the Telstar I and Explorer XV satellites.

out a magnetic shell labelled with L. Positively charged particles drift to the west and negatively charged particles to the east at a rate which depends upon the mass and energy of the particles, but which in all cases is very slow in comparison with the period of rotation of the particles around the magnetic field line and the bounce period of the particles between their mirror points.

All of the experimental results from the Telstar I and Explorer XV satellites have been obtained using semi-conductor p–n junction particle detectors. (BUCK et al., 1964; also BROWN et al., 1963). One of these devices is shown in Figure 2. The active volume is a disc shaped space charge region about 2 mm in diameter and 0.4 mm thick. This region contains a high electric field which separates holes and electrons created in the silicon by an incident charged particle and produces a pulse which is proportional to the amount of energy deposited by the particle in the active volume. The proportional response of the device is important in distinguishing between protons and electrons due to their very different rates of energy loss in material. By altering the condition for pulse height discrimination in the detector, by changing the thickness of the space charge region, and by varying the geometry of the shielding which surrounds the detector it has been possible to study protons and electrons over a relatively extensive energy region.

The results from these experiments have been treated in a variety of ways, one of which is illustrated in Figure 3 and 4 (BROWN et al., 1963b). In Figure 3 an array of points is shown in R-λ space representing all points at which data was obtained from an electron detector in Telstar I during a particular five day period in August, 1962. These points trace out various classes of orbits of the satellite during this period and they are spread in the magnetic coordinate space because of the irregularities of the

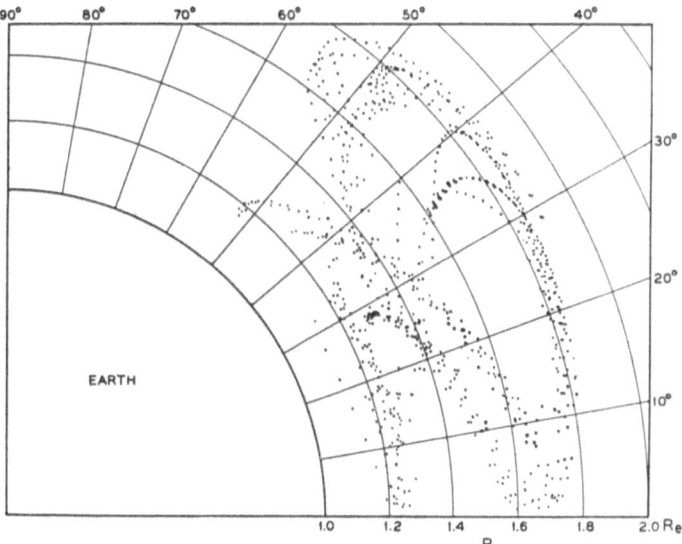

Fig. 3.   Points in R-λ space at which data from an electron detector on Telstar I was collected during a five day interval. Individual orbits and sets of orbits are clearly visible.

magnetic field. Figure 4 shows only those points from Figure 3 at which the counting rate of the detector and hence the flux of the particles incident upon it fell within a chosen range. Contours of equal flux have been drawn through successively selected collections of points such as these to produce maps of the flux distribution. Examples will be shown in following sections for both protons and electrons as measured by Telstar I. We can of course also examine the time dependence of the flux in a par-

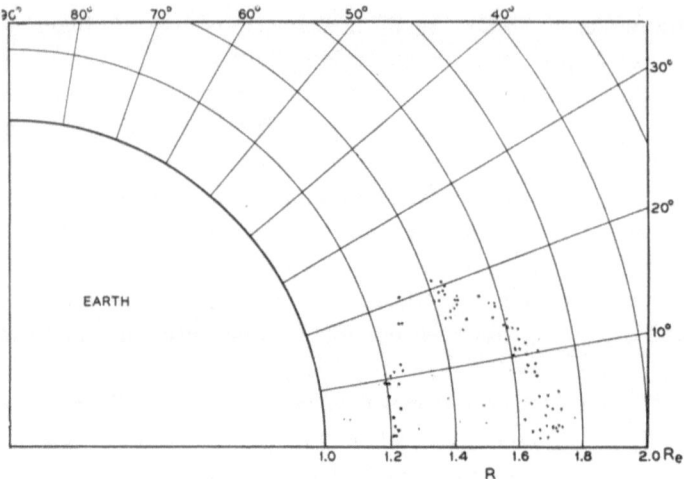

Fig. 4.   A. portion of the points in Figure 3 selected to fall in a particular counting rate range.

ticular region of space by sorting the data not on intensity as is the case in Figure 4, but on B and L or R and λ and displaying it against time. Examples of this sort will also be illustrated in connection with electrons measured by the Telstar satellite.

2. Protons

Figure 5 is a set of flux contours for a detector on Telstar I measuring protons between 26 and 34 MeV (BROWN *et al.*, 1963b). The contoures have been produced as described in the preceding paragraph and are drawn five to a decade. The logarithm of omni-directional proton flux is indicated on the contours, the highest corresponding to a flux of approximately 2.5×10^4 protons/cm^2 sec. This value agrees quite well with earlier measurements by VAN ALLEN (1959) and by MCILWAIN (1963). The flux distribution is characteristic of the inner Van Allen belt. It shows a single maximum at approximately 1.6 earth radii and on the equator. The contours are limited at about 1.9 to 1.95 earth radii by the apogee of Telstar I's orbit.

Figure 6 compares the equatorial flux profile as determined by the 26–34 MeV detector of Figure 5 with those from other proton detectors on Telstar I and Explorer XV. The very much higher apogee of Explorer XV is clearly evident. In all cases there is a single equatorial maximum, but there is a systematic increase of both

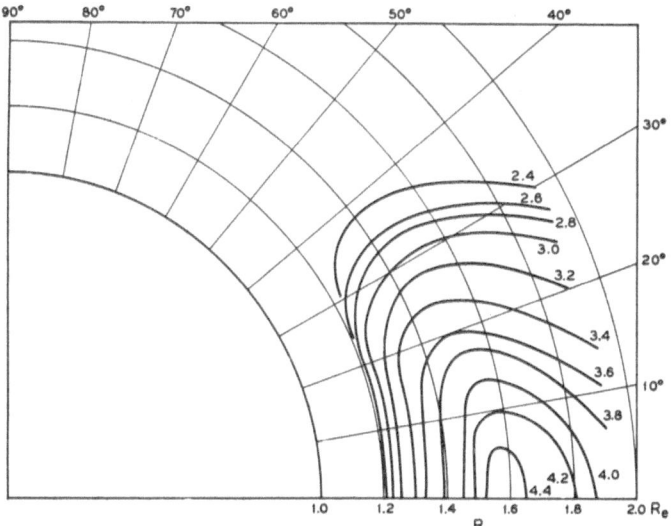

Fig. 5. Contours of equal flux of protons in the 26–34 MeV energy range as measured by Telstar I. The curves are labelled with the logarithm of the omnidirectional particle flux.

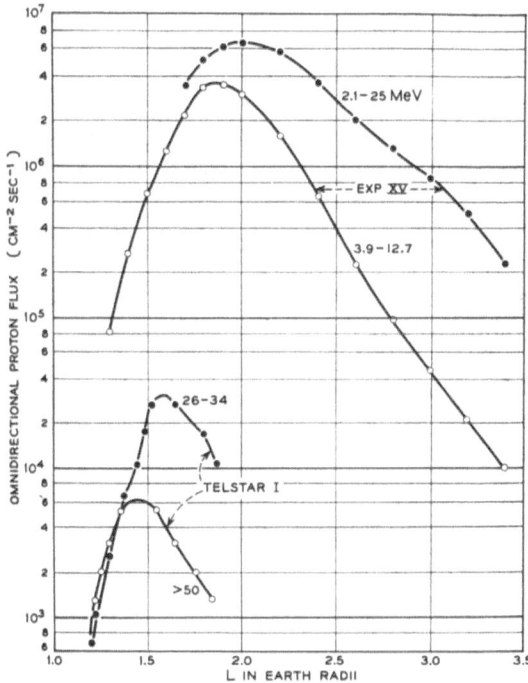

Fig. 6. The equatorial variation of omnidirectional proton flux from two detectors on Telstar I and two on Explorer XV.

maximum proton flux and the radial distance at which the maximum occurs with a decrease in proton energy. This same trend continues to still lower energies as measured by DAVIS and WILLIAMSON (1963). The distributions as determined by Telstar I and Explorer XV were quite stable over periods of several months. They represent thus an essentially steady state between particle source and loss mechanisms. There seems to be no doubt that a major source for the higher energy protons is provided by the decay of albedo neutrons produced by galactic cosmic rays and solar protons reacting with the earth's atmosphere (SINGER, 1960). There also seems to be no doubt that the atmosphere serves as the dominant loss mechanism controlling the very rapid fall off in the proton flux at low altitudes (RAY, 1960). The source for the low energy protons, however, and the mechanism which controls the upper altitude fall off in the flux are still uncertain. It has been suggested that the protons are lost above the flux maximum by scattering with hydromagnetic waves in the plasma (DRAGT, 1961), the waves perhaps arising at the magnetospheric boundary. It is also possible that acceleration mechanisms exist as a result of magnetic field fluctuations associated with boundary variations and that the lower energy protons of Figure 6 were once a part of the low temperature plasma. The importance of unravelling the details of such possibilities is quite evident.

3. Electrons

The distribution and stability of the electrons differ very markedly from the protons discussed in Section 2. Telstar I was observing the particles starting almost immediately after the United States high altitude nuclear explosion of July 9, 1962, the

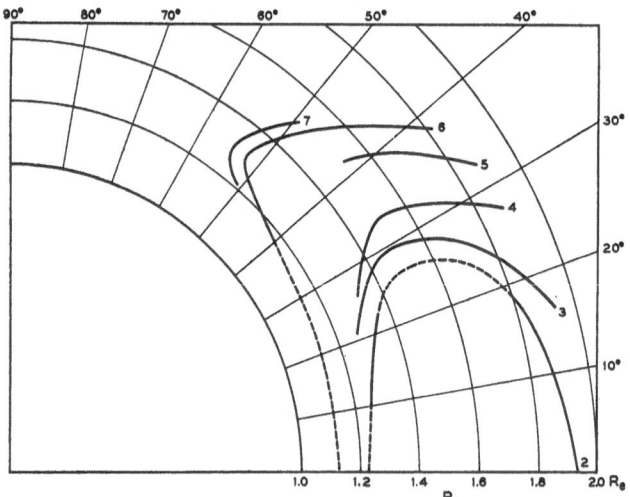

Fig. 7.   Contours of equal omnidirectional counting rate for electrons from Telstar I in the time period from Day 193–197, 1962. The contours are two to a decade as follows: $2 - 6.6 \times 10^7/cm^2$ sec; $3 - 2.1 \times 10^7$; $4 - 6.6 \times 10^6$, etc. The omnidirectional counting rate is related to the omnidirectional flux of particles by the efficiency factor of the detector.

Starfish event (*Collected Papers*, 1963). Explorer XV was launched just before the second of the group of three Russian high altitude nuclear tests that took place in October and November of 1962. A great many new electrons were added to the belts in these events and possibly even a redistribution of already existing electrons took place as well. Figure 7 is a flux map from Telstar I (BROWN *et al.*, 1963b) for the earliest time period in which there is sufficient data to construct a meaningful map. Starfish occurred on Day 191 and Figure 7 spans the five day interval from Day 193 through 197. The contours were constructed in the way described in Section 1 from data in this case obtained by a detector dominantly measuring electrons of about

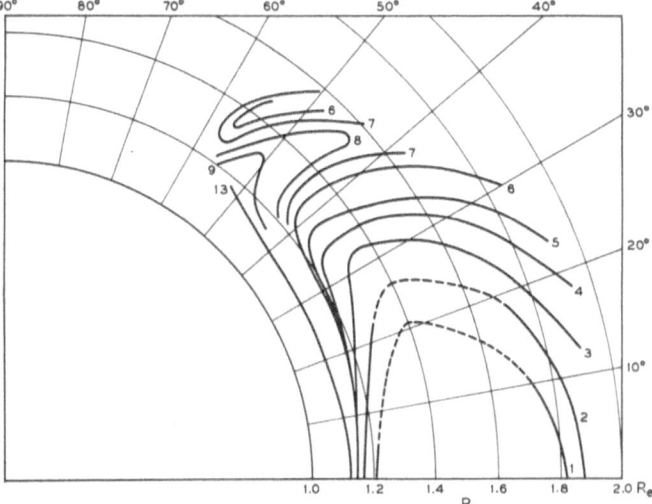

Fig. 8.   Contours of equal omnidirectional counting rates for Days 203–207, 1962. The contour labelling is the same as in Figure 7 except that Contour 1 with an omnidirectional counting rate of 1.5 × 10⁸ is now in evidence. This contour is slightly out of the normal two-to-a-decade order.

0.5 MeV. The contours are two to a decade, the highest shown, Contour 2, corresponding to an omnidirectional counting rate of approximately $7 \times 10^7/\text{cm}^2$ sec. The labelling of the curves has reserved Contour 1 for a still higher flux region not evident in this earliest case because of the orbit of the satellite. Notice that in Figure 7 there is only one maximum in the electron distribution up to dipole latitudes of about 50° and notice that Contour 2 crosses the $R = 1.8$ earth radii line at about 18°.

Figure 8 shows the situation about two weeks after Starfish during Days 203 to 207. Contour 3 now crosses $R = 1.8$ earth radii at 18° instead of Contour 2. There has thus been a decay in the observed particle flux in this region by about a factor of 3 in a ten-day period. A second maximum in the distribution is now evident as a result of an even more substantial loss of particles in the 40° to 50° region. This secondary maximum is the tip of the outer Van Allen belt, previously indistinguishable from the inner belt because of particles substantially filling the intervening space. Contour 1 is now observed at low latitudes, near crossings of the equator at both high and low altitudes. The satellite orbit has not yet precessed sufficiently to complete the contour as indicated by the dashed portions of the curve. Contour 1 has an omnidirectional counting rate of 1.5×10^8 counts/cm² sec. Because of the detector sensitivity as a function of electron energy the over-all efficiency of the detector in the presence of an electron spectrum produced by nuclear fission is approximately 0.2. The highest flux contour thus represents a total electron flux, if the spectrum is that of fission beta particles, of approximately 8×10^8/cm² sec.

Figure 9 shows the flux distribution in October just before the first of the Russian

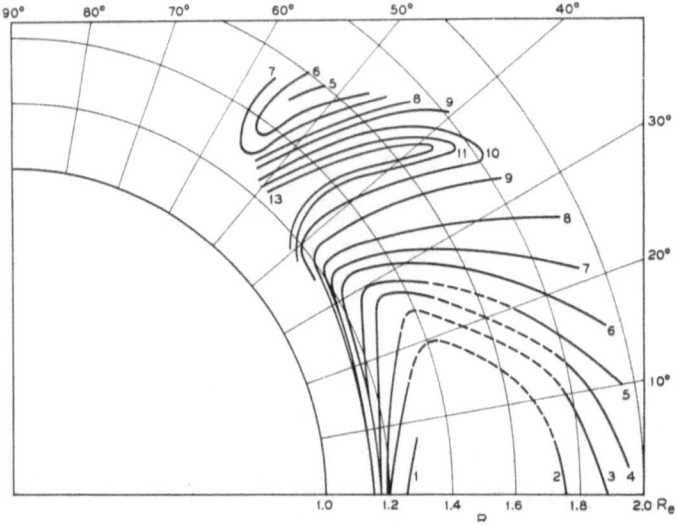

Fig. 9. Contours of equal omnidirectional counting rate for Days 288–294, 1962. The contour designations are as in Figures 7 and 8.

tests. The slot between the inner and outer belts is now extremely deep. The electron flux has decreased by about a factor of 300 at the deepest point. The position of Contour 2 at $R=1.8$ earth radii $\lambda=18°$ on Day 193–197 now falls between Contours 5 and 6 on Days 288–294, a decay of about a factor of 60.

Figure 10 shows the flux as a function of time in several small regions of space as indicated by the B-L values on the figure. The clusters of points along the lines reflect the times at which data was observed in each of these regions. On $L=2.5$ the decay is fast and over-all very large as observed in connection with Figure 9. On $L=1.7$ the decay is slow and only amounts to about a factor of 2 over the three month period. The decays are not exponential. If they were the curves would be straight lines on this semilogarithmic plot and would approach a final steady state value with a rather sharp corner. However, assuming the initial decay is approximately exponential, a time constant has been associated with the initial slope of these curves and others like them throughout the space in which Telstar I collected data. The time constants are themselves plotted in R-λ space in Figure 11. Notice that at the lowest altitudes the

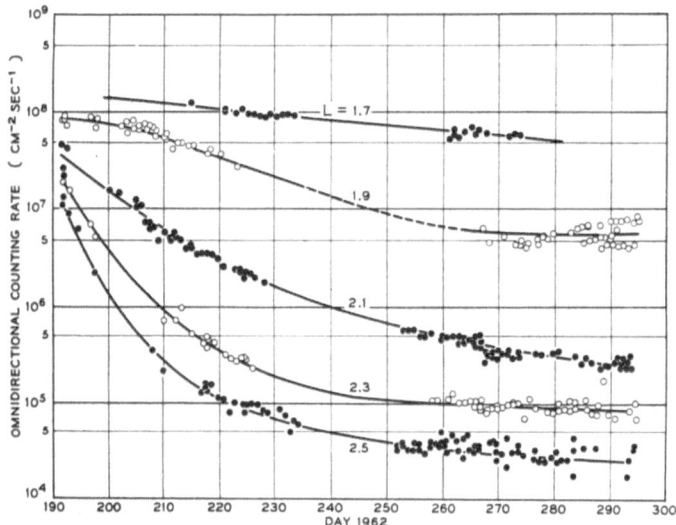

Fig. 10. The time dependence of the counting rate in several small regions of B, L space.

decay times are short as noted by VAN ALLEN *et al.* (1963). At intermediate altitudes near the equator the decay time is relatively long and at the highest altitudes and in the slot the time is once again short. The lightly dashed lines in the figure attempt to connect regions of approximately equal decay time. The shape of these lines is very roughly equivalent to the shape of the inner belt itself as seen in Figure 9. Clearly the rates of decay of the particles are closely related to the number of particles to be

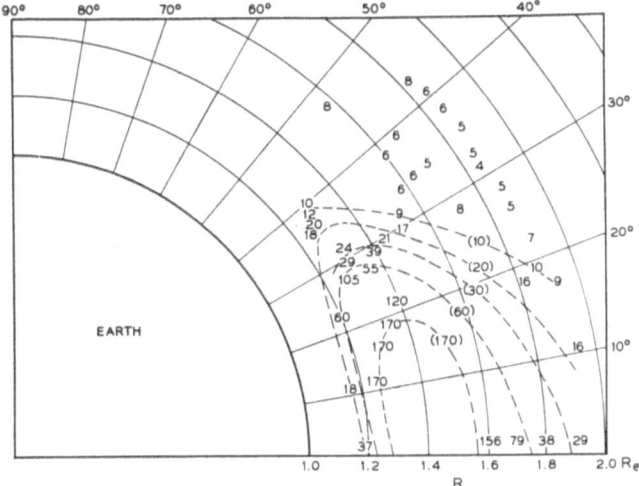

Fig. 11. The initial time constants from data such as in Figure 10 displayed in R-λ space. The decay times are in days. The dashed lines connect regions of equal decay times as indicated by the numbers in parentheses.

found in different regions. This is to be expected in a steady state case and in a case of a still decaying transient following a broad injection of new particles.

The question of the loss mechanism responsible for the results above is of very substantial importance, because these same mechanisms will be operative with respect to naturally occurring electrons as well as to electrons artificially introduced as in this case. At low altitudes (on L lines below about 1.3 earth radii) there seems to be no question but that loss is produced by interaction of the electrons with the earth's atmosphere (WELCH et al., 1963; WALT et al., 1963). Walt is calculating these decays in detail and is finding that his predictions agree quite well with the experimental observations. This correspondence is much more completely displayed with Van Allen's results from the Injun satellite (VAN ALLEN and LIN, 1964) than with results from Telstar I because of Injun's data coverage at low altitudes.

The atmospheric scattering process becomes less and less effective on higher altitude L lines as a result of the decreasing atmospheric density. On the other hand, Telstar I observations show the decay time becoming short again. Atmospheric scattering is certainly not producing the decay times of the order of a week, in the region of the slot. There has been considerable speculation concerning this process. One suggestion is that a Saturn-like ring of dust exists in space in the region of the slot and serves to remove electrons by scattering and energy loss. Such a dust ring would be tremendously effective in removing protons of a few MeV. The fact that the region of the slot is just where the maximum of the low energy proton distribution occurs (Figure 6) makes such a proposal exceedingly unlikely. It seems much more plausible that the loss is connected with some magnetic disturbance in this region of space. DUNGEY (1963) has just proposed the possibility that whistlers, interacting resonantly with the electrons in their cyclotron rotation around the magnetic field, are responsible for the rapid loss in the slot region. Whistlers are circularly polarized electromagnetic radiation produced by lightening discharges in the atmosphere. Dungey's proposal is an extremely interesting one which he will comment on in the discussion to follow. If this mechanism is correct it depends on the properties of the low energy plasma in the trapped particle space, because this plasma determines the propagation characteristics of whistler radiation.

On October 22, 28 and November 1 the Soviet Union carried out three nuclear tests which introduced new particles into the radiation belts in the region above $L=1.7$. Because of the character of the new particle distributions and the more extensive satellite instrumentation which was in space to observe them, these events added significantly in support and extension of information gained from the Starfish test. Figure 12 illustrates the observations of an instrument on the Explorer XV Satellite which measures electrons above 0.5 MeV. The data are for October 28, the day of the second Soviet test. The figure is a ϕ, L plot of the equatorial omnidirectional flux of particles, like Figure 6 for protons. The curve marked 1' was obtained on the second half of the first orbit of Explorer XV as the satellite returned in its highly elliptical, nearly equatorial orbit from apogee at an L of approximately 4.4. This pass crossed $L=2$ at 0407 on October 28. Curve 1' clearly shows the inner side of the outer electron belt, the slot, and the rise toward the inner belt maximum. Curve 2 is the outgoing half of the next orbit and the particle distribution has radically changed with the addition of new electrons above $L=1.8$. This orbit crosses $L=2$ about one hour after the orbit 1'. The initial transient of new particles is not yet complete at

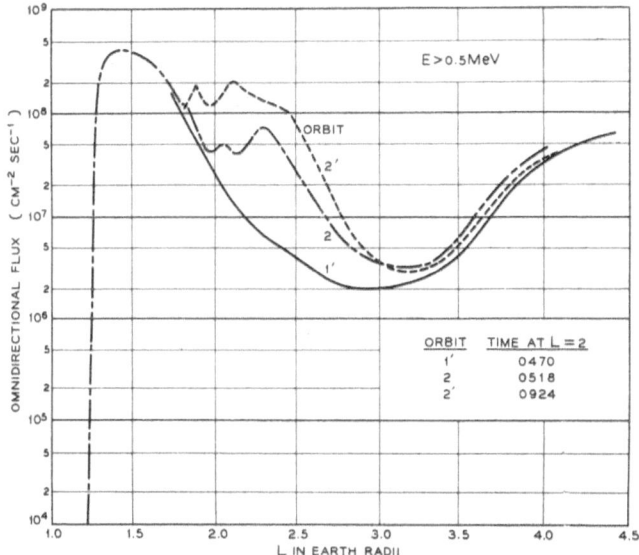

Fig. 12. The omnidirectional flux of electrons of > 0.5 MeV for Explorer XV for the early passes on October 28, 1962. The satellite is nearly equatorial and the figure thus gives a nearly equatorial trace through the particle distribution.

this time as evidenced by the further increases in flux seen on the returning half of the second orbit some four hours later, curve 2'. The fortuitous observation of this transient illustrates the longitudinal drift of the magnetically trapped particles mentioned in Section 1. At $L=2$ on orbit 2 the satellite is over the Atlantic and observing electrons which have drifted eastward around the world from their injection by the explosion over Asia. For 0.5 MeV electrons the drift rate on $L=2$ is approximately 5.7 degree/minute (WELSH and WHITAKER, 1959). For the 270° of longitudinal drift required, the corresponding time is 47 minutes. On $L=3$ the drift rate is 3.8 deg/min. During the time between its passage across $L=2$ and 3, the satellite is moving eastward at approximately 1 deg/min. On all of the significant L shells the particles are thus drifting faster than the satellite and catching up with it on their first transit around the earth. From the drift rates on different L shells it is possible to deduce the time at which the injection of new particles took place to be consistent with the observations. This time cannot be determined with great precision because of the energy dependence of the drift rate, but it appears to have been 0440±10 minutes. On orbit 2' of Figure 12, the electrons have drifted at least five times around the earth and the distribution in various L shells should be longitudinally uniform.

Figure 13 shows the early results for a second detector on Explorer XV, this time

measuring electrons above 1.9 MeV. On orbit 1' in this case there are two small spikes
in electron intensity at about $L=1.85$ and 2.0. These seem to be the remnants from
the first Soviet test six days earlier. There is no question but what much of the flux
between $L=2$ and 3 is also left over from that earlier test since the slot region is not
nearly as deep as it was observed by Telstar I to be just before the Soviet test series,

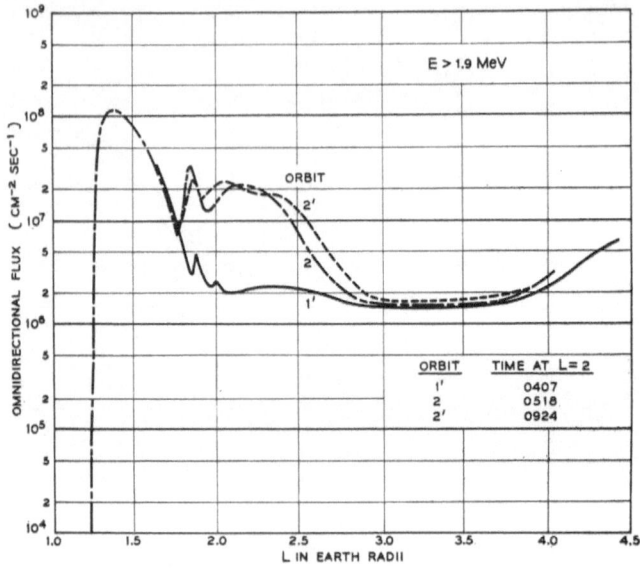

Fig. 13.   The omnidirectional flux of electrons of > 1.9 MeV Explorer XV for the same period as
that in Figure 12.

(Figure 9). In orbit 2 on Figure 13 the new electrons of greater than 1.9 MeV have
clearly already arrived. In fact in comparison with orbit 2' there are more electrons
seen earlier than later. This does not seem to be a decay phenomena, but rather a
decrease in flux associated with the longitudinal dispersion of the originally rather
well clumped group of electrons. The drift rate for 1.9 MeV electrons on $L=2$ is
approximately 20 deg/min. These particles are probably being observed on their
second transit around the earth. If they were in their first transit the injection time
would have to have been at about 0500, rather late for the lower energy particles
seen in Figure 12. The structure in the electron distribution as observed on orbits 2
and 2' is quite complex and can in principle at least be related to the motion of the
radioactive fragments carried in the expanding plasma of the nuclear explosion.

   In Figure 14, 15 and 16 results from three detectors measuring different energies
are shown together for the first passes of the satellite to illustrate the L variation in
the energy spectrum along the equator. The vertical spacings between these curves on
the semilogarithmic plot give the relative spectral hardness of the electron distribution.
From Figure 14 for orbit 1' the residue of the first Soviet explosion is seen at greater
than 1.9 and greater than 2.9 MeV, but is not distinguishable as sharp structure at
greater than 0.5 MeV. In the outer belt region the detectors indicate fluxes approxi-

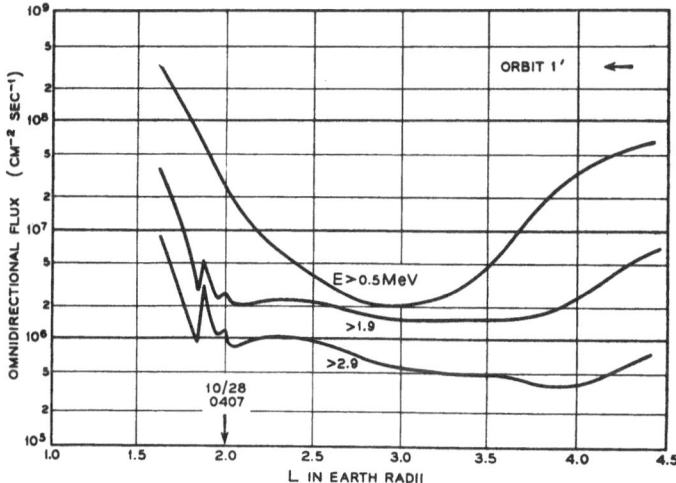

Fig. 14.   The omnidirectional flux from three electron detectors on Explorer XV for the first returning pass from apogee on October 28. The variation in the vertical spacing of the curves in the
semilogarithmic plot indicates the change in the spectral distribution of the electrons.

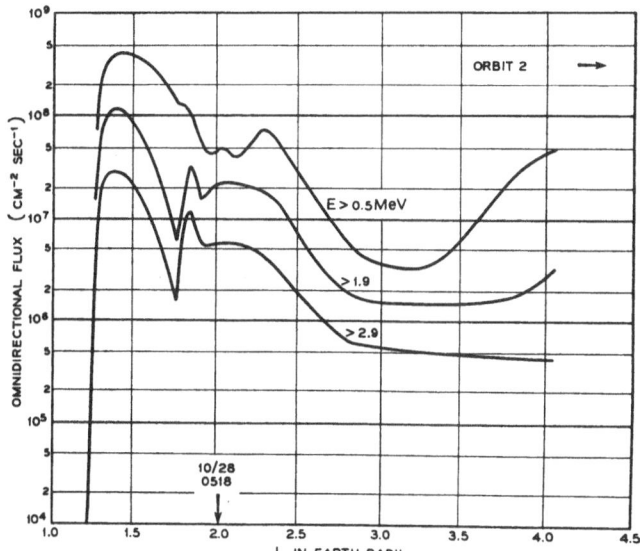

Fig. 15.   The omnidirectional flux from the detectors of Figure 14 as for the second outgoing pass
of Explorer XV. The time is very shortly after the second Soviet high altitude nuclear test.

mately in the ratio $1:0.1:0.01$. In the bottom of the slot at about $L=3$ the ratios are
$1:0.8:0.3$ a very much harder spectrum. In Figure 15, orbit 2, the peak of the inner
belt is seen to have ratios $1:0.3:0.09$. These are in quite good agreement with the
equilibrium spectrum of electrons created in fission beta decay. The electrons in the
inner belt peak are dominantly those produced by the U.S. Starfish test, and in the

peak region they have not decayed very much between July and the end of October as discussed in connection with Figure 11. Between $L=1.8$ and $L\simeq3$ the spectra in Figure 15 are confused by the transient of the newly added electrons in their drift around the world immediately after the second test. Figure 16 shows the electron distribution after it had time to disperse uniformly in longitude. Comparing these

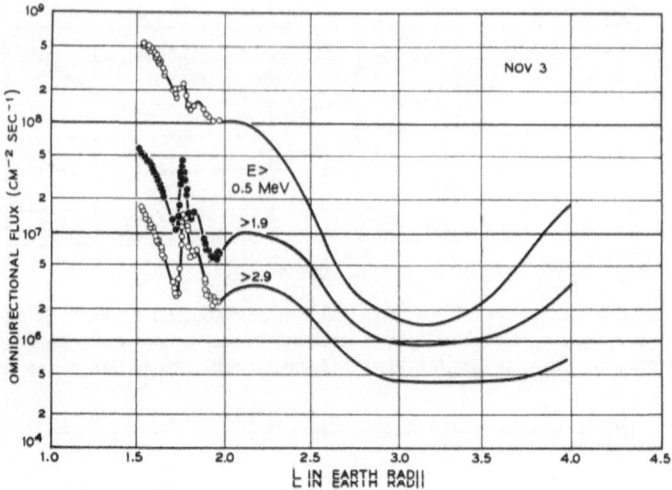

Fig. 16. The omnidirectional flux from the detectors of Figures 14 and 15 as for the second returning pass of Explorer XV. The newly injected electrons have now had time to disperse uniformly in longitude.

curves with those of Figure 14, it is clear that electrons of greater than 0.5 MeV have been added out to as far as $L=3.2$ or 3.3 whereas no significant number of the higher energy electrons has been added above $L=3$. This indication that the spectrum of added electrons is not everywhere the same is born out by comparisons over the whole L-region in which new particles have been added. Even at the maxima at $L=1.85$ and 2.15 in Figure 16 the spectra are different and in neither case are they what would be expected from β-decay of fission fragments. Somehow in the injection process the electron energies are reduced, possibly by the mechanism of Fermi deceleration in collision with the walls of the expanding plasma from the explosion as suggested by HESS (1963). Similar phenomena apparently occurred in Starfish as well, and served to produce a much softer spectrum of electrons at large L values than at small. Through such spectral differences and the difference in the detector response characteristics it is possible to account in a qualitative way for the apparent discrepancy between the Telstar I and Injun I measurements of the Starfish electron distribution in space (BROWN et al., 1963c).

Figure 17 shows the situation on November 3 following the third Soviet test on November 1. The actual data are presented in the region between $L=1.5$ and 2 to illustrate how narrow and well defined the spike of injection was at about $L=1.78$. Unlike the earlier tests, the third test seems to have added no significant numbers of electrons outside of the single narrow spike. At $L=1.85$ and above the electrons from the second test appear with decreased flux in comparison with Figure 16 because of particle decay over the week since their introduction. The decay of particles in the well-defined injection peaks from these two tests appears to occur by loss on the L shell, not by diffusion between L shells. That is, the peaks do not broaden substantially

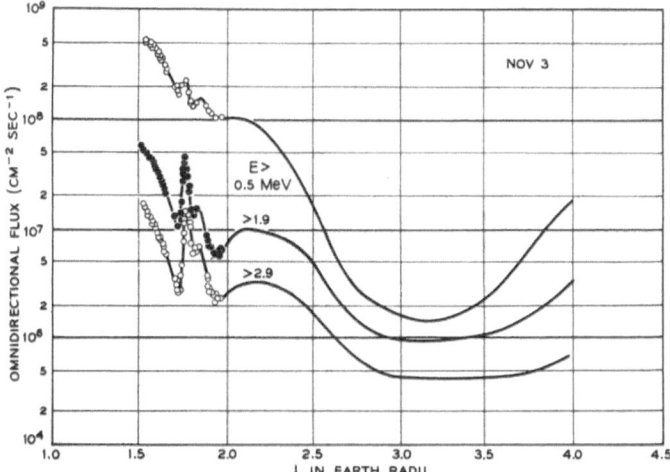

Fig. 17.   The omnidirectional flux of electrons from the detectors of Figures 14, 15, 16 on Explorer XV following the third Soviet test on November 1, 1962. The spike of newly added electrons at $L = 1.75$ is extremely narrow.

as they decrease in magnitude with time. This is to be expected by Dungey's whistler loss mechanism but perhaps not by a mechanism which produces a general mixing of the different field lines due to broad magnetic disturbances.

In Figure 18 the electron distributions are shown as a function of B, along the $L=1.75$ line for several intervals of time. The minimum value of B in the figure corresponds to the equator on this field line. The orbit of November 3 in Figure 17 contributes one of the near equatorial points in Figure 18. The electron flux decreases with increasing B in all cases in the figure. This is the same as saying the flux decreases with increasing dipole latitude (Figure 2). It is also equivalent to saying that the distribution of mirror points is peaked at the equator. The triangles in Figure 18 which fall in with the solid circles of the December 14–18 time interval are those points measured before the third Soviet test. This region at $L=1.75$ has been essentially unaffected by the earlier tests so the triangles give enough data points to define a "before" particle distribution on this field line rather well. The new particles added on November 1 not only increase the equatorial flux as indicated in Figure 17 but also drastically alter the distribution of electrons in B. The very flat uppermost distribution that is produced, gradually loses its anomalous shape over the next few weeks and reassumes the approximately linear dependence of log ϕ on log B it had before the test. The whole distribution then decays together. More rapid disappearance of par-

ticles at large values of B and the decay of the distribution as a whole at long times
are to be expected in processes controlled by diffusion of particle mirror points along
field lines with ultimate particle loss in the atmosphere. This is the situation found
by Hess (WELCH *et al.*, 1963) and by Walt (WALT *et al.*, 1963) for lower L values
where the atmosphere controls the whole process. It should also be expected in the

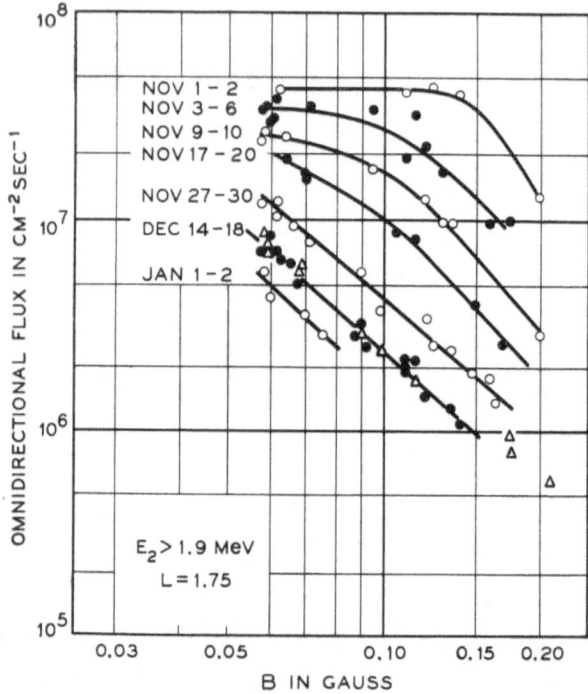

Fig. 18.   The variation of the electron distribution along the $L = 1.75$ line as it changes with time.
The triangles are points occurring before the third Soviet test on November 1.

whistler controlled diffusion of DUNGEY (1963). The fact that the flux in January in
Figure 18 has dropped below the data of the triangles for late October is presumably
due to a continuing decay of electrons either from the Starfish test or from the first
Soviet test.

   Figure 19 is analogous to Figure 18 but for $L=2.0$. There are now only two data
points (triangles) before the test on October 28. One of these corresponds to the outgoing, the other to the incoming pass of the first orbit of Explorer XV. The equatorial
point is that for orbit 1′ of Figure 14. The star marked "earliest point", is for orbit 2,
at which time the high energy electrons had not dispersed in longitude. At $L=2.0$
the ϕ, B variation shows a maximum off the equator, an even more anomalous
distribution than that on $L=1.75$. Such a distribution results from injection of new
particles far off the equator. This shape rapidly disappeared and the decay process
carries past the triangles of October 28 because of continuing decay of electrons from
the first Soviet test. Results for a third field line, $L=2.4$, are shown in Figure 20. On

this line the injection is much less anomalous and a steady state linear dependence of log ϕ on log B is rapidly reassumed. Comparisons of Figure 18, 19 and 20 show two particular features. First, the decay is more rapid on the higher L lines. This is in agreement with the results of Telstar I shown in Figure 10 and observed by Telstar I for the Soviet tests as well. Second, the slope of the log ϕ, log B lines decreases with

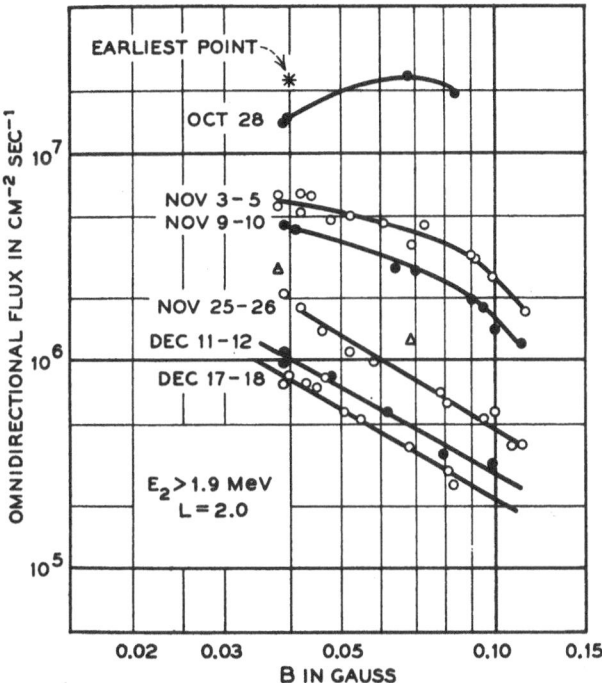

Fig. 19. The variation of the electron distribution along the $L = 2.0$ line as it changes with time. The triangles are the two points obtained before the second Soviet test on October 28. The star is the earliest point following the shot as in Figure 15.

increasing L. This is qualitatively consistent with the diffusion mechanism of particle loss because the ultimate sink for particles in the atmosphere is more remote for higher L lines.

Almost all of the discussion of electrons up to this point has been concerned with the distribution and the redistribution of the particles following artificially induced transients. In conclusion we will consider results for a region in which artificial effects have been unimportant in comparison with nature's own activity. Figure 21 shows the time record of greater than 0.5 MeV electrons from Explorer XV at $L=4$, in the outer electron belt. The two sets of points divide the data into near equatorial and off equatorial B-regions. There is very little B dependence as might be expected by extension of the decrease of the ϕ, B slope observed in Figure 18 through 20, for increasing L. Notice that there is a very rapid electron flux decline shown by the early data. It yields an approximately 5 day time constant. The two B-regions change

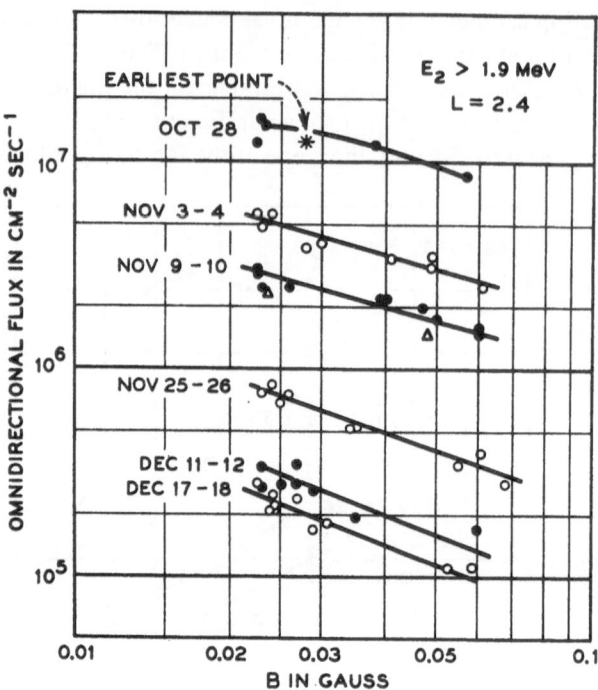

Fig. 20. The variation of the electron distribution along the $L = 2.4$ line as it changes with time. The triangles and the star have the same significance as in Figure 19.

together. Data is missing for about 12 days following day 314, but when the flux is measured on day 327, it is high again. There is a second decline with a similar but not identical decay time. Then on day 352, there is the start of a very large increase which in two days amounts to more than an order of magnitude. This phenomena occurs at the same time as SNYDER has reported observing very large disturbances in the plasma on the Mariner Spacecraft and large fluctuations in magnetic field on the earth (SNYDER *et al.*, 1963). On day 354, the flux reaches a peak and once again begins a rapid decline.

The situation for higher energy electrons is interesting in comparison. Figure 22 reproduces the curve drawn in Figure 21 to represent the shape of the 0.5 MeV data. This curve is superimposed on the greater than 1.9 MeV results with a factor of 10 scale change to make comparisons easier. The features of the two sets of data are similar, but note that the 1.9 MeV electrons rise before they decay in the approximately day 300 region and that the rise at approximately day 327 is visible after the data break whereas for the low energy electrons it was not. In the large rise on about day 352 the 1.9 MeV electrons again lag behind. The 0.5 MeV electrons have started to decline before the high energy electrons have reached their maximum. The increase for the 1.9 MeV electrons is about a factor of 40. These effects are certainly associated with plasma from the sun. Possibly magnetic disturbances initiated in interactions of the solar wind plasma with the earth's field are responsible. It is tempting to believe the observations are showing an acceleration mechanism in operation. It goes on for several days, increasing the energy of very low energy electrons until they are measured by the 0.5 MeV detector and then later by the 1.9 MeV detector. One might also interpret the results as due to a time varying source of particles outside the magneto-

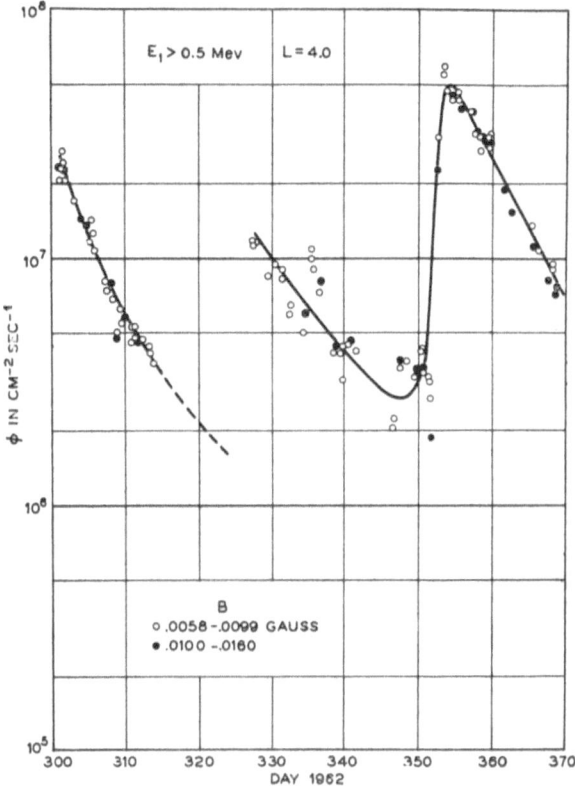

Fig. 21.   The fluctuations in the > 0.5 MeV electrons in the outer electron belt at $L = 4.0$ from October 28, 1962 through January 4, 1963.

sphere, the source producing more high energy electrons later. There are difficulties of course in carrying out injection of such electrons through the magnetospheric boundary and there is no evidence for such energetic electrons from the sun. The interpretation of these effects is presently unknown. It seems almost certain to require a mechanism that involves the plasma from the sun in interaction with the earth's field, a topic that falls naturally into a symposium on plasma space science.

4. Summary

In summary there is evidence that for both protons and electrons in the trapped particle region around the earth, plasma, the low temperature plasma inside the magnetosphere, the high temperature plasma arriving from the sun, and the manmade plasma of nuclear explosions, is important in mechanisms of injection, acceleration, and particle loss. A great many intriguing suggestions have been made as possible interpretations of the observed phenomena but many, if not most, of the questions that can be asked remain unanswered in any satisfactory detail. We should, however,

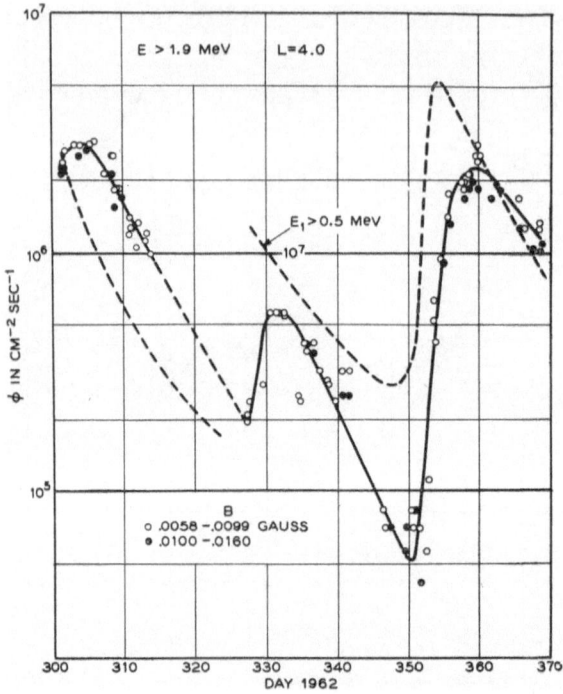

Fig. 22. The fluctuations in the > 1.9 MeV electrons in the outer electron belts at $L = 4.0$ for the same time period as Figure 21. The totally dashed curve in the figure is reproduced from Figure 21 with a change in vertical scale of a factor of 10.

expect major advances in our understanding in the next few years with the increasingly sophisticated experiments that are being prepared and launched and with the theoretical effort that is examining these complicated geomagnetic effects.

References

VAN ALLEN, J. A.: 1959, *J. Geophys. Res.* **66**, 1683.
VAN ALLEN, J. A., FRANK, L. A., and O'BRIEN, B. J.: 1963, *J. Geophys. Res.* **68**, 619.
VAN ALLEN, J. A. and LIN, W. C.: 1964, *Trans. Amer. Geophys. Union* **45**.
BROWN, W. L., BUCK, T. M. et al.: 1963a, *Bell System Tech. J.* **42**, 899.
BROWN, W. L., GABBE, J. D., and ROSENZWEIG, W.: 1963b, *Bell System Tech. J.* **42**, 1505.
BROWN, W. L., HESS, W. N., and VAN ALLEN, J. A.: 1963c, *J. Geophys. Res.* **68**, 605.
BUCK, T. M., WHEATLEY, G. H., and ROGERS, J. W.: 1964, Proceedings of the Ninth Scintillation and Semiconductor Counter Symposium, March 1964, to be published *IEEE Trans. Nucl. Sci.*
Collected papers on the Artificial Radiation Belt for the July 9, 1962 Nuclear Detonation, J. Geophys. Res. **68**, (1963) 619.

DAVIS, L. R. and WILLIAMSON, J. M.: 1963, *Space Science Review* (ed. by W. PRIESTER), Interscience Publishers Inc., New York.
DRAGT, A. J.: 1961, *J. Geophys. Res.* **66**, 1641.
DUNGEY, J. W.: 1963, *Plan. Space Sci.* **11**, 591.
HESS, W. N.: 1959, *Phys. Rev. Letters* **3**, 11 and 145.
HESS, W. N.: 1963, *Bull. Amer. Geophys. Union*, p. 76.
MCILWAIN, C. E.: 1961, *J. Geophys. Res.* **66**, 3681.
MCILWAIN, C. E.: 1963, *Science* **142**, 3590.
RAY, E. C.: 1960, *J. Geophys. Res.* **65**, 1125.
SINGER, S. F.: 1960, *Space Research 1* (ed. by H. KALLMANN-BIJL), North-Holland Publishing Company, Amsterdam.
SNYDER, C. W., NEUGEBAUER, M., and RAO, V. R.: 1963, *J. Geophys. Res.* **68**, 6361.
WALT, M., CRANE, G. E., and MACDONALD, W. M.: 1963, *Trans. Amer. Geophys. Union* **45**.
WELCH, J. A., JR. and WHITAKER, W. A.: 1959, *J. Geophys. Res.*, **64**, 909.
WELCH, J. A., JR., KAUFMAN, R. L., and HESS, W. N.: 1963, *J. Geophys. Res.* **68**, 685.

Discussion and Questions

J. W. Dungey: I would like to say briefly what I have done in considering electron loss produced by whistlers. A few years ago, I was looking at the effects of hydromagnetic waves on protons and found a resonance mechanism which changed their pitch angles in a random way. I have now done the analogous thing for electrons interacting with the circularly polarized electromagnetic waves propagating in the whistler mode. The frequencies in this problem, the gyro frequency of the electrons and the frequency of the electromagnetic wave, are of course much larger than in the proton case. The resonance condition depends on the dispersion equation for whistler propagation and the Doppler shift of the whistler wave as it is seen by the high velocity trapped electrons. The condition turns out to vary as B^3/n where n is a density of thermal electrons that enters the problem through the whistler propagation. Since n is believed to vary approximately with B the resonance condition varies as B^2 or as L^{-6}. I have only estimated the lifetime of trapped electrons due to this pitch angle scattering loss mechanism because it depends on the statistics of whistlers which are not very well known. The result is perhaps ten days, but this is only approximate. More striking than the absolute magnitude of the lifetime is its L dependence. Looking at the frequency spectrum of lightning, the source of whistlers, because the spectrum of whistlers is not as well known, one finds the maximum frequency between 5 and 10 kilocycles which in the extreme Doppler shifted limit corresponds to the gyro frequency of electrons on an L of approximately 2. With the L^{-6} dependence of the required resonance frequency one rapidly runs out of energy in the whistler spectrum at lower L's. The mechanism thus provides a very rapid variation of trapped lifetime with L.

W. L. Brown: I wonder if you can understand from this mechanism why one sees a relatively L independent lifetime through the bottom region of the slot. There is quite a large span in the values of L beyond about $L = 2.2$ in which the lifetime is the order of one week.

J. W. Dungey: The spectrum of whistlers has of course a very broad maximum, and what you are really doing in deducing the trapped lifetime is turning this spectrum upside down.

H. Alfvén: Yesterday I discussed the existence of electric fields in the magneto-sphere and observed that as long as the mean free path of thermal electrons in the plasma was much smaller than the distance which we called L, then the conductivity could be put equal to infinity and all the electric fields should be zero. That is, we have frozen in conditions. If the mean free path was long however compared to L the conductivity cannot be infinite and we can have electric fields different from zero. These could well be variable electric fields. The first situation obviously prevailed in the ionosphere and perhaps out into the magnetosphere. At some distance from the earth there should be a border between these two cases. It has been indicated in your paper and in that by Hess that inside an L value of about 1.7 the conditions are much more quiet, that is, the particle orbits are much more stable than outside. I should like to know whether you think this may be connected with the existence of variable electric fields. The mean free path of importance for the conductivity is of course not that for the high energy electrons but for the thermal electrons of the plasma.

W. L. Brown: We have speculated a little bit that non-neutral electrostatic regions would be nice scattering centers for electrons. It seems to me what you are proposing is a condition under which you might find non-neutral regions, regions having an electric field. Such regions would need to be relatively small of course or fluctuating rapidly to be very effective in their interaction with the high energy trapped electrons.

H. Alfvén: This is possible, but one could also think of large scale fields with large differences between the ionosphere and the equatorial region for large values of L.

W. L. Brown: These are gradually decelerating particles on each passage as they go back and forth?

H. Alfvén: And also transverse electric fields which cause the particles to move in and out across L shells.

W. L. Brown: I haven't considered such a process at all. It is an interesting idea.

C. Y. Fan: May I ask you what is the direction of the detector with respect to the magnetic field?

W. L. Brown: The Telstar I electron detector is taking a directional average of the particles in the field. It looks out normal to the spin axis of the satellite and averages over the electrons it sees, for a time large compared to the rotational period of the satellite. A correction has been made to the directional average for the angle between the satellite spin axis and the local magnetic field in order to obtain an omnidirectional flux. This correction is very simple and is assumed to be the same at all points on all L lines, but it provides a way of obtaining an approximate omnidirectional intensity which I believe is in no case in error by more than perhaps 30%.

L. Davis: You showed the time dependence of the low energy electrons only at $L=4$. I wonder if you could comment on the variation of time dependence with L.

W. L. Brown: As observed in the Telstar I case there is nothing but a decline in the electron distribution throughout the first three months of observation at L's between about 2 and 3. At high L values the flux does not always just go down how-ever, it goes up and down, as it does in a most spectacular way at $L=4$ as I have shown you. The time variability of the flux seems to get monatomically greater as one goes to larger L values. At $L=2$ in the Explorer XV data we see very little that we could interpret as fluctuations on the general flux decline. At $L=3$ they are quite evident and at $L=4$ they are very large.

E. Hones: I have a question regarding the mechanism that Professor Alfvén suggested might explain the very sharp boundary in the trapped lifetimes at $L=1.7$. I think McIlwain's data on protons showed long trapping lifetimes to distances well beyond 1.7, perhaps to $L=4$, and I wonder how this effects Professor Alfvén's inter-pretation.

W. L. Brown: Are you saying that he found the proton distribution very stable in time? If so, I agree that those were his observations.

E. Hones: Yes, what you have just shown is that at $L=4$ there are very rapid fluctuations in the population of electrons. McIlwain also showed this starting at about $L=1.7$, but he shows also that 40 to 80 MeV protons are very stable out to $L=4$. Now I should think that a mechanism involving electric fields beyond the $L=1.7$ border such as Professor Alfvén just mentioned would effect protons as well as electrons.

A. J. Dessler: May we wait until the round table discussion to take this question up. The existence of these electric fields is rather controversial at this point.

G. Sales: Concerning your proposal for the removal of electrons from high L shells due to magnetic disturbance it seems plausible from the data you presented to associate the disturbances with an injection of electrons. Removal was more common, or more normal, behavior when the magnetic variations were small.

W. L. Brown: I think of there being occasional large magnetic disturbances re-sulting in large increases of the particle flux. The particles are subsequently lost by a series of smaller and presumably different disturbances that go on over a more ex-tended period of time.

A. J. Dessler: I think it is particularly interesting to see the particles being ac-celerated by such a large amount during this very short time.

LOW ENERGY TRAPPED PROTONS AND ELECTRONS

L. R. DAVIS

Space Sciences Division, Goddard Space Flight Center, Greenbelt, Mld.

Abstract. The directional intensities of 100 keV to 10 MeV protons and the directional energy flux of 10 keV to 100 keV electrons have been measured on Explorers satellites XII, XIV, and XV. Preliminary results on the spatial distribution and spectra of the low energy protons and the temporal characteristics of the protons and electrons are presented. The total number of trapped protons having energy greater than 97 keV in the region from $L = 2.5$ to 7 is 4.4×10^{28} protons. The maximum proton intensity, 4.0×10^7 protons/cm^2 sec ster, occurs at $L = 3.5$. Below $L = 5$ the proton intensity is constant to within $\pm 25\%$ during magnetically quiet times for the period of observation. Low energy electrons in this region and above undergo abrupt increases of as much as a factor of ten at the beginning of small M-region magnetic storms.

This paper is a review of preliminary results on low energy trapped protons and electrons from measurements made with detectors flown on Explorers XII, XIV, and XV. Explorer XII was launched August 16, 1961 into an elliptical orbit with apogee at 13 earth radii geocentric; Explorer XIV was launched October 3, 1962 with an apogee of 17 earth radii; and Explorer XV was launched October 27, 1962 with an apogee of 3.7 earth radii. Nearly identical detectors were flown on all three satellites.

The detectors measured the directional intensity of 100 keV to 10 MeV protons and the directional energy flux of 10 keV to 100 keV electrons. Figure 1 is a schematic diagram of the detector which consisted of a thin (5 milligram/cm^2) zinc sulfide silver

ION ELECTRON DETECTOR DIAGRAM

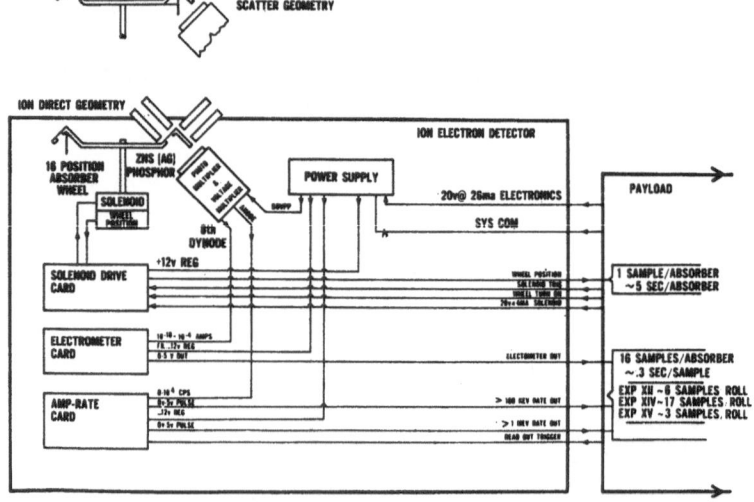

Fig. 1.

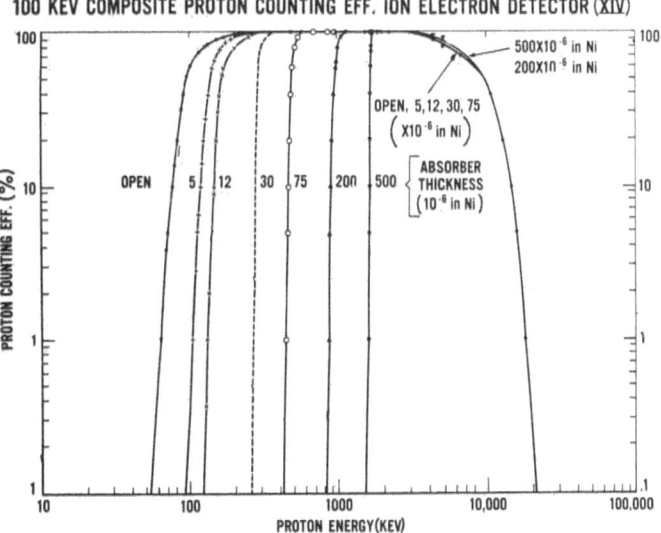

Fig. 2.

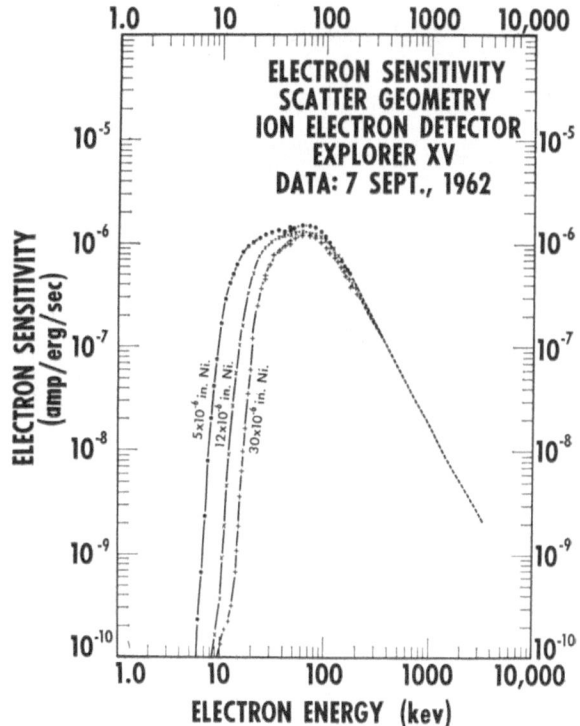

Fig. 3.

activated powdered phosphor scintillator, an absorber wheel, two entrance apertures, and the associated electronics. Both the phototube current and pulse counting rate were telemetered. The entrance aperture labeled "direct geometry" allowed particles to reach the scintillator directly after penetrating an absorber on the wheel. In this mode the phototube current was a measure of the total energy flux and the pulse count rate a measure of the proton intensity. The aperture labeled "scatter geometry" directed particles at a gold target and only scattered particles could reach the scintillator. The current measured in this mode was proportional to the electron energy flux.

Figure 2 shows the proton count rate efficiency as a function of proton energy for the six absorber thicknesses, which ranged from 5 to 500 millionth of an inch of nickel, and also for the open position of the wheel. As may be seen in this figure, the count rate measured the proton intensity integrated from an energy E_1 to about 10 MeV, with seven values of E_1 ranging from 97 keV to 1.7 MeV.

Figure 3 shows the electron sensitivity for the three absorber thicknesses employed in the scatter geometry. Only measurements from the 12 millionth of an inch nickel absorber will be presented in this paper. These measurements roughly correspond to the energy flux of 20 to 100 keV electrons. Responses of the detectors were essentially identical on the three satellites.

A typical picture of the belt as viewed with this detector is shown in Figure 4,

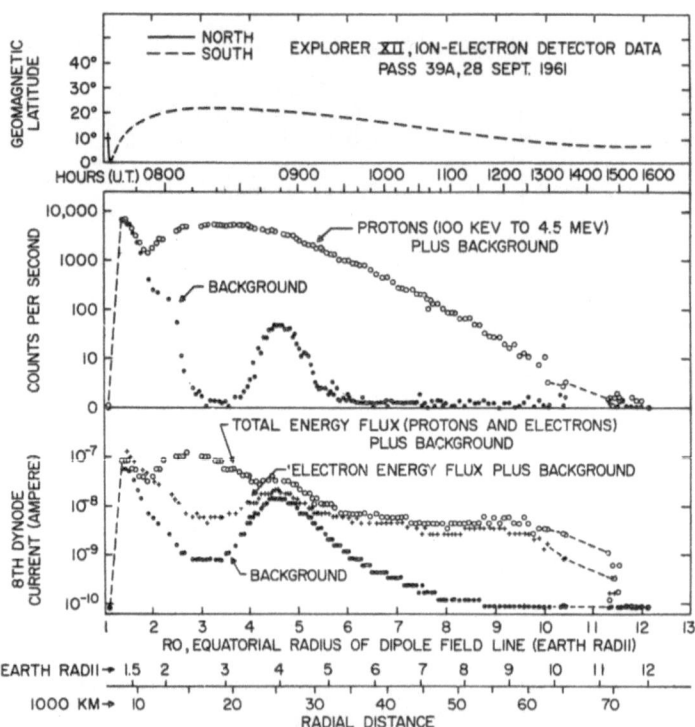

Fig. 4.

where the detector responses plotted are for locally mirroring particles and the thinnest absorbers. The detectors responses and in addition the satellite geomagnetic latitude are plotted as functions of R_0, the equatorial radius of the diple field line. The background counting rate and current, obtained on a wheel position which plugged both entrance apertures, are also shown. These backgrounds must be subtracted to obtain the true counting rate or current.

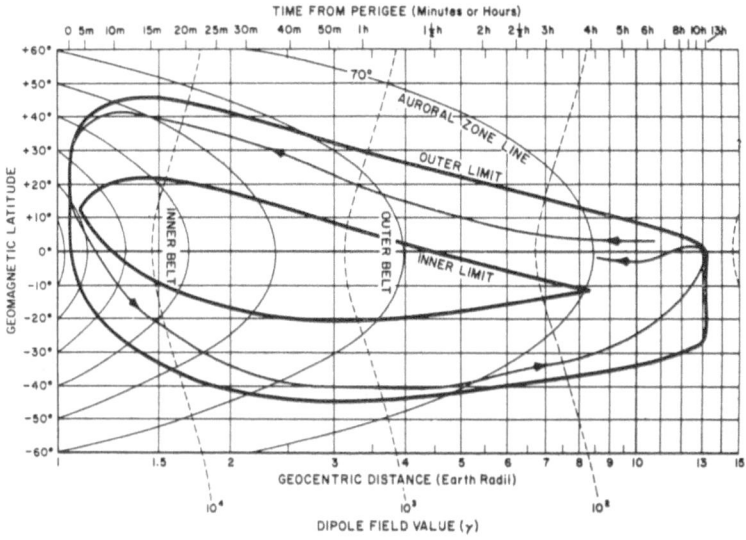

EXPLORER XII COORDINATE LIMITS AND SAMPLE ORBIT

Fig. 5.

The counting rate of 100 keV to 4.5 MeV protons shows the large belt of low energy protons extending from about 2 earth radii to at least 10 earth radii. One may note that the background counting rate over most of this region is one or two decades smaller than the proton counting rate. Below about 2.8 earth radii the background rate increases rapidly, presumably due to the penetrating protons of the inner belt. Because of this background no proton measurements are possible below about 2 earth radii. The background pulses appearing at 4 or 5 earth radii are apparently due to pulse-pile-up produced by the penetrating outer belt electrons, since the counting rate observed is proportional to the current cubed. The low energy protons also show up around 3 earth radii in the current responses as a difference between the total energy flux and the electron energy flux.

Low energy electrons are measured by the difference between the electron energy flux and the background current. They are present throughout most of the region above 2 earth radii. To date, our studies have been directed toward obtaining a detailed map of the proton intensities and spectra and studying time variation of the protons and electrons.

Figure 5 shows the region of space on a magnetic meridian plane traversed by the

Explorer XII satellite. A representative orbit and the outer and inner limits of the region sampled by the satellite on successive orbits are also shown. As may be seen there is a rather large region near the equator between 1.4 and 4 earth radii which was missed by Explorer XII. A similar situation existed with Explorer XIV, but Explorer XV conveniently filled in this region of space.

We have chosen to use an L parameter-equatorial pitch angle (EPA) coordinate system. Our detector, with the aid of an optical aspect sensor on the satellite, measured the directional intensity of particles in inertial coordinates. We then determined the local pitch angle (LPA) of the particles being measured, using a computed value of the field direction. The EPA was then determined using the following relation:

$$\sin^2 \text{EPA} = \sin^2 \text{LPA} \times B_0/B = B_0/B_m$$

where B is the field intensity at the satellite and B_0 the minimum intensity on the field line. The relation to the mirror point field intensity, B_m, is also shown in the above equation. The JENSEN and CAIN (1962) model of the geomagnetic field has been used to compute EPA and L.

This coordinate system adequately orders the proton data in the region below about 5 earth radii as we shall show in more detail. Above 5 earth radii the data are not well ordered. While we have not as yet made a detailed study of this region it appears that the disordering is due both to time variations of the proton intensities and to a distortion of the geomagnetic field by external current sources. Figure 6 illustrates the effect of one possible source, the cavity distortion. The figure shows two dipole field lines as dashed curves and the same two lines emanating from the earth as they would be distorted by confining the geomagnetic field to a spherical cavity of 10 earth radii radius. As can be seen the directions of the distorted field line and the dipole field line differ by as much as 30° in the higher latitudes traversed by the satellite. Thus one could make errors of this size in computing the LPA of the particles being measured and therefore produce disordering of all data obtained on a

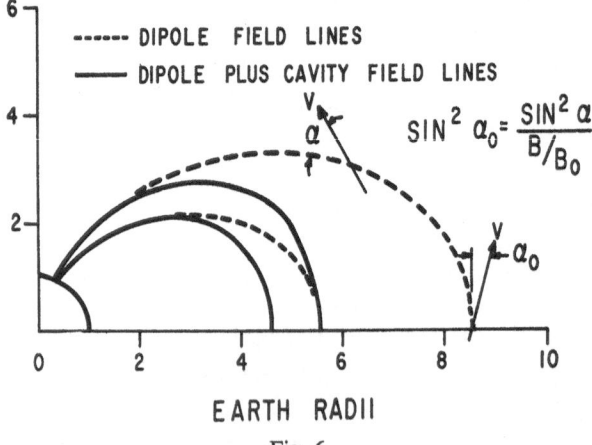

EARTH RADII

Fig. 6.

given pass through the shell. Measurements made at high and low latitudes on the same shell as indicated by L computed for the dipole field line would, if the distortion existed, have been obtained on widely different shells.

To illustrate how well ordered the data are in the region below five earth radii,

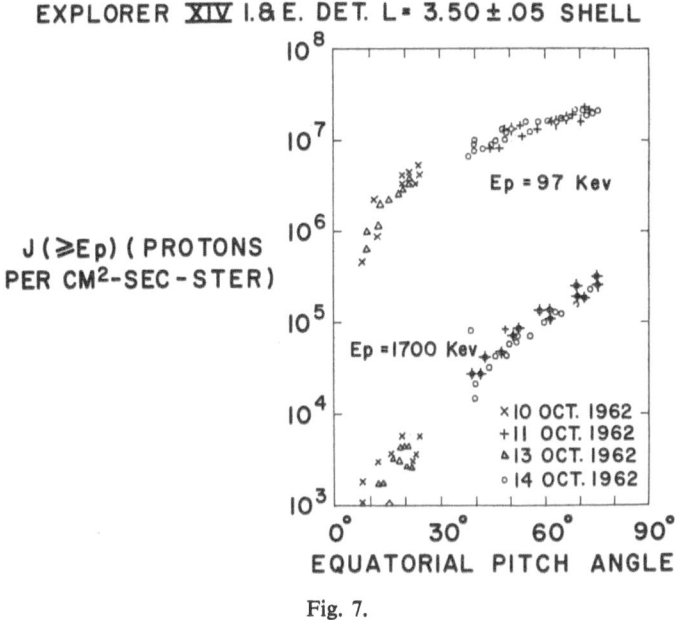

Fig. 7.

pitch angle distributions of measurements obtained on four successive passes of Explorer XIV through the $L=3.5$ earth radii shell are shown in Figure 7. The October 10, and October 13, 1962 passes were at high latitude with B/B_0 equal to about 10. The October 11 and 14 passes were nearly on the equator with B/B_0 equal to 1.2. As may be seen, data taken from a single pass as well as the entire set of data from the four passes form a well ordered pitch angle distribution. Figure 8 shows data obtained on three of the passes crossing the $L=6.0$ shell. As can be seen the data obtained in a single pass are not as well ordered as they were on $L=3.5$ shell. The data obtained on different passes show even larger discrepancies particularly in the higher energy protons. In this particular case the disordering of the pitch angle distribution of protons having energy greater than 495 keV is due both to a decrease of the intensity with time and a distortion of the geomagnetic field.

Detailed comparison of measurements of the low energy proton intensities in August and September of 1961 with those measured in October, November, and December 1962 in the region $L=2$ to 5 show that the proton intensities during magnetically quiet periods were constant to within $\pm 25\%$ and the data were well ordered using the JENSEN and CAIN (1962) model of the geomagnetic field. Because of this we have been able to combine data obtained on Explorer XIV in early October 1962,

(Explorer XIV satellite precessed during a large portion of its life. In early October the satellite was not precessing and good directional information could be obtained) with Explorer XV data obtained in December of 1962, when the high background produced by the Russian bomb tests had decayed away, to produce a general mapping

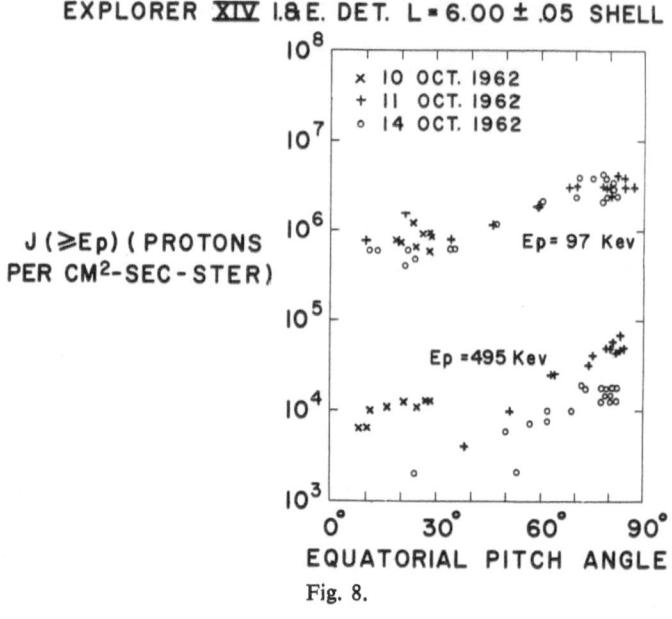

Fig. 8.

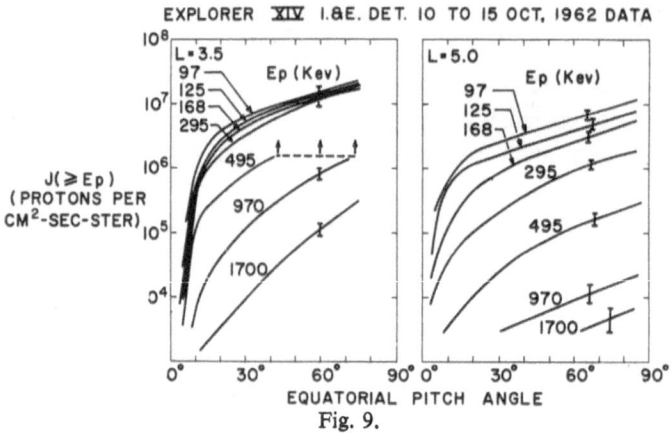

Fig. 9.

of the low energy protons in the region 2 to 5 earth radii. This distribution we believed to be typical of the magnetically quiet times throughout this period.

Figure 9 shows pitch angle distributions for each of the seven energies measured at $L=3.5$ and 5.0. The pitch angle distributions steepen as one moves from low to high energies, showing that the proton spectra soften as one moves from large to small EPA.

Figure 10 shows the consistency of the measurements when plotted as a function of L. The intensity corresponding to the background counting rate is plotted as open

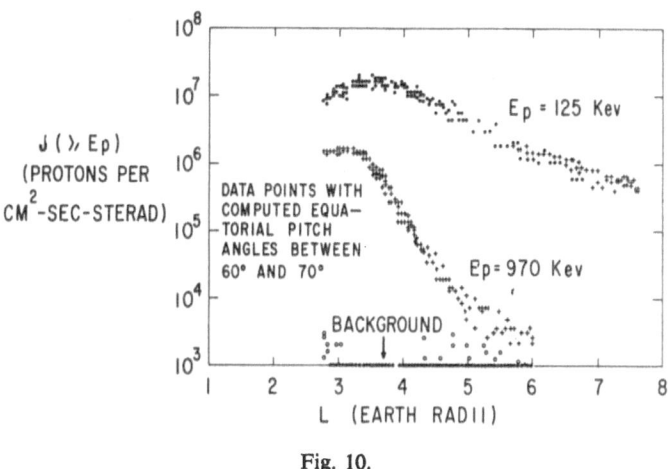

Fig. 10.

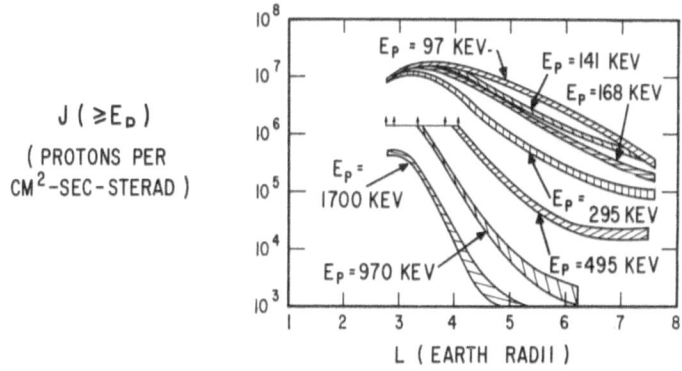

Fig. 11.

circles at the bottom of the graph. Background values equal to or less than 10^3 are plotted at the bottom extremity of the graph. As may be seen the background counting rates were insignificant on these passes. There were two each of the 5, 12, and 30 millionths of an inch nickel absorbers on the wheel, with telescope factors that differed by a factor of 200 to increase the dynamic range of the intensity measurement. The upper curve in Figure 10 shows data obtained with the two apertures covered with 5 millionth of an inch thick nickel foils, the data from the small aperture as dots and the large aperture as crosses. The intensities measured with the two apertures are in complete agreement showing that the proton measurements were not affected by the pile-up of pulses produced by low energy electrons.

Figure 11 shows the proton intensity as a function of L for protons having EPA values between 60° and 70° for each of the seven energies measured. (The curve labeled 141 keV should read 125 keV.) The detector was in saturation in the region 2.5 to 4.5 earth radii on the 75 and 200 millionth of an inch nickel foil wheel positions. This may be seen in Figure 11 where the curves for protons of energy greater than 495 keV and 970 keV show only a lower limit for the intensity in this region. Figure 12

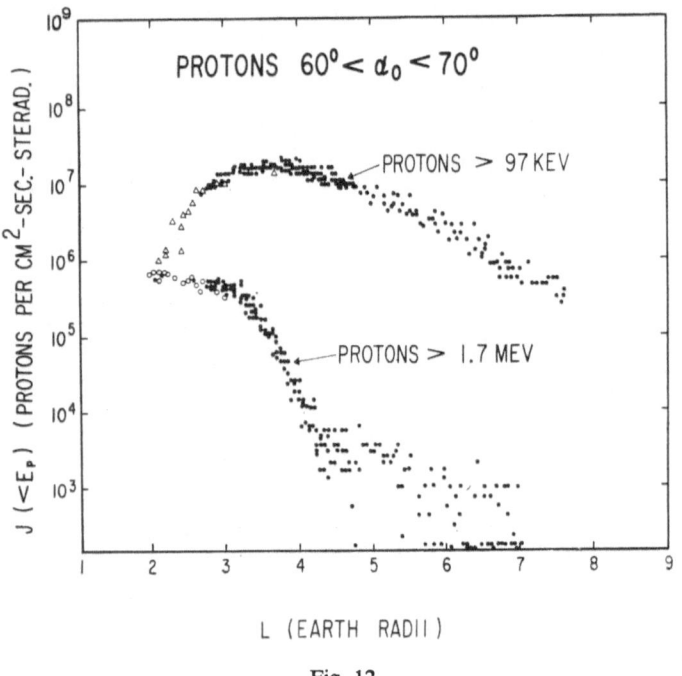

Fig. 12.

compares the proton intensities measured on Explorer XIV, plotted as dots, with the proton intensities measured on Explorer XV, shown as open triangles and circles. The good agreement seen here is typical of all the data.

As has been previously reported (DAVIS and WILLIAMSON, 1962), the proton spectra are better approximated by an exponential than a power law form. Some representative spectra are shown in Figure 13 for protons having equatorial pitch angles between 60° and 70°. As can be seen in this figure at low L values, (for example $L=3$), one has only four closely spaced measurements at low energies and a measurement at

1.7 MeV, the detector being in saturation at 495 keV and 970 keV. Thus one cannot determine the form of the spectra here. If we assume that the spectrum is exponential, the three lowest points would indicate that there is some flattening of the spectrum below about 200 keV. In the region around $L=4$ the spectrum is fairly well represented by an exponential for all of the seven energies measured. At higher L values, as the $L=5.0$ curve illustrates, the spectra show a definite flattening with increasing energy

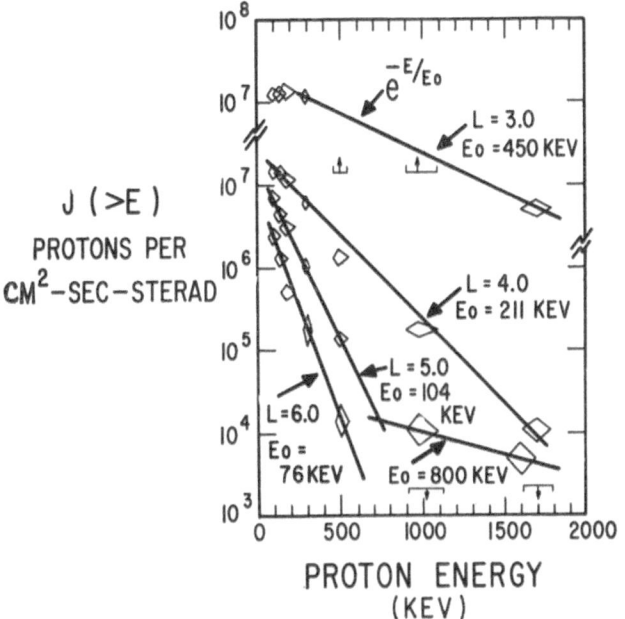

Fig. 13.

on the semilog plot. Nevertheless one may represent a large portion of the low energy protons by fitting exponential spectra to the data, favoring the lower energy measurements where necessary, as is shown in Figure 13. Such a mapping is shown in Figure 14. In the upper part of the figure are contours of constant proton intensity $(E_p \geqslant 97 \text{ keV})$ plotted in the equatorial pitch angle-L plane. The absolute accuracy of the contours are about $\pm 25\%$. In the bottom of Figure 14 are contours of constant e-folding energy plotted in the same coordinates. As is noted during the earlier presentation of raw data, the proton spectra soften as one moves to higher L values or to lower pitch angle values. The e-folding energy contours are accurate to $\pm 50\%$.

With such a set of contour maps one can determine the total number of protons trapped, the total energy of the trapped protons, and the number and energy densities

CONTOURS

OF CONSTANT

PROTON INTENSITY

($E_p \geq 97$ KEV)

CONTOURS

OF CONSTANT

e-FOLDING

ENERGY

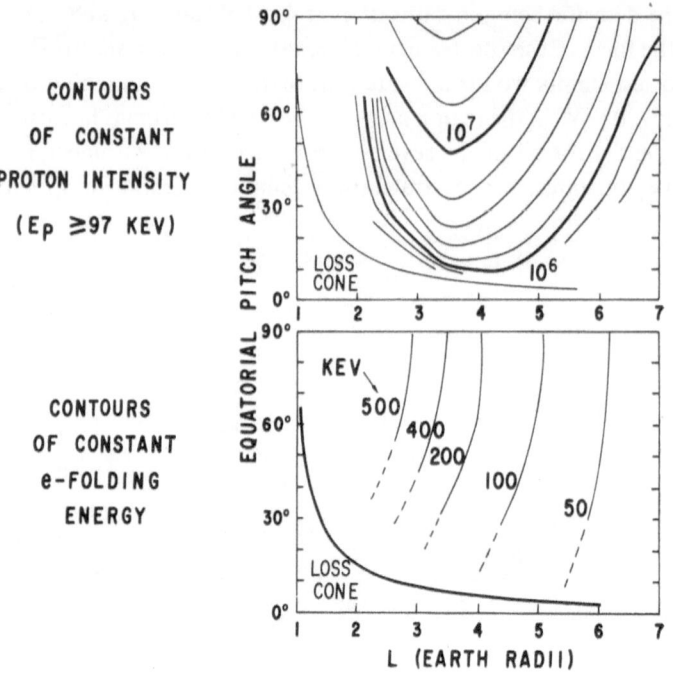

Fig. 14.

ENERGY

DENSITY

(EV/CM^3)

ON

EQUATOR

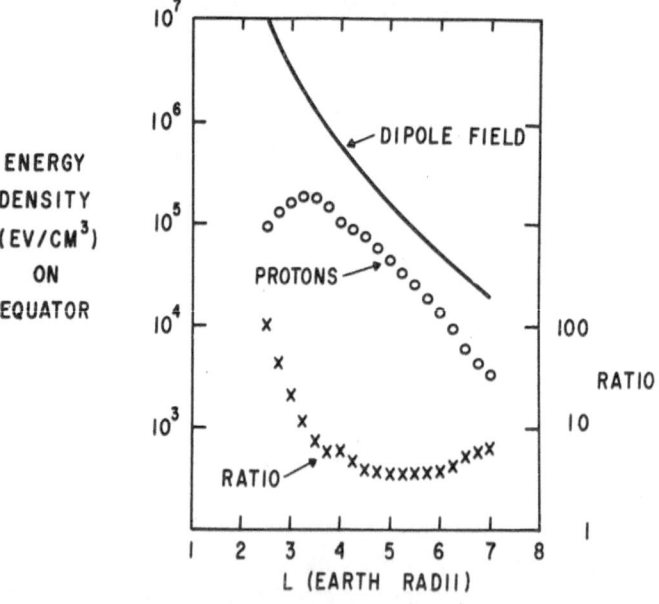

Fig. 15.

as a function of L. Integrating from $L=2.5$ to 7 we find the total number of trapped protons with energy greater than 97 keV is 4.4×10^{28} and the total kinetic energy is 1.5×10^{15} ergs. The data show that the kinetic energy density of the low energy protons on the equator decreases with increasing distance as L^{-6}. This is displayed in Figure 15, where the solid line at the top shows the energy density as a function L for the dipole approximation of the geomagnetic field. The open circles indicate the kinetic energy density of protons having energies greater than 97 keV as computed from the mapping shown in Figure 14. The curve at the bottom of Figure 15 is the ratio of the dipole field energy density to the proton kinetic energy density. Starting with a ratio of about 100 at $L=2.5$, the ratio rapidly decreases with increasing L up to about 3.5 earth radii and then remains nearly constant on out to 7 earth radii

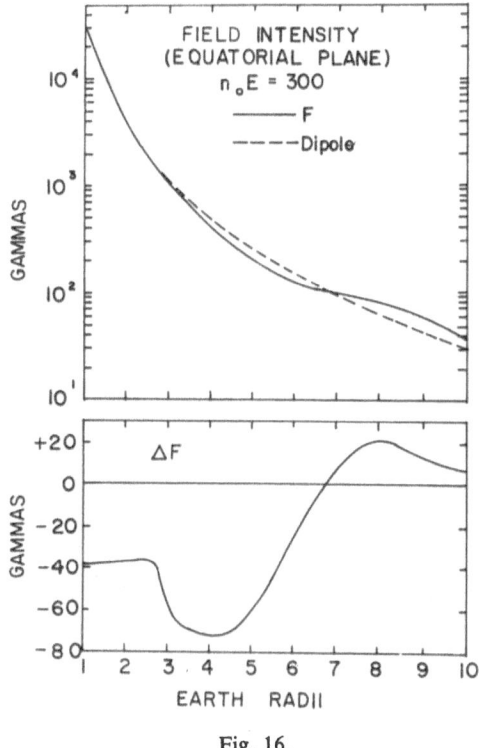

Fig. 16.

with a value ranging from 4 to 7. This perhaps implies a limiting effect by the geomagnetic field on the intensity of particles that can be stably trapped in the field.

The magnetic field disturbance produced by the trapped protons have been computed by AKASOFU, CAIN, and CHAPMAN (1962) for a model proton belt which roughly approximates the map shown in Figure 14. Their results are shown in Figure 16. The bottom curve shows the field disturbance as a function of radial distance. At the top of the figure the undisturbed dipole field is plotted as a dashed line and the dipole plus ring current field is plotted as a solid line. The results predict a 40 gamma depression of the field on the earth's surface at the equator.*

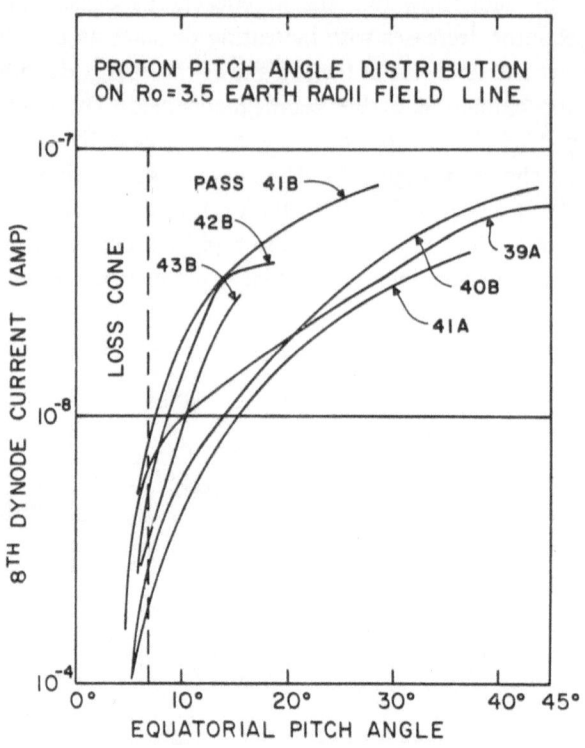

Fig. 17.

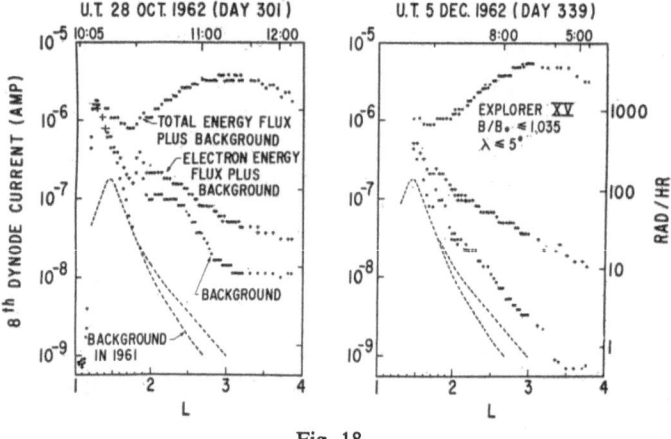

Fig. 18.

* HOFFMAN and BRACKEN (1964) have recently made a more accurate computation of the disturbance field produced by the quiet time proton belt based on a more complete mapping of the low energy protons. Their results show that the surface depression of the field is only about a fourth the value computed by AKASOFU, CAIN, and CHAPMAN (1962).

Enhancements of the protons trapped below 5 earth radii have been observed during magnetic storms on two occasions on Explorer XII. Figure 17 shows the increase observed during the 30 September 1961 storm. The pitch angle distributions labeled pass 39a, 40b, and 41a were obtained on the days just prior to the sudden commencement which marked the beginning of this storm. The curve labeled pass 41b was obtained 17 hours after the sudden commencement and passes 42b and 43b on subsequent days. The enhancement observed during both of the storms amounted to an increase of about a factor of three and was confined primarily to the lowest energy protons. On both occasions the satellite crossings of this region were at geomagnetic latitudes greater than 30° during the periods following the sudden commencements. Thus only protons having equatorial pitch angles of 34° and less were sampled. Calculations by AKASOFU, CAIN, and CHAPMAN (1962) of the field depression which would be caused by the enhanced proton belt, assuming that the proton flux was enhanced at all pitch angles, and the energy spectra remained the same, indicate that the storm time decrease of the field may have possibly been produced by these low energy protons. The D_{st} values ranged from 30 to 60 gamma during the period of enhanced proton flux.

Explorer XV, whose five hour orbital period provided relatively frequent sampling

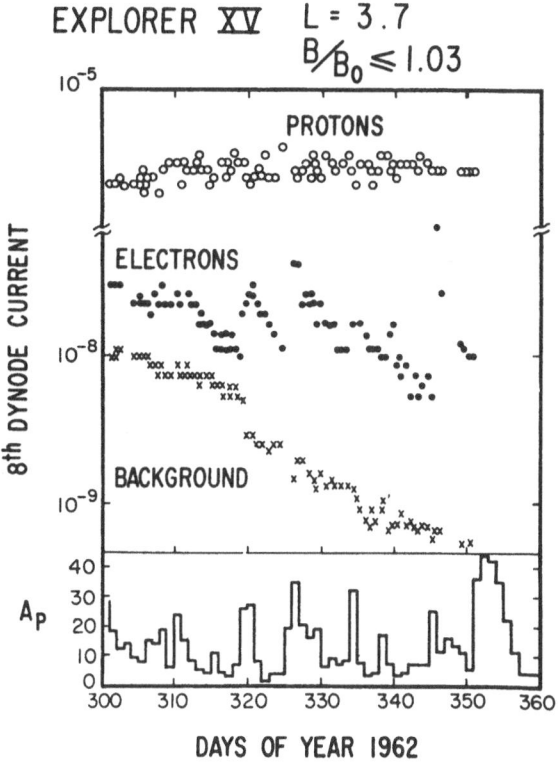

Fig. 19.

of the outer belt, has provided data which are well suited for studying the time variations in the low energy protons and electrons. The data obtained in the early part of the satellite's life are of somewhat less value because of the high background which resulted from the artificially injected electrons from the United States and Russian series of high altitude nuclear tests. On the left hand side of Figure 18 we see the total energy flux, the electron energy flux, and the background measured on October 28, 1962. Also shown as a dashed curve is the background measured on Explorer XII in 1961. The artificially injected electrons show up as a large increase in the background measured in 1962 as compared to that measured in 1961. The electrons from the United States tests appear at about $L=1.25$ and the electrons from the Russian tests appear as a sharp peak at about $L=1.9$ and with some enhancement extending on out to about $L=3$. The total energy flux measured on December 5, 1962 is plotted in the right hand side of Figure 18, and shows by comparison, that the background has almost decayed back to the 1961 value. The background current provides us with a crude measure of the radiation dose behind one to three grams of aluminium and plastic, for which the values are indicated on the right hand side of the figure.

Figure 19 contains the time history of protons and electrons mirroring on the equator at $L=3.7$ earth radii, the background current measured by the detector, and at the bottom of the figure, the planetary index A_p. As can be seen the protons remained essentially constant to within the scatter of the data ($\pm 25\%$) throughout this 50-day period. The electrons, however, underwent frequent and rapid changes in intensity characterized by sharp, almost instantaneous, increases which occurred at the beginning of small M-region type magnetic storms as indicated on the A_p plot. Following the rapid increases the electrons decayed away with e-folding times of about 5 to 10 days. The modulation of the electrons extended down to about $L=3.0$.

Acknowledgements

The detector was developed by J. M. Williamson and the author. Mr. Williamson was responsible for the Explorer XIV and XV detectors, including their excellent calibration and near perfect operation throughout the satellites' lives. We thank Dr. R. A. Hoffman for helpful discussions about our findings and for the use of his results on the storm time enhancements of the protons. The higher energy proton calibrations were performed on the 5 MeV Van der Graph at the Naval Research Laboratory and we thank Dr. Eligius A. Wolicki for making this possible.

References

AKASOFU, S.-I., CAIN, J. C., and CHAPMAN, S.: 1962, *J. Geophys. Res.* 67, 2645.
DAVIS, L. R. and WILLIAMSON, J. M.: 1962, 'Low-Energy Trapped Protons', *Space Research* 3, 365.
HOFFMAN, R. A. and BRACKEN, P. A.: 1964, 'Magnetic Effects of the Quiet Time Proton Belt', *Goddard Preprint Series* X-611-64-186.
JENSEN, D. C. and CAIN, J. C.: 1962, 'An Interim Geomagnetic Field', *J. Geophys. Res.* 67, 3568.

PRELIMINARY RESULTS OF
MAGNETIC FIELD MEASUREMENTS IN THE TAIL OF
THE GEOMAGNETIC CAVITY*

LAURENCE J. CAHILL, JR.**

Headquarters National Aeronautics and Space Administration, Washington 25, D.C.

Exposition

A triaxial flux gate magnetometer was included in the Explorer 14 payload. This instrument was similar to the one carried on Explorer 12 (CAHILL and AMAZEEN, 1963). The range of magnetic field measurement along each axis of the Explorer 14 magneto-meter was nominally ± 500 gammas. Explorer 14 was launched on 2 October 1962 into an elliptical orbit with apogee near 16 earth radii, inclination 33° and initial angle between the orbit major axis and the earth-sun line 70°. The initial spin rate of the satellite was 10 rpm. The satellite also had a precession motion of initial half-angle about 5°, this had decreased to 3° during the period 10–20 October. In November the precession cone opened up to 40° half-angle decreasing to 1° or less in early January 1963. To date we have analysed the magnetic data from a number of passes in October and in January when the precession cone was small. The October data are consistent with a geomagnetic cavity boundary, often beyond 16 earth radii, near 90° to the earth-sun line. Similar observations were obtained with Explorer 12 near the same region of the cavity one year earlier; Explorer 12 results indicated that the boundary was often beyond 12 earth radii.

The data obtained in January 1963, when the angle between the major axis of the orbit and earth-sun line was between 140° and 150°, are quite different from that ob-tained by Explorer 12 and form the basis of the present preliminary report. Some pre-cautionary comments about these data are necessary. There are still some questions not completely answered about (1) the spin axis orientation of the satellite; (2) possible zero level drift and sensitivity changes of the magnetometer. While investigation of these problems is proceeding, this preliminary report is submitted in view of interest expressed by several individuals.

Figure 1 shows measurements obtained on 9 January. This is a period of very low magnetic activity on earth; preliminary K_p indices for 9 January do not rise above 1^+. Thus this record represents the conditions in the tail of the magnetic cavity during very quiet conditions.

The magnitude of the field is close to that predicted until 10 earth radii. Beyond this it is higher than predicted and continues between 30 to 50 gammas out to apogee

* Supported by NASA Contract NASw-155.
** Permanent address: Department of Physics, University of New Hampshire, Durham, New Hampshire.

at 16 earth radii. On this particular pass the field appears to reach a minimum near 12 earth radii and to increase to 40–50 gammas beyond 14 earth radii. This is not a typical feature on other records in this period; it may be a time fluctuation in the magnitude of the field. Also at 10 earth radii the direction of the field commences to depart from that of a dipole field. The most apparent change occurs in the spacecraft

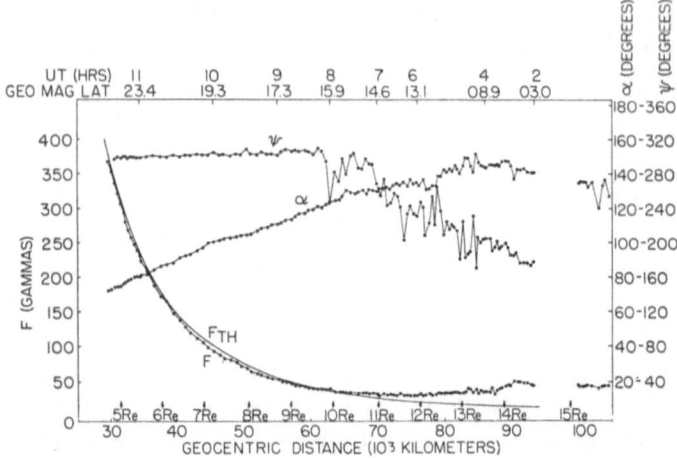

Fig. 1.   The magnetometer record for the inbound pass of January 9, 1963, measured magnitude of the earth's magnetic field and the field direction in spacecraft direction angles α and ψ. The points plotted are 16 measurements average where the measurements are taken 3 times a second.

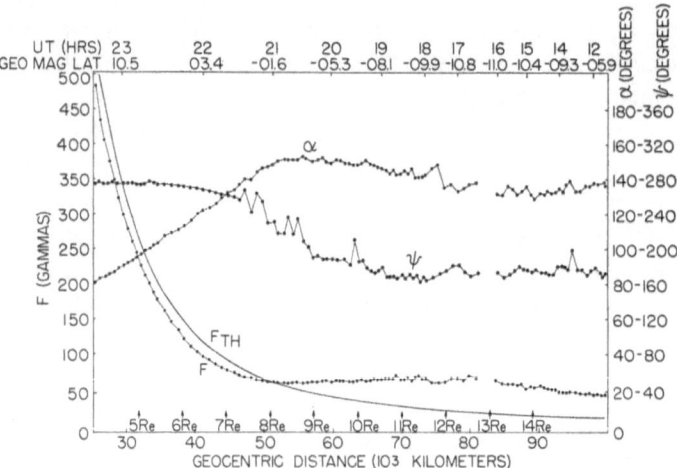

Fig. 2.   The record for the inbound pass of January 7, 1963.

direction angle ψ. The record of ψ becomes more irregular beyond 10 earth radii in addition to the slow change in the average value from about 300° at 10 earth radii to 180° at 14 earth radii. After a gap in the record near 15 earth radii the field is seen to continue similar in direction and magnitude. The angle α deviates from that expected for the earth's dipole field but by a less important amount.

Figure 2 shows a similar record two days earlier. Here the magnitude of the field becomes higher than expected at about 8 earth radii and the change in angle ψ commences there. Deviation in the angle α is also more apparent in this record. The direction angles α and ψ are somewhat different between 5 and 8 earth radii than in Figure 1, presumably because the geomagnetic latitude is different. The negative departure of the field below 8 earth radii may or may not be significant. Further discussion of this feature will be reserved until a more complete assessment of the data has been made. Also apparent in this record is the observation that the field assumes an approximately constant average direction and magnitude beyond 11 earth radii. The same feature was suggested in the 9 January record but beyond 14 earth radii. In the 7 January record the magnitude of the field is somewhat higher than on 9 January, remaining above 50 gammas to 14 earth radii then decreasing slowly to 40 gammas.

Interpretation

These two records, and a few other partial records now available during this period, allow some preliminary statements to be made about the nature of the tail of the magnetic cavity. The magnetic field remains essentially that of a centered dipole, with minor distortions, out to a distance of 8 to 10 earth radii. This distance apparently varies from day to day. Beyond this distance the field remains at a level of 30 to 60 gammas with relatively small short term (period less than one hour) fluctuations in magnitude. Slow time variations may be present with changes of 20–30 gammas occurring in periods of several hours although these cannot be distinguished from spatial variations along the satellite orbit. The direction of the field changes from the dipole direction at 6–10 earth radii to a direction, identified in spacecraft coordinates, as $\psi \simeq 180°$, $\alpha \simeq 140°$ within a transition distance of 2 to 4 earth radii. It appears that when the transition starts at a closer distance as on 7 January the changes occur more rapidly. Once the field direction has reached the values mentioned above the average direction and magnitude remain quite constant out to apogee beyond 16 earth radii.

These results are to be compared with those obtained by Explorer 10, also launched into the tail of the magnetic cavity (HEPPNER et al., 1963). Explorer 10 encountered a departure from dipole field direction that occurred gradually between 5 and 10 earth radii. Beyond 10 earth radii the field direction remained relatively constant until the abrupt change beyond 22 earth radii that has been interpreted as first penetration of the cavity boundary. The Explorer 10 trajectory traversed the tail of the cavity considerably south of the geomagnetic equator, apparently close to the edge of the cavity, and near the 2100 local time meridian.

The Explorer 14 satellite passes being discussed, are contained within local time meridians 2300 and 0200. The satellite orbit inclination was 33° and the period under discussion was 2–3 weeks after winter solstice. The satellite was close to the geomagnetic equator during these passes but below the ecliptic plane. Considering theoretical models of the geomagnetic cavity when the earth's dipole is not perpendicular to the earth-sun line, the measurements were perhaps taken near the lower edge of the cavity.

The Explorer 14 measurements are consistent with those of Explorer 10 and indicate a cavity field in the tail stretched away from the sun and from the earth.

The departures from dipole field occurring as low as 7–8 earth radii in January 1963 are to be compared to the SUI charged particle measurements in early December 1962 on the same satellite (FRANK *et al.*, 1963). At that time, also magnetically quiet, the flux of trapped electrons of energies >40 keV showed an abrupt (−1000 km) decrease near 35000 km (5 earth radii). This decrease closely resembled the abrupt decreases in trapped electron flux seen in Explorer 12 data (ROSSEN *et al.*, 1962). The decreases were coincident with penetration of the geomagnetic boundary as observed in the magnetic measurements from Explorer 12 and were observed on the sunlit side of the earth. Trapped electron and magnetic field measurements from Explorer 14 are not yet available on the same pass and we can only point out that the distortion of the magnetic field observed in January occurs at a somewhat greater distance than the trapped radiation boundary observed in December. The abruptness of the decrease in trapped particles is certainly not similar to the slow changes in field direction. These occur in a distance of 10000 to 20000 km. Since the abrupt particle cutoff does not appear to be related to an abrupt change in the local magnetic field, it may be that the cutoff is caused by some large scale characteristic of magnetosphere. The recent work of HONES (1963) may provide an explanation of this cutoff.

References

CAHILL, L. J. and AMAZEEN, P. J.: 1963, 'The Boundary of the Geomagnetic Field', *J. Geophys. Res.* **68**, 1835–1843.

FRANK, L. A., VAN ALLEN, J. A., and MACAGNO, E.: 1963, 'Charged Particle Observations in the Earth's Outer Magnetosphere', *SUI Research Report* 63–10, unpublished.

HEPPNER, J. P., NESS, N. F., SCEARCE, C. S., and SKILLMAN, T. L.: 1963, 'Explorer X Magnetic Measurements', *J. Geophys. Res.* **68**, 1–46.

HONES, E. W.: 1963, 'Motions of Charged Particles Trapped in the Earth's Magnetosphere', *J. Geophys. Res.* **68**, 1209–1220.

ROSSEN, W. G. V., O'BRIEN, B. J., VAN ALLEN, J. A., FRANK, L. A., and LAUGHLIN, C. D.: 1962, 'Electrons in the Earth's Outer Radiation Zone', *J. Geophys. Res.* **67**, 4533–4542.

Discussion and Questions

H. Alfvén: When you measure the field in the sunward direction, you sometimes observe that the direction changes by 180 degrees. Does it stay at 180 degrees, that is antiparallel to the earth's magnetic field?

L. Cahill: Not always.

H. Alfvén: How much could it vary? I'm speaking of a case when you have a 180 degree change plotted at a sharp boundary.

L. Cahill: There's a tremendous variety of variations. Sometimes it just oscillates about this 180 degree direction although there's often a skew (change in the angle α) after the initial change of 180 degrees.

H. Alfvén: A difference of 45 degrees in the direction?

L. Cahill: Yes, and sometimes it returns to the approximate dipole direction. Sometimes it points back toward the sun.

H. Alfvén: When you go in the sunward direction, are there also cases when you don't observe the 180 degree change? When the direction remains parallel to the earth's magnetic field even outside what you call the boundary?

L. Cahill: Well, in that case I wouldn't call it a boundary because I use this change in direction to indicate where the boundary is.

H. Alfvén: The point where the variation stops. Are there cases when the field has the same direction and magnitude while the radiation stops?

L. Cahill: I'm not aware of any. There may be some and I can't claim that we have made a day by day comparison of trapped radiation boundaries and magnetic field boundaries but I am not aware of any case where the field is in the dipole direction and the trapped radiation boundary occurs. Except that October 1, 1961 case, but on that day the boundary was moving. Are you aware of any Leo (Davis)? Do you see a boundary considerably before we see the magnetic field boundary?

L. Davis: No.

H. Alfvén: I think the conclusion from your diagram is that the boundary is a rather exceptional case and that your data are not in agreement with a closed model. It is much more in agreement with an open magnetosphere model as I have suggested.

L. Cahill: In the rear of the magnetosphere?

H. Alfvén: Yes, it isn't necessary at all you see that there is a boundary. You always assume that there must be a boundary but that is not a necessity. It is only in the case of a specific and at present a very popular model that there should be a boundary.

L. Cahill: I agree that the data don't support a teardrop boundary in the rear, at least not to within 16 earth radii. The data doesn't indicate one way or the other whether there is a boundary in the rear.

R. Lehnert (chairman): 'Are there any more questions', I would like to say, before we go into more detailed discussions, if these pertain to the discussion between the speaker and Prof. Alfvén I think we did better postpone this and take this up in the round table discussion.

E. Schmerling (NASA Headquarters): Instead of attempting to correlate the distance of the boundary index, I wonder if you'd consider looking at whistler data. Now whistler dispersion is, very roughly speaking, the electron density integrated along the field lines. Consequently if one goes to middle high latitudes whistler dispersion can measure the fractional extension of field lines which one might think would be fairly closely correlated with distances to the boundary.

L. Cahill: That's a very good suggestion, but I don't know if any comparison has been made. That should be done.

J. Dungey: I spoke with L. Frank (SUI) and he was discussing the distortion of magnetic field as a possible explanation of the trapped particle boundaries in the rear. He thought it wouldn't be sufficient with the data he had at that time.

L. Cahill: The distortions that I showed in the magnetic field are considerably thicker, take longer to occur and are, in general, further from the earth than the quite abrupt boundaries of the trapped radiation so I don't think that the local magnetic field determines the boundary but rather the general shape of the magnetosphere determines the trapped radiation boundary.

J. Dungey: Well, can you say that the distortion of the field is sufficient to produce the observed distortion of the electron boundary?

L. Cahill: The local distortion of the field I don't believe would do that. I think you would have to consider the magnetosphere as a whole.

OBSERVATIONS OF THE GEOMAGNETIC
CAVITY BOUNDARIES

NORMAN F. NESS

Space Sciences Division, Goddard Space Flight Center, Greenbelt, Mld.

Abstract. This paper reviews the plasma and magnetic field data (BONETTI *et al.*, 1963; HEPPNER *et al.*, 1963) obtained in March 1961 from the space probe Explorer X. Particularly emphasized is the strong correlation of these separate results when viewed with respect to multiple crossings of the geomagnetic cavity boundary on the dark hemisphere of the Earth. In addition, the magnetic field data have been subjected to detailed analyses of its spectral characteristics for fluctuations of small amplitude (less than several gammas) and low frequency (less than 0.003 cps). Comparisons of the experimental data with theoretical models of the cavity reveals certain gross similarities but there exist substantial discrepancies which have yet to be explained. It is concluded that Explorer X traversed the cavity boundary on a number of occasions and tentatively established the experimental fact that the cavity tail is flared outward from the antisolar direction to a distance of at least some 280 000 km at an angle of approximately 20°.

Introduction

A review is presented of the plasma and magnetic field data obtained in deep space from the Explorer X spacecraft launched March 25, 1961. The primary objective of the satellite was to investigate the magnetic fields and low energy proton fluxes in cislunar space. The power supply of the Explorer X satellite was silver cadmium batteries which limited the lifetime of the satellite to 55 hours during which all but the last 3 hours of telemetry transmission yielded calibrated scientific data. During this interval the satellite trajectory was oriented generally at 135° to the Sun–Earth line.

The two principal experiments carried on board the spacecraft were a Rubidium vapor and fluxgate magnetometer experiment prepared by the Goddard Space Flight Center and a Faraday Cup Plasma Probe prepared by Massachusetts Institute of Technology (M.I.T.). Figure 1 illustrates the various locations of different sensing elements. The Faraday Cup looked out normal to the spin axis of the satellite with a directional response characteristic that was cylindrically symmetrical about the normal to the cup face. The half-angle acceptance cone was 63° with a minimum positive flux sensitivity equivalent to 4×10^6 protons/cm²/sec. Two monoaxial fluxgate magnetometers were mounted at the extreme ends of two oppositely located structural booms so as to remove the sensors from the magnetic contamination associated with the spacecraft and its electronics. The two fluxgates are identical in construction with slight differences in the calibrations and were designed with a small dynamic range (± 20 gammas) to provide high sensitivity (± 0.25 gammas) in order to map definitively the weak magnetic fields which were anticipated. The Rubidium vapor magnetometer which completed the instrument repertoire was mounted on a support structure along the positive spin axis. This instrument measures the total magnitude of the scalar magnetic field and is insensitive to the direction of the field. Superimposed

Chang & Huang (eds.), Proc. Plasma Space Sci. Symp. All rights reserved.

on the ambient unknown magnetic field was a magnetic field created by a set of
current carrying coils wound on the surface of the spherical ball at the top of the
support structure. The addition of a known vector field to the unknown scalar field
allows mathematical analysis to determine the vector characteristics of the unknown
field.

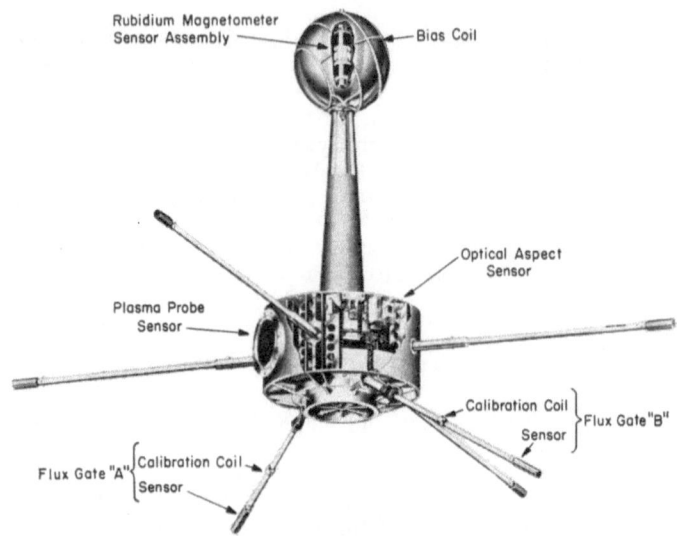

Fig. 1.   Explorer X Satellite (after HEPPNER *et al.*, 1963).

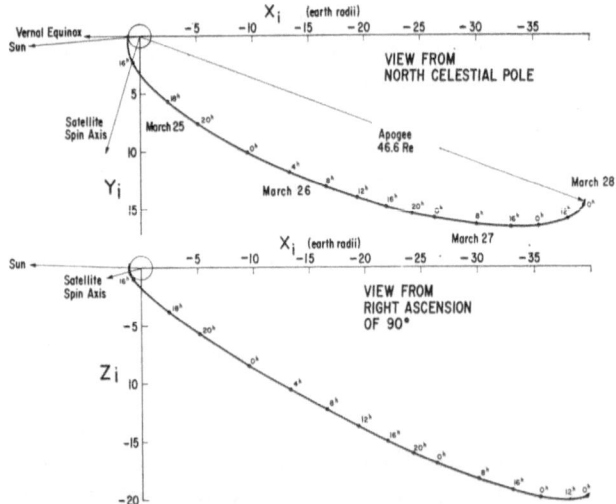

Fig. 2.   Trajectory in celestial inertial coordinates (after HEPPNER *et al.*, 1963).

Geomagnetic Field Results

The trajectory of the satellite in celestial-inertial coordinates is shown in Figure 2. The projections of the Earth-Sun vector and the probe spin axis vector are shown as well as the trace of the trajectory on the three planes formed by the coordinate system. Accurate scientific data were obtained over the distance range 1.8 to 42.5 R_e (Earth Radius = R_e), and at positions which were below the equatorial plane of the Earth and behind the Earth relative to the Sun. Figure 3 summarizes the total field data obtained with the Rubidium vapor magnetometer over the distance 1.8 to 6.6 R_e and is seen to decrease as approximately $1/r^3$ over this region. In order to accentuate

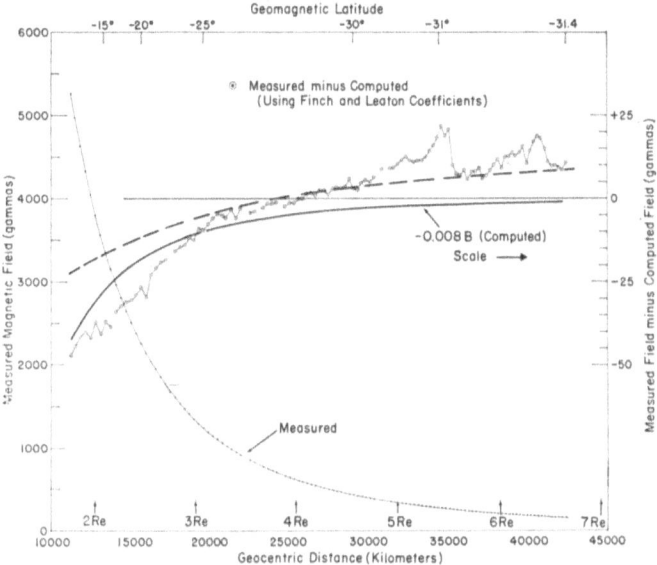

Fig. 3.   Total magnetic field data from 1.8 to 6.6 R_e (after HEPPNER *et al.*, 1963).

the minor differences between observed and theoretical fields extrapolated by spherical harmonic coefficient sets from terrestrial surface measurements the difference of these two fields has been superimposed on this same scale. The data points indicated are taken at one minute intervals and represent averages of the continuously operating Rubidium vapor magnetometer. In the region 3 to 4 R_e it is seen that the departure of the observed magnetic field relative to that computed displays adjacent negative and positive anomalies. A possible explanation of this dipole-type anomaly is that it is due to the trapped particles in the radiation belts and leads to the formation of effective ring currents analogous to those studied some years ago in connection with theories on the Earth Storm. These data however, do not substantiate the existence of large scale ring currents found in the vicinity of 6 to 10 R_e by the Explorer VI satellite. Deviations in this region are smooth and do not show the dipolar charac-

teristic which is only seen on Explorer X at a distance of 3.8 R_e (when the satellite
was passing through the region of maximum intensity in the radiation belts).

Figure 4 illustrates the magnetic field data from approximately 4.1 to 12.3 R_e.
In this presentation the data is referred to a payload coordinate system in which
alpha (α) is the polar angle between the spin axis and the magnetic field vector
and psi (ψ) is the azimuthal angle in the equatorial plane of the satellite between the
projections of the satellite-sun vector and the magnetic field vector. The observed
magnetic field is presented as averages at 1 minute intervals on the diagram and

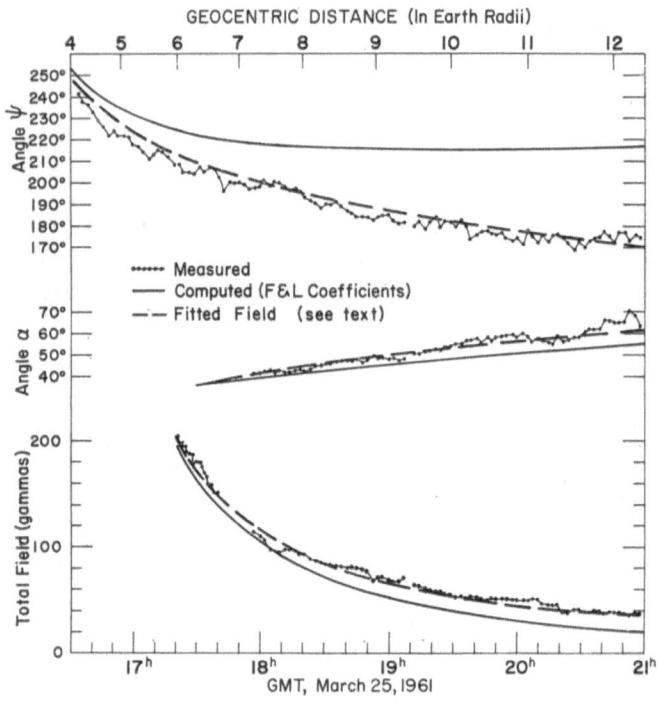

Fig. 4.   Vector magnetic field data from 4.1 to 12.3 R_e (after HEPPNER *et al.*, 1963).

compared with the theoretical fields obtained by use of the Finch and Leaton coefficient set for the Earth's magnetic field.

Unfortunately the operation of the Rubidium vapor magnetometer ceased to be
continuous at this point along the trajectory of the satellite. Passive thermal control,
implemented through the use of specially prepared surfaces on the magnetometer
sphere, failed to correctly maintain the proper operating temperatures for the instrument. Outgassing of the third stage nose-faring during the initial launch maneuver
contaminated the surface of the sphere and modified the absorptivity and emissivity
characteristics, so that the temperature rose above the proper operating temperature.
Rubidium vapor magnetometer data from 6.5 R_e out were continuous only at brief

and short time intervals. Fortunately the Earth's magnetic field was below the maximum dynamic range of the fluxgate magnetometers at approximately 1800 UT March 25 and the utilization of the intermittent Rubidium vapor magnetometer data allowed an accurate calibration of them at this time. Since the fluxgate magneto-meters are relative devices with no absolute zero scale, it is extremely important that some in-flight means be provided for determining the accuracy of their data.

It is to be noted in Figure 4 that there is an increasingly large discrepancy between the measured magnetic field magnitude and more importantly the directional charac-teristics when compared with the theoretical magnetic field. The departure of the observed field from the theoretical field, however, is a smoothly varying function increasing slowly and in such a manner that the interpretation of data obtained on this satellite at considerably larger altitudes is particularly relevant in the proper understanding of the departures at low altitudes.

The data presented thus far have been referenced to coordinate systems associated with the Earth and the payload. The reason for doing so was that in this region of space geomagnetic fields dominated the character of the measured magnetic field. During this time no plasma was detected which can be associated with a flux of low energy particles directly from the Sun. Shortly after injection the Plasma Probe did show small signals briefly but the interpretation is that "velocity scooping" of a relatively stationary plasma by the satellite during its high initial velocity led to such a response.

Geomagnetic Cavity Boundaries

The remainder of the data obtained from the Explorer X satellite is presented in a coordinate system which reflects the interplanetary plasma or "solar wind" aspect of the data more appropriately. A solar-ecliptic coordinate system is defined centered at the Earth in which X_{se} is parallel to the Earth–Sun direction, Z_{se} is normal to the ecliptic plane in the same sense as the spin axis of the Earth, and Y_{se} is chosen to form a right-handed coordinate system. Figure 5 illustrates the trajectory of the Explorer X spacecraft in an isometric projection referred to these solar ecliptic coordinates. The trace of the trajectory is projected on the X_{se}–Z_{se} plane, illustrating how far behind and below the Earth the satellite was relative to the solar direction. The numbers adjacent to the trajectory indicate the times at which the satellite passed through the points during the interval March 25th through March 27th.

The magnetic field data presented for the rest of the trajectory are presented as a magnitude and two angles, theta (ϑ) being the latitude of the magnetic field vector relative to the ecliptic plane, and phi (ϕ) being the longitude of the magnetic field vector as projected on the ecliptic plane. Theta is positive above the ecliptic plane, and phi is positive in a direction anti-parallel to the Earth's orbital motion in the plane of the ecliptic (to the west of the sun). An anti-solar direction would be defined by $\vartheta = 0°$ and $\phi = 180°$. This is the most appropriate coordinate system to discuss the observations of both the streaming solar plasma and the interaction of this plasma with the geomagnetic field. Figure 6 presents the results of the magnetic field experiment over the distance 12 to 21 R_e in which the data are obtained from the fluxgate magnetometers. Included in this Figure is the representation of the theoretical geomagnetic field magnitude.

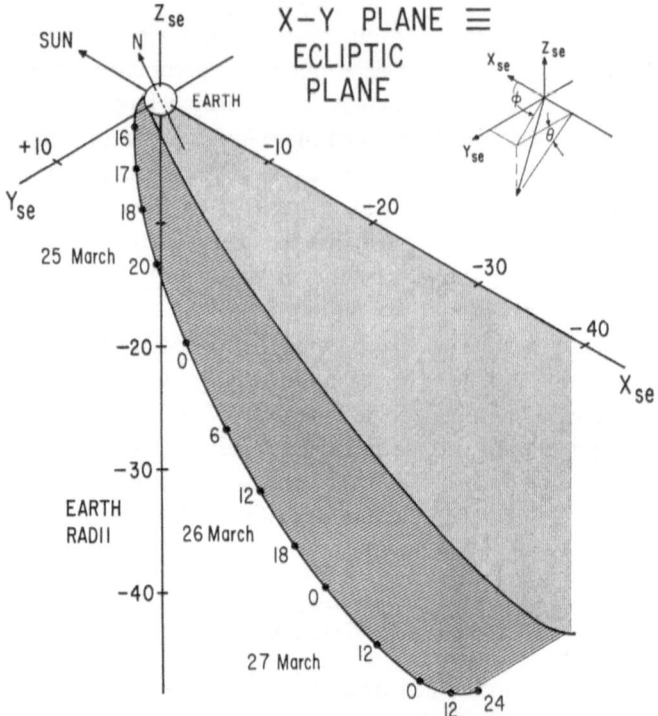

Fig. 5.   Isometric projection of trajectory in solar-ecliptic coordinates.

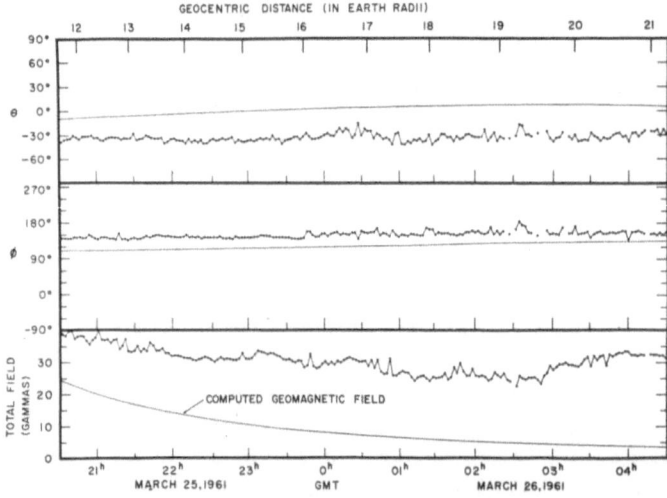

Fig. 6.   Magnetic field data from 11.7 to 21.3 R_e (after HEPPNER *et al.*, 1963).

The sensitivity of the fluxgates, $\pm\frac{1}{4}$ gamma is important in the analysis of the data from this portion of the flight through to the termination of successful transmission of scientific data. The contamination of the magnetic field measurements due to fields associated with the spacecraft was measured before flight and found to be less than 1 gamma. The calibration of the fluxgate magnetometers through the use of direct comparison with the absolute Rubidium vapor magnetometer during its intermittent operation yielded an accuracy of $\pm\frac{1}{4}$ gamma. Although the calibration of the fluxgates was possible over a limited time interval past experience has shown that these sensors do not exhibit zero level drifts or sensitivity changes over short time intervals when a constant temperature is maintained on both the sensor and the electronics.

The total field is larger than the theoretical magnetic field by approximately 20 gammas at large distances from the Earth. At approximately $19.5\,R_e$ the measured total field actually increases rather than decreases to a value of 32 gammas. In addition, the direction of the magnetic field is considerably different from that predicted by extrapolation of the surface geomagnetic field. The direction is such that the magnetic field vector is 40° below the plane of the ecliptic and about 145° away from a solar direction. This large distortion of the magnetic lines of force makes the field vector point in approximately an anti-Earth direction along this portion of the trajectory. During this interval no plasma was observed by the M.I.T. detector. The data points on this plot are separated at time intervals of 147 seconds, which correspond to the programmed sequence of telemetry transmissions for the fluxgate magnetometers and plasma probe. The measurements of the magnetic field and the

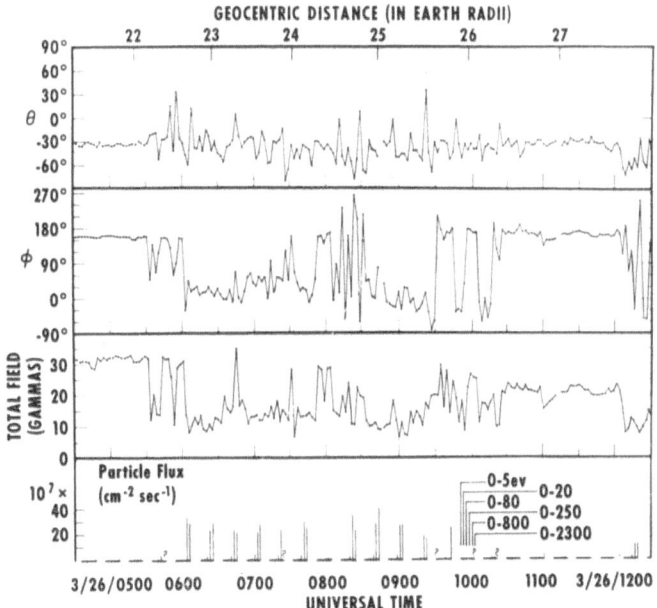

Fig. 7.   Magnetic field and plasma data from 21.3 to 28.0 R_e (after BONETTI *et al.*, 1963).

plasma were immediately adjacent to each other so that in spite of the large time interval separating successive measurements of either the magnetic field or the plasma, the measurements of both phenomena were almost simultaneous.

Figure 7 summarizes the magnetic field data in solar ecliptic coordinates and the plasma intensity measurements over the distance range 21 to 28 R_e. Up to the geocentric distance of 22.3 R_e the magnetic field maintains its steady but highly distorted configuration and its large value of 30 gammas. At 0530 on March 26th, however, the character of both the magnetic field data and the plasma data changed and from thereon the inter-relationship of these two phenomena was quite intimate. At this point on the trajectory the satellite was 22.3 R_e from the Earth at approximately a Sun–Earth-probe angle of 140°. The magnetic field abruptly changed to a low level at the same time as a substantial flux of plasma was first observed. Figure 7 also illustrates a unique characteristic of the magnetic field: its alternation from a stable configuration identical to that measured before 0530 UT to a variable orientation observed frequently thereafter. It is also noted in this Figure that when plasma is not observed the magnetic field returns to an orientation very close to that observed before 0530 UT.

The programming of measurements of the different energy levels for the plasma probe was such that it took approximately 15 minutes for a complete spectral scan. This is indicated in the Figure by the equally spaced markers corresponding to a separation of 147 seconds and the identification of the energy levels as shown. The energy levels chosen were 200, 300, 500, 800, 1500 and 2300 volts. Measurement of the plasma is effected by modulating the voltage on a grid in front of a current collector and measuring the resulting current modulation arising from the streaming positive ions. The energy of the ions is known from the corresponding modulated voltage, but the mass distribution is unknown. The measurement of the flux is an integral measurement of all ion species whose energies are equal to or less than the grid potential. In this Figure is seen a typical case in which the total flux measured at the 2300 volt level was less than that measured in the preceding telemetry transmission corresponding to the 800 volt level. The only possible interpretation is that the intensity of the flux changed on a time scale of 147 seconds since the integral measurement of the plasma probe requires that the 2300 volt measurement be equal to or greater than the 800 volt level. The fluxes are indicated in units of 10^7 particles/cm^2/sec and assumes a single ionic species of protons. It is seen from the data that fluxes of 20 to 40 in the units previously defined, represent a good average so that the data indicate an average flux of several times 10^8 protons/cm^2/sec.

It is important to note the detailed correlation of the plasma data and the magnetic field data. When the magnetic field is approximately 30 gammas, stable in configuration with a latitude of approximately $-30°$ and longitude of 140°, no plasma is detected. However, when the magnetic field fluctuates rapidly both in magnitude and direction strong plasma fluxes are observed coming from the solar direction.

In Figure 8 the magnetic field and plasma data obtained over a distance scale 28 to 33 R_e are presented. The correlation between the plasma and the magnetic field is striking, when the magnetic field returns to the characteristic direction of $\vartheta = -30°$ and $\phi = 140°$, plasma is not observed. However, the total intensity of the magnetic field is only 20 gammas, as opposed to the previous interval in which plasma was not seen when the magnetic field strength was 30 gammas. The magnetic field is not as variable as was observed previously, and in addition there now appears a second

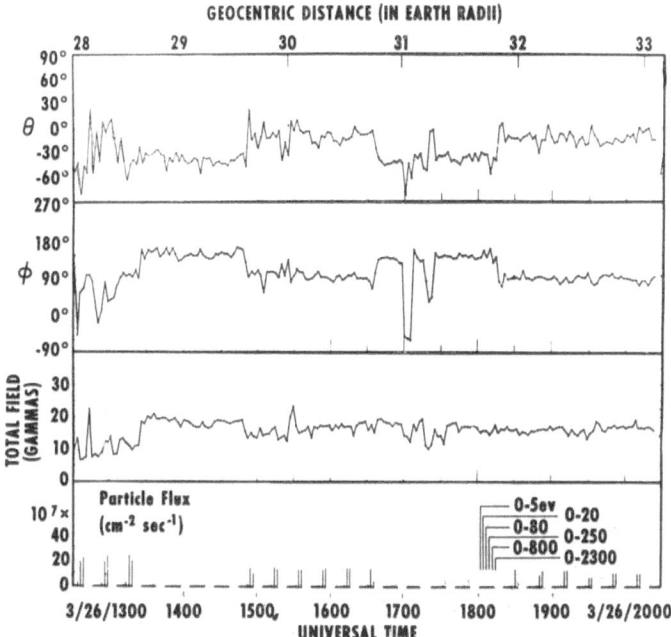

Fig. 8.   Magnetic field and plasma data from 28.0 to 33.2 R_e (after BONETTI *et al.*, 1963).

stable magnetic field configuration corresponding to $\vartheta = 0°$ and $\phi = 90°$ during which plasma is observed. Thus there are two orientations of the magnetic field vector for which the magnetic field is stable in magnitude and maintains its direction for time scales up to several hours. Plasma is observed only during one of these stable configurations however, although plasma is always observed whenever the magnetic field fluctuates rapidly in magnitude and direction.

These three general classifications of the correlated plasma and magnetic field data are sufficient for the remainder of the flight in cataloging the various characteristics of the data. It is not possible to unique'y catalog all of the data obtained into these three classes, but approximately 90% of the data fall into these three regions. Between them are transitions which have yet to be explained satisfactorily; one of the major problems being the inadequate time sampling on the extended scale of 147 seconds.

In Figure 9 is shown the results of the plasma and magnetic field data over the distance 33 to 37 R_e. Again the correlation of the plasma and magnetic field data is clearly seen as well as a rather gross change in the direction of the field around

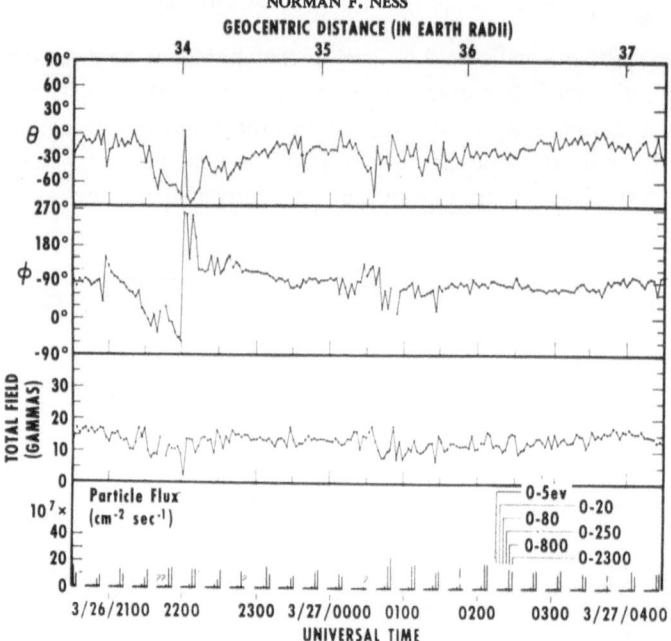

Fig. 9.   Magnetic field and plasma data from 33.2 to 37.3 R_e (after BONETTI *et al.*, 1963).

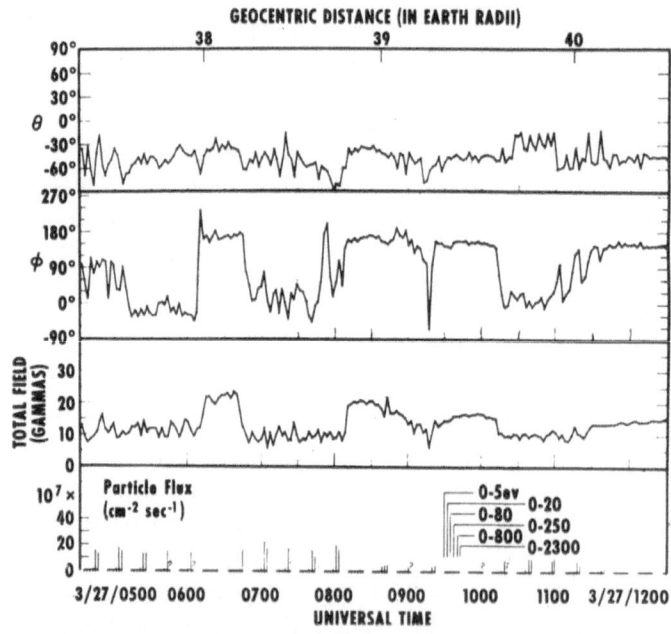

Fig. 10.   Magnetic field and plasma data from 37.3 to 40.4 R_e (after BONETTI *et al.*, 1963).

2200 UT on March 26th. Figure 10 illustrates the magnetic field and plasma data over the distance scale 37 to 41 R_e, and very strikingly illustrates the separation naturally of the data into the three regions previously mentioned. It should be noted that during the second stable configuration of the magnetic field during which plasma is observed, the magnetic field intensity in general is approximately 10 gammas. Figure 11 illustrates the magnetic field and plasma data over the distance 41 to 42 R_e. One item to be noted in this presentation of the magnetic field data is that in general the direction was such as to always be below the plane of the ecliptic, although there are a few occasions on which the field was pointed above the ecliptic.

Figure 11 also includes plasma and magnetic field data which were measured during a sudden commencement geomagnetic storm, observed terrestrially at 1503 on March 27th. This storm followed a class 3 solar flare which occurred at 1005

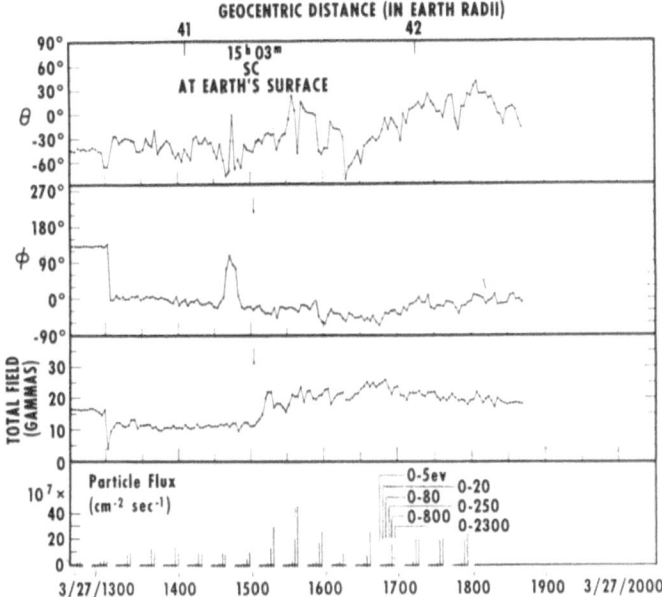

Fig. 11. Magnetic field and plasma data from 40.4 to 42.4 R_e (after BONETTI *et al.*, 1963).

on March 26th. It is desirable to attempt an interpretation of the magnetic field and plasma data on Explorer X from the viewpoint of identifying at what instant the sudden commencement occurred at the satellite. Unfortunately the character of the magnetic field and plasma observed in space is by no means similar to that observed on the surface of the Earth. A rapid direction change in the magnetic field occurred some 15 minutes preceding the 1503 sudden commencement, but no magnitude change. A sharp magnitude change occurs approximately seven minutes after the sudden commencement is observed at the surface of the Earth. Either one of these changes could be interpreted as corresponding to the sudden commencement on the

Earth's surface. Which of these two one chooses, however, strongly effects the interpretation of the character of the magnetic storm in interplanetary space when unmodified by the propagation of the disturbance to the Earth's surface. The satellite data are insufficient in solving this problem.

Interpretation

The attempt in the presentation of the data has been to classify the magnetic field and plasma correlation into three definite and separate regimes. Data of the regions classified as A are typical of the magnetic field data when no plasma is observed during the entire lifetime of Explorer X. The direction of the magnetic field is away from the Earth for this region and the interpretation of the data is that it represents the geomagnetic field distorted by the streaming solar plasma forming a magneto-spheric tail of flux lines interior to which there is no direct plasma penetration. A most unique feature of the Explorer X data and trajectory is that the trajectory appears to have approximately paralleled the boundary separating the geomagnetic cavity interior and exterior. Regions B and C correspond to intervals during which plasma is observed, and the interpretation is that the satellite was outside the geo-magnetic cavity boundary, and thus was able to observe plasma directly although under two separate magnetic field conditions. A definitive explanation for the alter-nate appearance and disappearance of plasma and the changing but highly correlated magnetic field data on this basis has not thus far been presented. It is proposed in this paper that either directional and/or intensity changes in the streaming plasma altered the geometrical position of the cavity tail relative to the satellite trajectory.

The character of the magnetic field during the region classified as B was as stable but weaker in magnitude than A: approximately 10 to 15 gammas in the plane of the ecliptic but pointing in the Y_{se} direction. A relatively weak flux of plasma was always observed. This direction is in the same sense as would be expected for solar magnetic field lines stretched out by the highly ionized streaming plasma in an Archimedean spiral configuration. The angle of the field, however, is too small for nominal values of plasma velocity assumed in various theoretical models. Interpretation of this dis-crepancy is that the presence of the geomagnetic cavity alters the interplanetary magnetic field, and distorts it so that it is not truly representative of either the inter-planetary magnetic field magnitude or direction.

In Region C the magnetic field points back towards the sun and well below the plane of the ecliptic. Plasma flux during the classification C is generally more variable than in the previous interval B. In Figure 12 is summarized the magnetic field data without the plasma data but including the classifications A, B and C with different shading codes. This illustrates a forty-eight hour segment of data compared on a twenty-four hour time scale to illustrate the effect of the rotation of the Earth and the tilt of the dipole magnetic field. Clearly, region A intervals tend to be periodic on a time scale of twenty-four hours with respect to their magnitude behavior. It is seen that the character of the magnetic field changes cyclically from region to region, in a

fashion which strongly suggests an interpretation on the basis of motion of the geomagnetic cavity boundary. Note that the three classifications which have been presented do not cover the entire set of observational data. The additional data that
remains unclassified consists of transitions of various types between the three regions
considered.

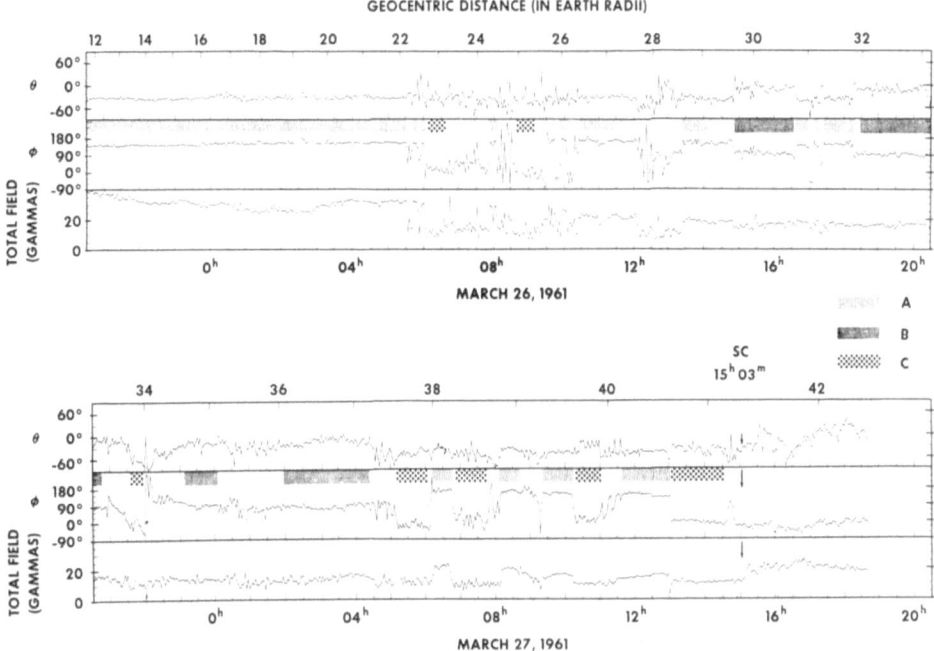

Fig. 12.   Summary of Magnetic field data.

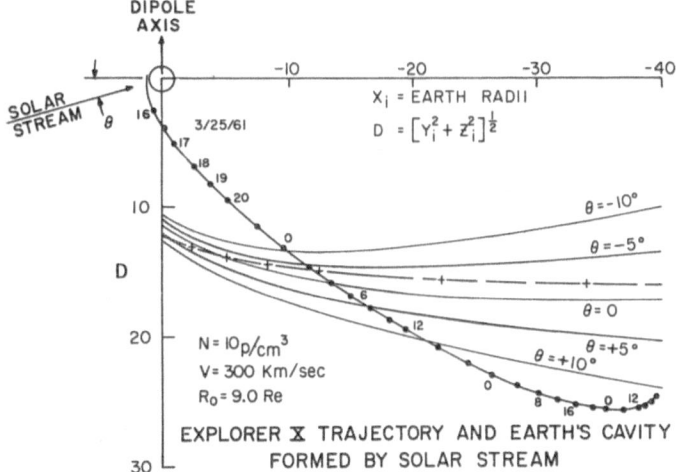

Fig. 13.   Explorer X trajectory and theoretical geomagnetic cavity for $V_p = 300$ km/sec.

At the present time it is possible to speculate upon the true nature of why the data indicate a continuing change in relative position of the satellite and the geomagnetic cavity boundary. The main feature to be explained is the positional instability of the cavity boundary as being due to either an intensity change and/or a directional change in the plasma flux. In Figure 13 is presented the trace of the geomagnetic cavity as theoretically computed on the basis of specular reflection and no interaction with the direct flux of the plasma reflected from the geomagnetic cavity boundary. The traces are in the meridian plane containing the magnetic dipole and the direction of the plasma flux is assumed to be from the sun. The different traces at varying angles indicate the variability in the direction of the cavity tail behind the Earth when the direction of the incident flux changes relative to the direction of the magnetic dipole. Superimposed on these traces is the projection of the Explorer X orbit. The dash line indicates the position of the cavity trace in the equatorial plane. It is assumed that the cavity tail is approximately cylindrically symmetrical about the Sun–Earth line.

For values of the plasma flux a proton density of 10 particles/cm³ with a velocity of 300 km/sec, were chosen as being representative of the values obtained from the Explorer X Plasma Probe. Superimposed on the trajectory is the identification of the satellite position at 0530 on March 26th. It is seen from this Figure that the position of the geomagnetic cavity tail boundary is extremely sensitive to the incident solar stream and directional changes of only a few degrees could strongly effect the relative position of the Explorer X satellite and the cavity boundary. A notable feature of this interpretation is that the data obtained on Explorer X is inconsistent with this theoretical model since the cavity boundary does not exhibit an outward flaring feature on the dark side. This must be assumed from the direction of the field in region *A* and the trajectory which paralleled a conical surface the half angle of which

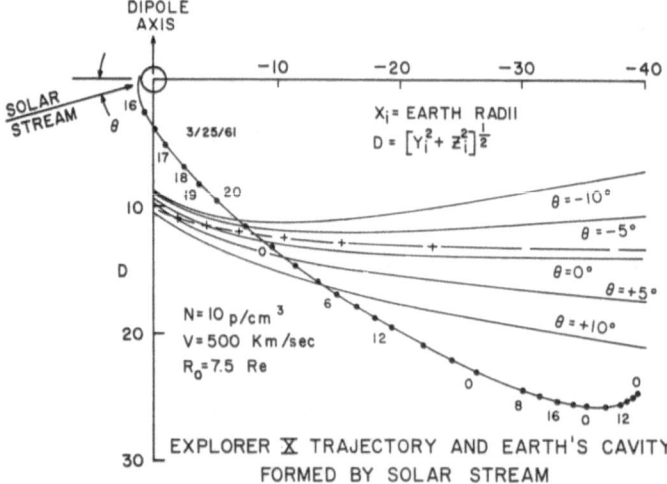

Fig. 14. Explorer X trajectory and theoretical geomagnetic cavity for $V_p = 500$ km/sec.

was approximately 15° to 20°. All theoretical computations thus far have neglected the effects of either a plasma with a finite temperature and an imbedded interplanetary magnetic field. These theoretical results have indicated that the cavity tail should not open out, but indeed should close off and terminate at some distance.

The distance on the Sunlit side of the Earth to the geomagnetic cavity boundary for the balance of magnetic and particle pressure for the assumed plasma flux is $9 R_e$. The actual data obtained from the Explorer X plasma detector indicated that during regions A and B, plasma flux corresponded to a streaming velocity of 300 kilometers per second with particle densities ranging from 4 to 12 with an average of $10/cm^3$. In Figure 14 is shown a similar presentation of the geomagnetic cavity but one in which the particle velocity has been increased to 500 kilometers per second. In this case the cavity is considerably compressed relative to that shown in Figure 13, although similar effects are illustrated showing the dependency of the cavity tail position to the direction of the incident plasma flux. In summary, these two Figures illustrate how both directional and intensity changes can lead to a reasonable, but not completely consistent explanation of the alternation of the satellite position between inside and outside the cavity boundary.

Cavity Boundary Instabilities

There is a third possibility for the explanation of the boundary traversals observed on Explorer X. This is related to temporal instabilities on the cavity surface which would propagate somewhat similar to surface waves and would be associated with the response of the cavity and its boundary to temporally varying plasma pressure. In an attempt to investigate the possibility of such instabilities, the energy spectrum of the magnetic field orthogonal components and the spectrum of the total field has been computed for the time interval from 0530 UT on March 26th, to 1830 UT on March 27th. The magnetic field data for this study has been selected to correspond to the regions A, B and C and their associated transition periods and the data were eliminated when no plasma was observed inside the cavity boundary prior to the first boundary traversal. Figure 15 presents these energy spectra and the 5% confidence limits for the analysis in which there were 18 degrees of freedom. The significance of the 5% limits is that it is indicative of the statistical confidence that an observed spectral peak is real and is not associated with the finite sample of data available. It is seen that there are significant spectral peaks in the three components and the total magnetic field at periods of several minutes to hundreds of minutes.

The discrete nature of the data utilized in the analysis and the correspondingly long time interval between successive samples do not allow a unique determination of the frequency spectrum however. This is due to a phenomena which in the theory of discrete time series is referred to classically as aliasing of spectral components.

Associated with the periodic sampling of the magnetic field data, is the non-uniqueness in the identification of what spectral component is actually observed. In Figure 16 is illustrated graphically the effect of aliasing on the interpretation of

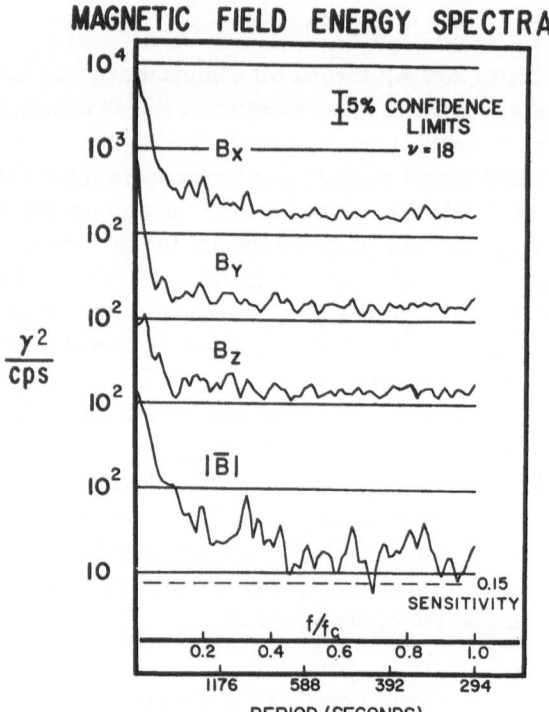

Fig. 15.   Magnetic field energy spectrum.

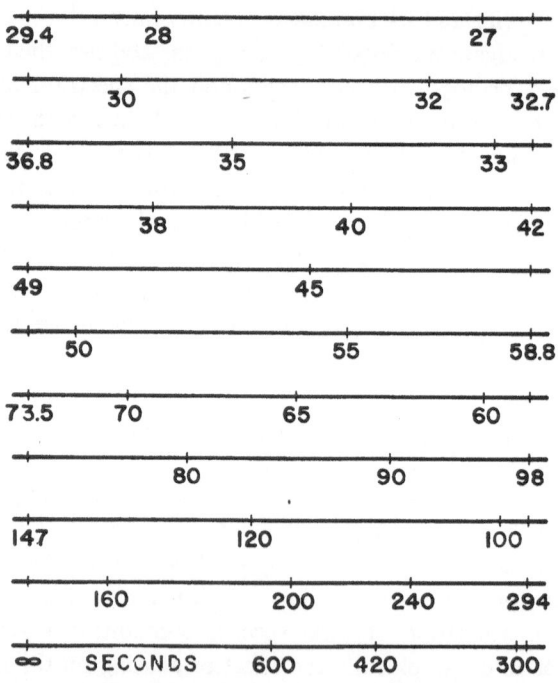

Fig. 16.   Aliased periods of magnetic field data.

the magnetic field spectra. This diagram illustrates that the fundamental frequency or period (which occurs along the lowermost horizontal line) possesses aliases which are found by inspection vertically of the remaining portion of the spectrum. Thus a fundamental period of 550 seconds has a first alias of 200 seconds, and a second alias of 60 seconds. Because of the inability of discretely sampled data to uniquely specify the frequency spectrum, those spectral peaks which have been observed are not capable of being identified at a particular frequency within the pass band of the sensor (0 to 10 cps).

The transmission of data from the individual fluxgate sensors was continuous over a time interval of some 3 seconds. During this time interval it has been possible to study the fluctuations through a detailed numerical procedure which allows a determination of the vector magnetic field. In Figure 17 the magnetic field is seen

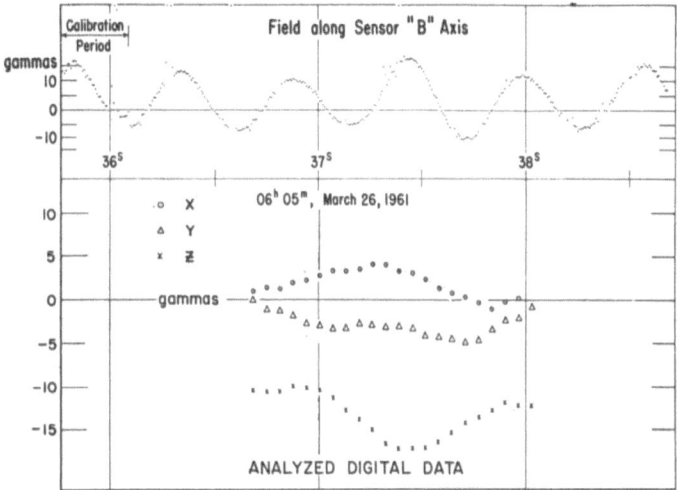

Fig. 17.   Analysed digital data for single telemetry transmission.

to vary on a time scale of seconds indicating that the aliasing problem previously discussed is a real problem in this data sample. Certain features in the detailed analysis are important to note. The modulation of the basic signal is dependent upon the component of the field perpendicular to the spinning satellite. The periodic excursions of the data shown as sharp spikes are associated with noise induced by the receiving antenna in response to the transmitting antenna pattern and the rotating spacecraft.

In Figure 18 is shown an illustration of a magnetic field fluctuation observed on a considerably larger time scale when the satellite was within the geomagnetic cavity. The bars indicate the variation within 3 seconds associated with the determination of the magnetic field for each data point. An apparent damped sinusoidal fluctuation with a period of twenty minutes was observed when on the interior of the cavity boundary. It is not known at this time whether this periodic variation is indicative

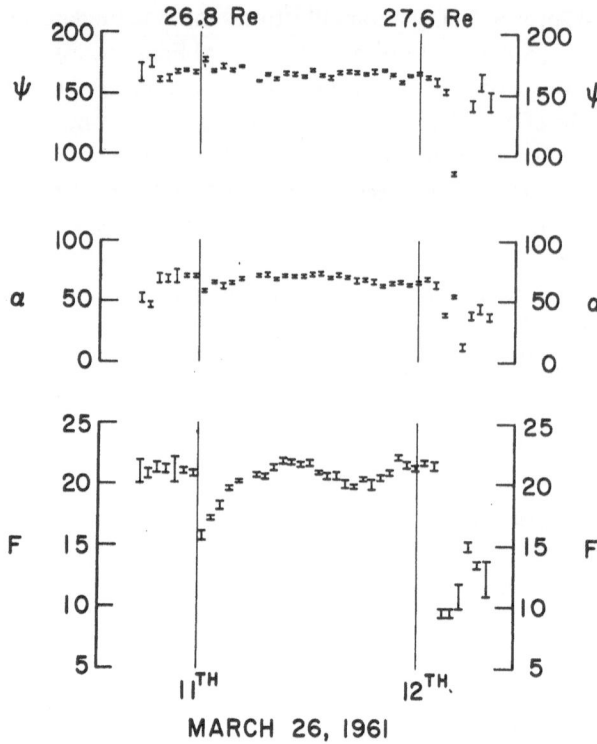

Fig. 18.   Time variation of vector field data at cavity boundary.

of the temporal instability of the cavity boundary, or whether there is a periodic spatial gradient of the magnetic field on the interior of the cavity boundary. If it is assumed that the constancy of direction is indicative of the mode of propagation of the disturbance, then a compressional surface wave propagating on the geomagnetic cavity boundary has been detected.

The interpretation of the Explorer X magnetic field and plasma data is limited by the presently available theoretical treatments of the interaction of an ionized gas with a dipole magnetic field. Satisfactory results are obtained only for regions of space corresponding to the subsolar point on the cavity boundary while the dark side of the Earth and the associated magnetic "tail" are as yet incapable of being analytically treated. More importantly, the topology of the magnetic field lines and the distribution of the intensity is unknown so that a direct comparison between a theoretical model of the geomagnetic cavity and the Explorer X data cannot be made.

Finite Cylinder Approximation

A recent analysis (PARKER, 1962) is pertinent for it has treated the problem of a dipole enclosed within a cylindrical cavity plugged at either end. The condition that no magnetic lines of force traverse the boundary and that the dipole field corresponds to that of the Earth at the origin of the coordinate system forms the mathematical model of containment of the geomagnetic field. This is illustrated in Figure 19 and includes the dimensions chosen for the detailed computations. The purpose of this investigation has been to determine the topology of field lines for physically real and mathematically soluble problems. It is not the intent to directly compare the results of the computation with the Explorer X satellite, since the flaring out of the cavity is not predicted. The computations will however, indicate the characteristics of the

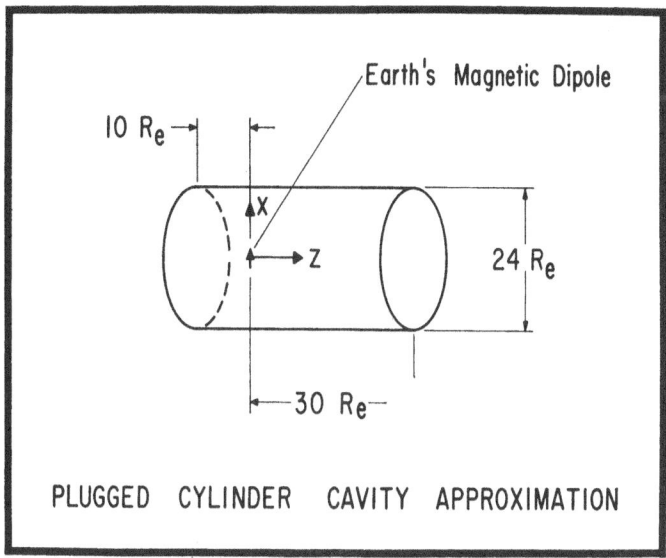

Fig. 19. Plugged cylinder cavity approximation.

field lines within such an enclosure. In Figure 20 field lines in the meridian plane associated with a magnetic dipole in free space (dashed lines) and for the dipole in a plugged cylinder approximation (solid lines) are shown. A compression of the field lines in the Sunlit hemisphere has been intuitively assumed and analytically found in all schematic representations of the cavity and the field line topology. An important feature noted in this diagram is that on the dark hemisphere (the portion of the cavity behind the Earth) the field lines are also compressed. The topology of the field lines is strongly dependent upon the particular boundary which exists. In this case the termination of the boundary on the dark side has led to a compression of the field lines in spite of the fact that the termination occurs at some $30\,R_e$ as opposed to $10\,R_e$ of the front side.

A second feature which is of equal importance is the distortion of the magnetic meridian planes which is associated with absence of axial symmetry of the cavity with reference to the magnetic dipole axis. Figure 21 gives the projection of magnetic field lines on the equatorial plane and indicates the distortion of the magnetic meridian planes at successively higher magnetic latitudes. It is this distortion of the magnetic

PLUGGED CYLINDER CAVITY APPROXIMATION

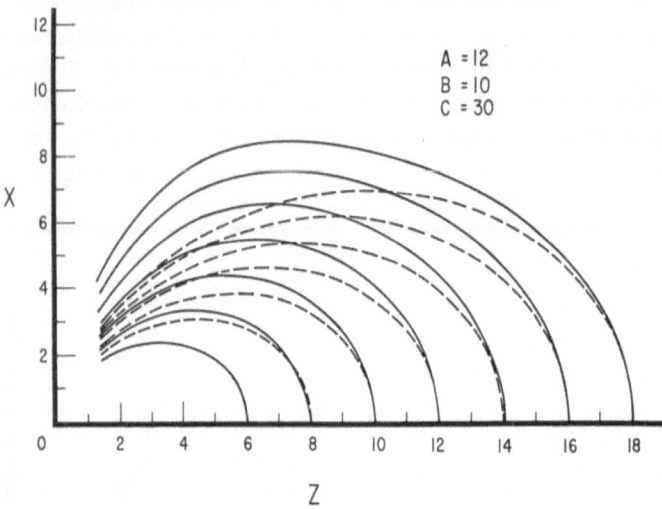

Fig. 20.   Distortion of dipole field lines by cavity containment (meridian plane).

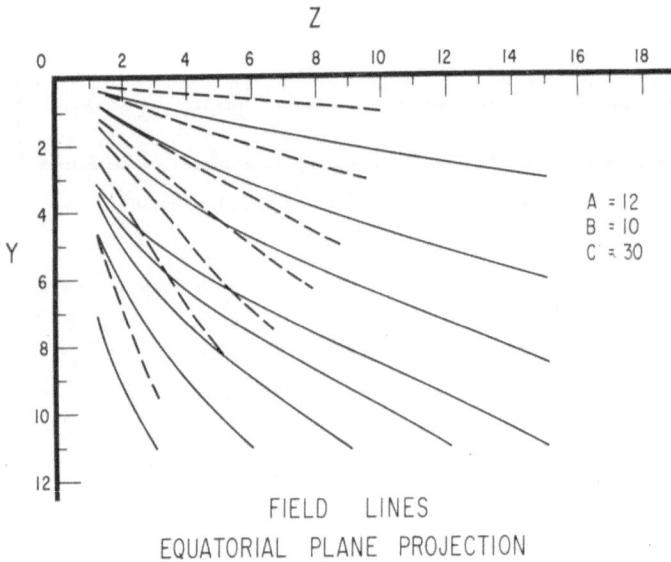

FIELD   LINES
EQUATORIAL PLANE PROJECTION

Fig. 21.   Distortion of dipole field lines by cavity containment (equatorial plane).

meridian planes which led to the incorrect conclusions for the Explorer VI magnetic field experiment as demonstrating the existence of a ring current at $10\ R_e$. Such ring currents were not detected on the Explorer X satellite in a similar region of space, and has not been seen on the Explorer XII, or XIV satellites. In Figure 22 is shown a

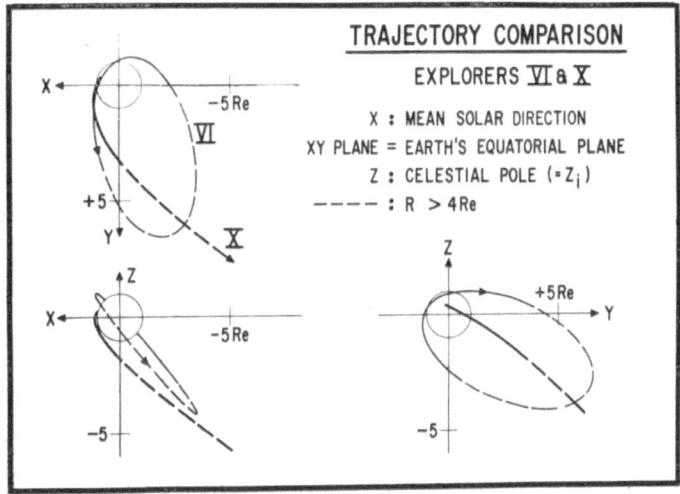

Fig. 22.   Trajectory comparison of Explorers VI and X.

comparison of the trajectories for Explorer VI and X, satellites which presented the
first data on the distortion of the geomagnetic field by the solar plasma flux.

Acknowledgements

The discussions of the Explorer X data with co-experimenters from GSFC and MIT
is gratefully acknowledged. Miss M. B. Ball performed the cylindrically contained
cavity programming for the IBM 7090.

References

The reader is referred to the source publications by the GSFC and M.I.T. experimenter groups
for detailed discussions of the separate experiments. In addition to the references cited therein should
be added the following.

PARKER, E. N.: 1962, 'Dynamics of the Geomagnetic Storm', *Space Sci. Rev.* **1**, 62–99.

SMITH, E. J.: 1962, 'A Comparison of Explorer VI and Explorer X Magnetometer Data', *J. Geophys.
Res.* **62**, 2045–2049.

BLACKMAN, R. B. and TUKEY, J. W.: 1958, *The Measurement of Power Spectra,* Dover Publications.

BLOCK, L. and HERLOFSON, N.: 1956, 'Numerical Integration of Geomagnetic Field Lines', *Tellus*
8, 2, 210–214.

BONETTI, A., BRIDGE, H. S., LAZARUS, A. J., ROSSI, B., and SCHERB, F.: 1963, 'Explorer-X Plasma
Measurements', *J. Geophys. Res.* **68**, 4017–4064.

HEPPNER, J. P., NESS, N. F., SCEARCE, C. S., and SKILLMAN, T. L.: 1963, 'Explorer-X Magnetic Field
Measurements', *J. Geophys. Res.* **68**, 1–46.

ROUND TABLE DISCUSSION ON THURSDAY

Participants. A. J. Dessler (chairman), H. Alfvén, D. Beard, W. Brown, C. C. Chang, S. Forbush, W. Hess, U. Liddel, G. Mead, K. Ogilvie.

Prof. H. Alfvén gave a short talk at the blackboard first which started things off. Then there were questions.

E. Hones (IDA): My only remaining question is that in the model you can have a situation where higher rigidity particles are trapped at greater L values than lower rigidity particles. This is obtained for the case where you've got a uniform field in one direction, either aligned or opposed to the direction of the dipole. This is probably a difficult model of electric and magnetic fields to conceive of, because the outer magnetosphere is more or less random and time varying. It is hard to see how you extrapolate from this particular configuration to that more complex one.

H. Alfvén: This is quite right. This is an extremely simplified case. However, an electric field in this direction may be produced if you have a magnetized plasma wind which gives an electric field which in part goes into the magnetosphere. Of course this wind varies both spatially and temporally. But there should certainly be cases when the time variation is so slow that this is a reasonable approximation. This is just one approach which may give the essentials of the phenomena.

D. Beard (University of California): I'd like to say that in the steady state problem it would be hard to conceive how an electric field could be permanently perpendicular to magnetic field lines, since the conductivity along the magnetic field lines is very high and there would be some grounding at the ionosphere; i.e., a short circuiting across the field lines.

H. Alfvén: The essential part of my lecture yesterday was to show that this indeed is the case in the part of the magnetosphere where the mean free path is small compared to the dimensions of the magnetosphere. However we cannot be sure that this is so in the outer magnetosphere, because there the mean free path is long compared to the dimensions of the magnetosphere. And if that is true we could not put the conductivity equal to infinity, because the medium does not act as a perfect conductor. It is much more like a dielectric, because it has the possibility to have an electric field parallel to the lines of force under the conditions that we have a magnetic gradient, and a temperature which is not zero.

A. Dessler: I would like to say that this is a controversy which has gone on for years and we have a much shorter time. So we will take one more remark on this, then we will go on to something else.

C. Roberts (Bell Telephone Laboratories): As I understand the electric field Prof. Alfvén was discussing yesterday, the one that would cause fluctuations in the distributions along the magnetic field lines, this would give rise to an **E** field in the north-south direction on the earth. I think he said this might explain how particles can drift into and out of the slot. Now if you have an **E** field predominately north-south and a **B** field with north-south and radial components, the $\mathbf{E} \times \mathbf{B}$ drift will be east-west, in other words around the world. You need a horizontal electric field to produce an in-out drift, so I fail to see how these fields along the lines can cause particles to move into and out of the slot.

H. Alfvén: The only point I would like to make is that in the outer magnetosphere, we cannot say that there is a short circuit along the lines of force, so that we may very well have an electric field along the lines of force. Thus in the equatorial plane we may have electric field in any direction.

A. Dessler: The essence of the controversy, which I think is still a controversy, is the point which Dr. Parker and Dr. Beard brought up. It is hard to understand how one can have an electric field in a region of high conductivity. I think we should just say that we don't understand and change the subject to some of the other papers.

D. Beard: I would like to comment briefly on the results of some considerations I have given to the presence of an interplanetary magnetic field. Consider first the case where the weak interplanetary field lines are parallel to the velocity vector of the solar wind. The forward momentum of the wind is large and predominates over any magnetic field pressure which might exist in the interplanetary medium. A detailed analysis of the collisionless individual particle orbits yields the result that if the magnetic lines in the interplanetary plasma encounter any obstruction such as the magnetosphere, these lines become plastered against the edge of the magnetosphere. The particles move adiabatically around on the field lines except for a small region near the subsolar point (the stagnation point) where the particle motion would not be adiabatic. At the subsolar point there would be no magnetic field exterior to the magnetosphere, just a particle pressure maintaining a gradual curvature for the magnetic field lines.

If the interplanetary field is not homogeneous, turbulence would occur. In the case of a homogeneous magnetic field, a plot of magnetic field versus geocentric distance would fall off smoothly with height, decreasing abruptly at the magneto-pause to approximately one-half the value interior to the magnetopause, and con-tinuing roughly constant for several earth radii beyond the magnetosphere before decreasing to the interplanetary field value. If the interplanetary field is not homo-geneous, the field intensity beyond the magnetosphere would fluctuate in magnitude between zero and the intensity interior to the magnetopause. The direction of the compressed field adjacent to but exterior to the magnetosphere must in all cases be tangential to the magnetopause, but otherwise unrelated to the direction of the field interior to the magnetopause.

If, as is very likely, the interplanetary field has a component perpendicular to the velocity of the solar wind, the compressed field exterior and tangential to the boundary will be weaker than if there were no perpendicular component, since the particle motion and the field are tightly coupled in some directions in this case. This increases the particle pressure perpendicular to the field lines.

Unidentified: You say that the particle pressure is very strong and that it plasters the magnetic field lines along the surface of the magnetosphere. Well, when this happens the magnetic field pressure is not strong, and must be comparable to or equal to the particle pressure. I would think that in this process that a number of particles will be reflected.

D. Beard: The interplanetary magnetic field becomes compressed and is large at the surface of the magnetopause. What happens is that as long as the motion is adiabatic the particles just move around along the field lines to the antisolar side of the earth. When I consider the particle pressure at the boundary, I am considering the pressure due to the perpendicular motion of the particles, and this is always very small as long as the motion is adiabatic. The sum of the magnetic pressure and the particle pressure perpendicular to the magnetic field is equal to the pressure inside the boundary.

W. Hess: I would like to change the subject and ask Walt Brown to tell us more about these large time variations seen on days 350 and 330, and ask if he can give a description of the relation of the particle population and what was going on in the field. For example was there a sudden commencement or what?

W. Brown: K_p was 5 and rose suddenly on both occasions, and was high for about 5 days. It was on both of these occasions also that Snyder saw large increases in the velocity and intensity of the plasma from Mariner II.

W. Hess: Was the increase in K_p at the same time as you started to see an increase in the low energy particles?

W. Brown: Yes.

W. Hess: And did it stay up during the time the high energy particles were responding?

W. Brown: Yes, it stayed high for a period of several days. Now the high energy electrons seemed not to reach their maximum for a period of nearly a week, and that begins to stretch the time that K_p stayed up.

W. Hess: So that a model one might make of this would be a particle acceleration and a later time variation of the field.

A. Dessler: I have a question along the lines of the talk. You have a plot of proton energy as a function of density showing harder spectra inside and softer as you go out. Dragt predicted, in his theoretical analysis of the termination of the proton belt due to scattering by hydromagnetic waves, that the maximum energy would fall off as R^{-11} (*J. Geophys. Res.* **66** (1961) 1641). Have you tried to check that?

W. Brown: We haven't done it, but there is no reason why we couldn't do it for these points. From $L=1.5$ to $L=2.0$ the peak energy does not change rapidly enough.

A. Dessler: One other question then. You showed that the particle density decayed very rapidly at $L=4$, and that beyond this the particles were accelerated very rapidly during magnetic storms. What do you think will happen at sunspot minimum when there is only one sunspot every four months?

W. Brown: I don't know.

A. Dessler: Well, I think it is going down nice and straight.

W. Brown: You might think it will go down curved and low. (Laughter.) On the other hand why are those particles moving anyway? If things are very, very quiet you have a question of what is there to take them away as well as put them in, and I don't have a strong feeling on this.

Unidentified: Can you really typify any of the events mentioned as a storm? My understanding was that small disturbances, so-called polar type events, occurred throughout the solar cycle.

W. Brown: You are asking whether this class of disturbance will occur at solar minimum. That's a good point. It may.

A. Dessler: Well, in addition to the fact that there are few onsets of sudden storms, there is also only a small variation in magnetic ranges during the day during a sunspot minimum. The amount of variation you get even without storms shows a strong sunspot cycle.

S. Bowyer (NRL): Dr. Chang, in the discussion of the geomagnetic cavity, I saw three different mechanisms for pumping up the energies. Was that correct, and would you discuss their relative importance?

C. Chang: We are talking of the steady case where you get this flapping at the polar region. The period is between 200 and 10 seconds as detected in the polar region. The other thing is that since the solar wind is unsteady according to Dr. Snyder's lecture, the magnetic field is unsteady, and the shape of the cavity changes, as shown in many papers. May I take the floor for just a moment to answer Dr. Alfvén's question about the geomagnetic boundary? The boundary we calculate is based on the assumption that there are no trapped particles. We assume that the outside solar wind is steady. In the actual case this is not true for many reasons, particularly in the neighborhood of the neutral point. Because the magnetic field is so low, if you have any trapped radiation you would get a demagnetization. So I feel that this model is far from the actual case.

H. Alfvén: You see it is a very important question whether we have a closed magnetosphere or not. In some directions there are some very sharp boundaries such as the Cahill discontinuity. In other directions there are apparently boundaries where the radiation drops, but then it comes back again; i.e., there is no sudden transition at all. Now the sharpness of the border between the magnetosphere and interplanetary space depends very much on the existence of large scale electric fields. If we have large scale electric fields, plasma can drift out and in. The final decision about the existence of electric fields must obviously come from measurements, and there are quite a few ways of measuring them. This is a very important thing to do.

C. Fan: I should like to change the subject back to the trapped radiation. I just want to make one comment. On Explorer VI and Pioneer V the University of Chicago has put a proportional counter telescope which detects protons exceeding 100 MeV. The decrease of proton density as a function of L is not as fast as the theory indicates. What is worse is that it has double peaks. The first peak is at $L=1.5$, which is in close agreement with Dr. Brown's measurements. Then it has a minimum at $L=1.75$ and a second maximum at $L=1.82$.

W. Brown: I wonder if it is related to the maximum that McIlwain sees in these 40 MeV protons at $L=2.2$. There is a secondary hump there. You are at 100 MeV?

C. Fan: Yes, so he is seeing the low energy tail of our proportional counter distribution.

W. Hess: Along this line there is a very interesting feature of this double peak which does seem to be true now. One has to get rid of the electrons and high energy protons in the outer belt. There are various ways to do it, but it looks as if we need two of them operating simultaneously rather than just one of them. The point is that you do have two places where the particles go away. You have one at $L=1.5$ and one a little past $L=2$. So it seems to me that one cannot hold on very strongly to any particular process as, for example, the Dragt scattering process, which would have them go away at one place. You just couldn't have anything after that. Well, one of the ideas that has been put forward recently is just to calculate what the trapped Störmer orbits are in the real field, and not the dipole field. It turns out that you get rather low limits for trapping energies. You get 300 MeV at 1.5 R_E. It might be that you have a process like this operating to limit particles in one place and some second process, as for example scattering, limiting them in another place. If this is the case then the innermost process has to operate less efficiently than the outermost process. This introduces a new complexity.

A. Dessler: Dr. Dungey has just the answer for that with the electrons, and it just might work for the protons too (*Planet. Space Sci.* **11** (1963) 591). If there were some oscillations in the geomagnetic field in some narrow spectral range, wouldn't this pick a particular range of L's? You have the whistlers creating the slot in the electrons because the whistler frequency spectrum is peaked near 5 or 10 kc. If you did this with some lower frequency electromagnetic waves near 1 cps, it might produce a narrow slot in the protons. How about that?

W. Hess: You are going to whistle out the outer protons?

A. Dessler: No, I would scatter them out.

W. Hess: Well, you don't want to scatter them out because you still find some there.

A. Dessler: Well, do it slowly, say you put some in and scatter them out slowly. This is only a relative minimum, not a zero.

J. O'Hara (North American Early Warning Service): I thought this might contribute to the understanding of the level of the magnetic storms that have occurred. Since the first of November there has been only one sudden commencement type storm and that was about January 28. The others have been the more gradual recurrence type that have extended over a period of 6 or 7 solar rotations. During that time, the whole day K index has never exceeded, I believe, about 40, which is about a K of 4.5 on the average day. There has been one $K=6$ and several values of $K=5$ in the past 6 or 7 months.

W. Hess: Talking about K_p again, I have a question for Walt Brown. Did the protons do anything at the time of the two events for the electrons?

W. Brown: We were only looking at low energy protons in Explorer XIV, and we didn't see anything in this region.

Unidentified: I was wondering if any of the microflares on the sun have produced solar cosmic rays? Does anyone have any data on this?

W. Hess: By microflares do you mean the thing that Lindsey and coworkers have seen on OSO? You don't mean class 1⁻, do you?

Unidentified: Well, class 1⁻ also. I am concerned primarily with the small ones that are not normally observed.

W. Hess: During the history of OSO, if I remember correctly, there were no solar protons. There were a few 1⁻ flares and a substantial number of these micro-flares. Am I correct?

Unidentified: The most reliable thing that correlates with solar proton events is Type IV radio emission.

Unidentified: I wonder if we can get back to the Frank, Van Allen, and Macagno story here which shows the peculiar contours on the back side (*J. Geophys. Res.* **68** (1963) 3543, Figure 10). I wonder whether the various members of the panel or the audience would like to take a stab to put some intuitive understanding into the way in which these patterns arise, and particularly to come up with a void region and then another large counting rate as you go out to further distances.

D. Beard: If you are asking for suggestions I have thought about that. The particles which are trapped along the magnetic field lines are compressed while flowing around the magnetic field lines and into the tail. These magnetic field lines have quite a bit of inertia because they are carrying particles, and thus they don't turn corners very easily. They tail off in the back, and it might be this that causes the dense counting region. The void would be inside that. This is just a suggestion, I haven't calculated this.

A. Dessler: There would be a difficulty with that because in Explorer XII you very seldom see particles accelerated. You should think then that any particles that were accelerated and expanded into the tail where the field flared out would certainly be no more energetic. Even if the energy remained constant, the particles they are measuring are over 40 keV particles, and there is no evidence of 40 keV particles just outside the tail.

Unidentified: I'd like to ask Dr. Beard what he thinks of this standing shock wave upstream from the magnetosphere. Because presumably this would be a col-lisionless shock and thus form an interesting experimental object.

D. Beard: I don't see any reason for it. I am not saying it doesn't exist, but I don't see any mechanism by which it can exist because the particles are not reflected into it. They go around on back except for a very small spot in the front.

W. Hess: I talked this same thing over with Petschek from AVCO, whom I'm afraid may have left. Let me say a few things which he said. I hope I repeat them correctly. He wanted to have the best half of two arguments simultaneously. The things he would like are one, to have a detached shock form just as you would have it in fluid mechanics; but then to have the shock of appropriate thickness to have a collisionless shock, and the characteristic thickness for this is just the ion cyclotron radius. So you get something which is rather thin, and in that sense magnetic properties dominate it, but it is at a place and of a shape so that the fluid dynamic properties control it. Thus you are winning on both sides, which seems unfair but may be correct.

C. Chang: I would like to add a few words to this. The collisionless shock wave is of the order of 10 Larmor radii. If you have that size you automatically have a compressed magnetic field. Now if you have particles coming back from reflection into this region, it is very hard for them to come back out, so that they are going to accumulate in this region. Also, as I mentioned before, if the traveling signal velocity or the Alfvén wave velocity is one order of magnitude less than the solar wind velocity, then the particles cannot know what is going on at the boundary. This is why you get this shock wave picture.

A. Dessler: I can see one more point bearing on the shocks in the paper Dr. Snyder gave, where the plasma density, energy and velocity exhibited a very sharp discontinuity in interplanetary space. This may also be evidence of a shock wave. We have run out of time so I think we should adjourn. Thank you very much.

PART IV

MISCELLANEOUS GEOPHYSICAL AND LUNAR PHENOMENA

THE RING CURRENT, GEOMAGNETIC STORMS
AND THE AURORA

SYDNEY CHAPMAN

Geophysical Institute, University of Alaska, College, Alas. and
High Altitude Observatory, Boulder, Colo.

Abstract. During the main phase of geomagnetic storms, an additional magnetic field (DR) is superposed on the normal field of the earth. The DR field is nearly uniform over the earth, and is along the direction of the dipole axis of the earth. It is ascribed to the growth of a ring current flowing in the region of the Van Allen belts. This hypothesis is borne out by magnionic theory (namely that of the motion of charged particles in magnetic fields), and by satellite observations of enhanced population of the Van Allen belts.

Throughout the storm there may be recurrent growth and decay of a distinctive electric current system (DP) in the ionosphere, especially in the polar regions; its magnetic effects are called a DP substorm. The development of the main phase and of the DR field is correlated with the DP activity (in number and intensity).

Quiet auroras can appear during periods of magnetic calm, during magnetic storms, and even during lulls in the DP activity. In the latter case the latitude of the aurora seems to be connected with the ring current. During a DP substorm the aurora becomes much more active, in ways recently studied by Davis and by Akasofu, using all-sky camera films obtained in the IGY.

These related events are manifestations, long observed at the earth's surface, of phenomena occurring in the Van Allen region. Some tentative interpretations of their interconnections can be given, but much remains to be discovered.

I begin by indicating what I think are the suggestions offered by the time incidence of magnetic storms as to the manner in which the sun ejects plasma. Figure 1 shows the plasma stream whose continued flow is inferred from the tendency of the magnetic disturbance to be repeated after one solar rotation. This implies a fairly narrow beam of emission which continues (sometimes intermittently) for weeks, sometimes for months, and after each solar rotation when it strikes the earth there may be a recurrence of magnetic disturbance. Solar flares occur at different points on the sun's

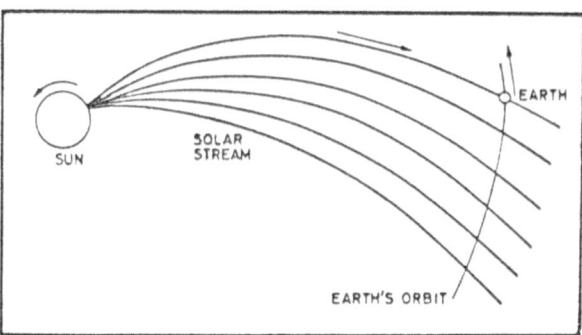

Fig. 1. Sketch illustrating a solar stream of particles moving with a speed corresponding to a time of travel of 36 hours from the sun to the earth.

Chang & Huang (eds.), Proc. Plasma Space Sci. Symp. All rights reserved.

surface, mainly within 20 degrees of the equator, and the great ones are almost always accompanied by magnetic storms on the earth. But there are many weaker ones which don't have any clearly distinguishable effect. At the same time as a flare is observed, there is an ionospheric disturbance and a small magnetic disturbance over the sunlit side of the earth (Figure 2), and then after about a day there often occurs a magnetic disturbance, beginning simultaneously all over the earth within about a minute

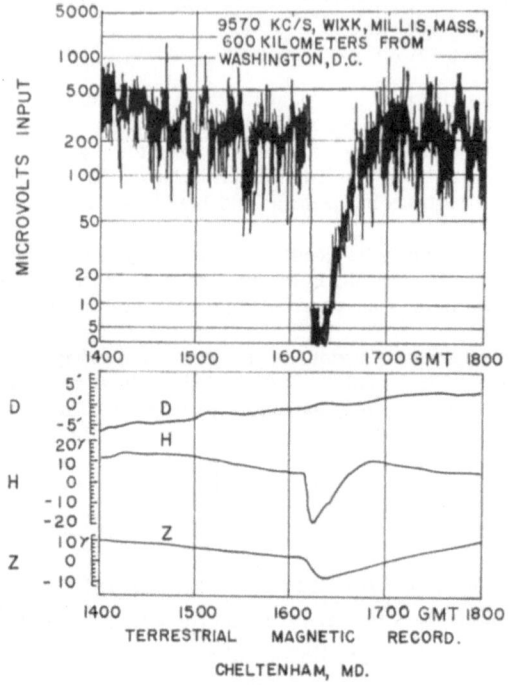

Fig. 2.   The radio fade-out (low signal intensity) and magnetic disturbance on November 26, 1936 (after J. H. DELLINGER).

(Figure 3). The fact that flares at the very edge of the sun can affect the earth in this way, when they are great flares, suggests that the flares eject solar plasma for brief periods in the form of a cone of wide angle; Figure 4, which is diagrammatic, shows part of a spherical shell of gas proceeding from a flare; the semiangle of the cone here shown is 80 degrees; it may be 80, it may be 90, or it may be less in other cases. As time goes on the shell travels outwards, and owing to the probable differences in speed of its particles it thickens, extending radially. If finally it crosses a planet, such as the earth, a magnetic storm will follow. The interval during which the earth is in the plasma will depend on the radial thickness of this shell, or on the breadth of the solar stream (Figure 1). The solar streams proceed from particular regions not easy to identify; Bartels called them M-regions. Besides solar plasma shells from flares and the streams from M-regions, there is in addition a general outward flow of plasma, as spoken of by Dr. Parker two days ago. It may be that the correlations noticed by Dr. Snyder between the events observed on the Mariner when it was far

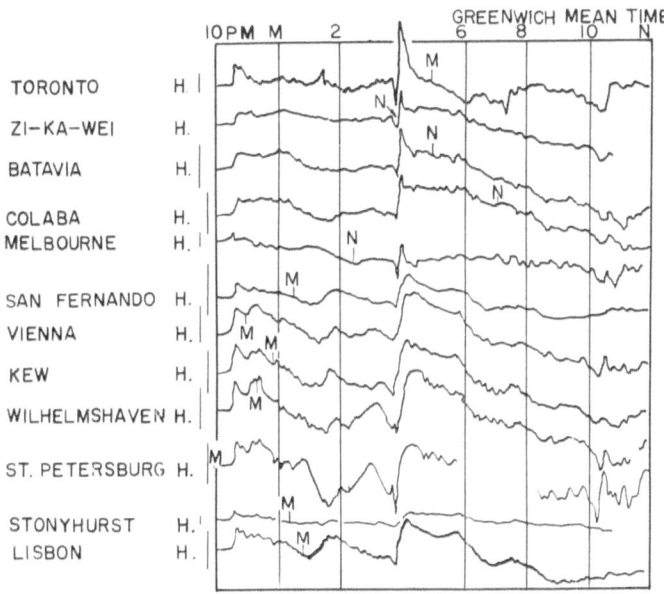

Fig. 3. Records of the variation of the horizontal magnetic force during the magnetic storm of June 24, 25, 1885. The scale of force varies from one station to another; it is indicated by the small vertical scale lines, which in each case represent 0.001 gauss. The time interval represented by the adjacent vertical lines is 2 hours. The letters N and M indicate the epochs of local noon and midnight.

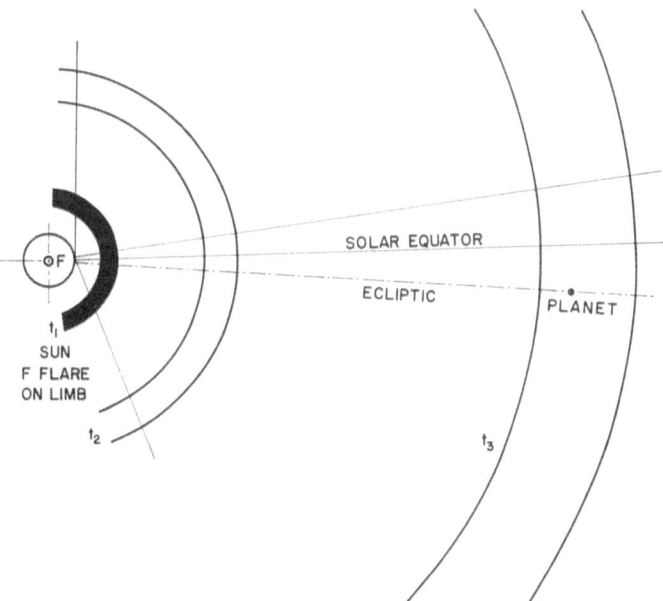

Fig. 4. Part of a spherical shell of gas proceeding from a solar flare.

from the earth were due to the passage across both the earth and the Mariner of some weak flare plasma shells; otherwise, one would not expect the sun to affect both simultaneously. Figure 5 presents the record of one of the three great storms of the IGY, the 13th of September 1957, in which the aurora was seen as far south as

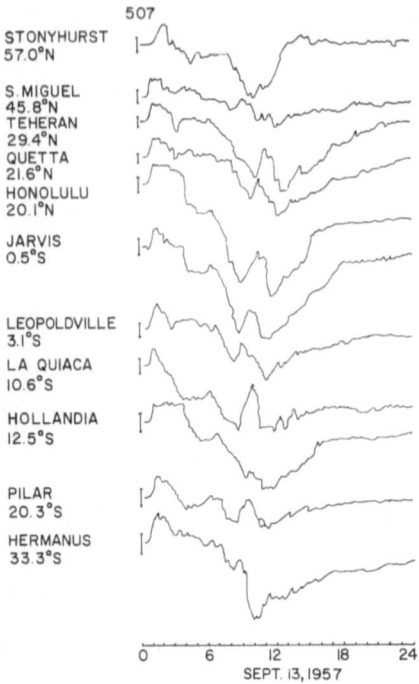

Fig. 5.   Magnetic record of storm of September 13, 1957. The small vertical scale lines at each station represent 0.001 gauss.

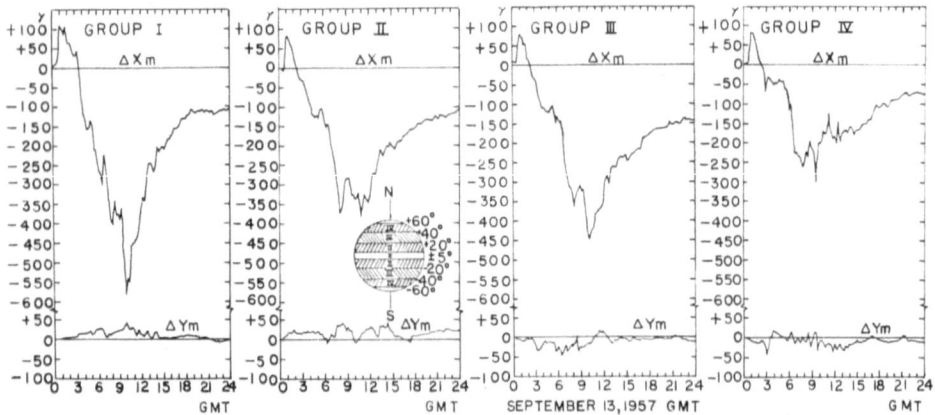

Fig. 6.   The Dst (X_m) and Dst (Y_m) curves for September 13, 1957, for four zones bounded by circles of geomagnetic latitude (indicated in the figure). The SSC occurred at 00h 47m GMT on September 13, 1957. Note the rapid growth and decay of the ring current.

Mexico City. Magnetic records are shown in Figure 5. Figure 6 shows these variations averaged around four belts of latitude. We see that even when thus averaged, though there are some irregularities in the curves, in the first phase of the storm there is an increase of horizontal magnetic force, followed by a much larger decrease. These vector changes of magnetic intensity lie on the average in dipole meridian planes, and can be attributed to the immersion of the earth in a changing field parallel to the

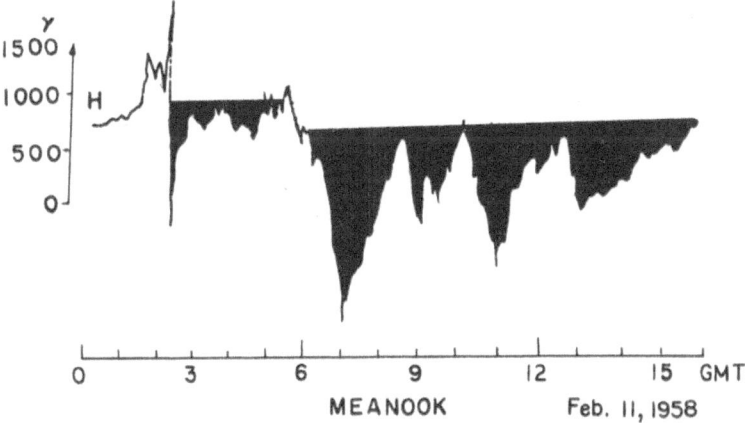

Fig. 7.   Magnetic record from the high latitude station Meanook of the storm of February 11, 1958.

dipole axis: first northward, second southward, but modified by the field of the induced currents within the earth.

The generally adopted explanation of the first phase is that it is due to the impact of plasma on the earth's magnetic field. Whether the plasma comes in the form of a stream or shell or a general solar wind, makes little difference, because the earth is so small compared with the scale of the plasma in all three cases. The plasma beats on the earth's field and it compresses and confines it, thus it enhances the field (along the northward direction) throughout the space occupied by the earth, in a fairly uniform way over a region on the scale of the earth. One may look at this enhancement in another way as due to currents in the surface of the hollow formed in the plasma, can be caused by slightly different speeds of the protons and electrons in the surface. These particles in the stream are deflected backwards or sideways. The protons are always deflected a little more to the right than the electrons, and this constitutes in the equatorial plane an eastward electric current. Now we come to the problem of the main phase, in which the field is decreased. The picture I have shown of the magnetic changes in the great storms has so far been confined to moderate and low latitudes. In higher latitudes there is often very much greater disturbance, and there one cannot easily recognize the first phase and the main phase. Figure 7 shows a record from the high latitude station Meanook of the storm of February 11, 1958; it does not show an orderly increase and then a decrease. It shows several great decreases starting from and returning to a normal base. Each of these disturbances

lasts for perhaps one or two or three or four hours. They may be called polar substorms occurring within the main storm, which itself may last for a day, or for
weaker storms, two or three days. Figure 8 shows the electric current system in
the ionosphere which is believed to be the cause of the polar substorms, and which
causes the irregularities in the main sweep of the curves (Figures 3 and 5) for lower
latitudes. Here we are looking down on the northern hemisphere; the noon meridian
extends downwards from the center, the midnight meridian is directed upwards, and
the dawn and twilight meridians are to the left and right. There is a very intense
current system along the auroral zone, so intense and laterally limited that it is
called an electrojet or electrical jet stream. It is mainly westwards over the night

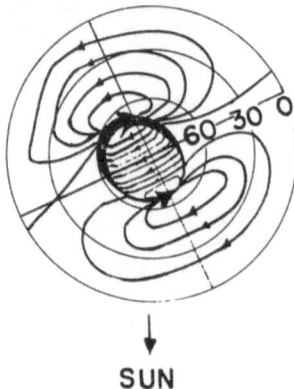

SUN

Fig. 8.   Electric current system in the ionosphere, looking down on the northern hemisphere. The
noon meridian extends downward from the center, midnight meridian is directed upwards, and the
dawn and twilight meridians are to the left and right.

hemisphere (in the upper half of the diagram), and the path is closed mainly over
the auroral cap within the auroral zone, but also partly over the great belt between
the two auroral zones. The fluctuations of this current system without much change
of pattern cause the polar substorms I have spoken of. This system as a whole
first increases and then dies away, sometimes with moderate changes of orientation;
this produces the jags in the curves. This polar current system is closely associated
with the active stage of auroras. The main phase of the storm has long been thought
to be due to westward electric currents above the earth. In the past some people
(including myself in 1918) have thought that these westward currents are located in
the ionosphere. In 1918 I gave a theory of magnetic storms which Lindemann criti-
cized in 1919. Like many earlier theories, it was based on solar particles of one sign
only. Lindemann proposed that the stream is ionized but neutral, and this is now
generally accepted. Figure 9 shows the auroral zone and the lines of equal frequency
of auroral visibility outside it, due to Fritz, who published it about 1873. The auroral
zones were among the features which Birkeland tried to explain by a theory that
magnetic storms and auroras are due to charged particles shot out from the sun.
Birkeland made laboratory experiments which showed that a dipole field would

deflect electrons polewards. Störmer worked out the theory of the motion of a single charged particle in a dipole field. He was one of the pioneers in developing what may be called the science of magnionics, the motion of charged particles in a magnetic field. At a later stage Prof. Alfvén added greatly to this theory by finding one of the invariants of the motion, and by introducing the conception of the motion of a

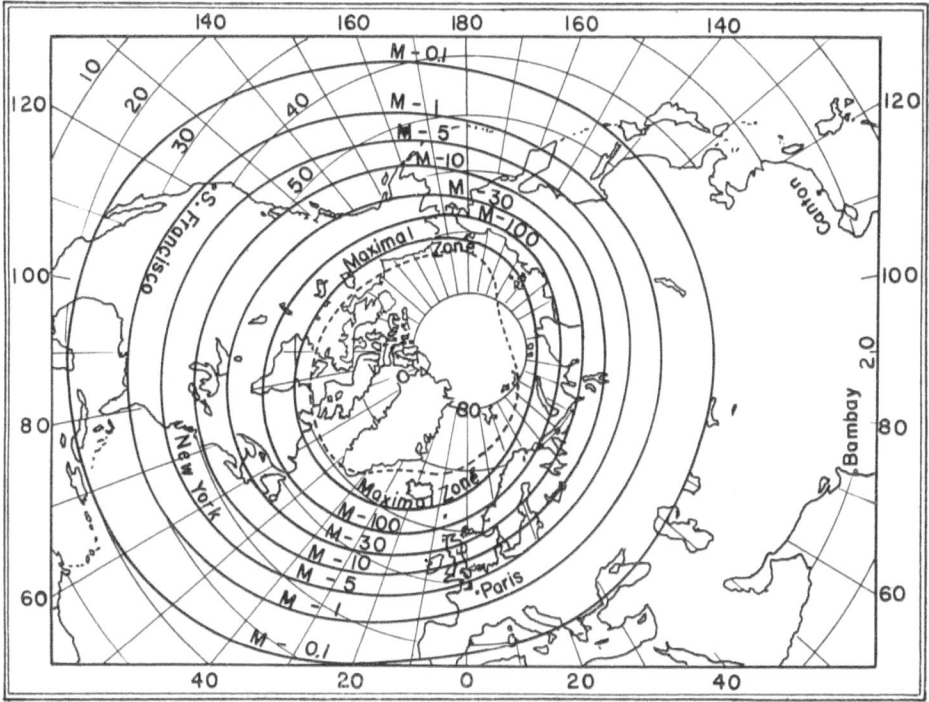

Fig. 9.   The distribution of isochasms, or lines of equal auroral frequency, in the northern hemisphere, according to Fritz.

guided center (which I think should be called the Alfvén center) about whose path the particles spiral. Later invariants were discovered by Rosenbluth and Longmire and by Northrop and Teller. Störmer could not explain by his theory why the charged particles, which he supposed to be electrons of a certain velocity corresponding to a certain depth of penetration, appeared in the atmosphere especially in the auroral zone. The electrons sweep around the earth in the eastward direction; this corresponds to a partial westward current. Störmer conceived the idea that this general sweep of the electrons might explain the main phase of magnetic storms, and that it might also draw the auroral zones down from his calculated latitude to the observed one. He calculated the field of such a stream, taking it, for mathematical simplicity, to be a circular current, instead of the unclosed current of the electrons. That was the first appearance of what may be called a ring current in the literature. But his current cannot explain magnetic storms and auroras. Although his theory might explain certain features of auroras, its basic assumption of particles of one sign only is untenable. But his calculations were eventually valuable in the theory of cosmic rays.

In 1930–1931 Ferraro and I showed that a plasma beating on the earth's field would confine the field within a cavity around the earth. We estimated the size of this cavity in the plasma, from the magnitude of the first phase of the storm, as 5 to 10 earth radii. We were also inclined to put the ring current in this hollow, at some distance of about 4 or 5 earth radii. We took its form to be toroidal. This is not correct, and Singer in 1957 was the first to suggest that the ring current might be a distribution of energetic charged particles trapped in the earth's field, of the same form and in much the same location as was later proved to exist by van Allen and later satellite workers, when the magnetosphere began to be explored by the satellites.

Prof. Alfvén in 1940, when he extended the science of magnionics, showed that charged particles in a dipole field will drift partly due to the gradient of the field, and partly to the curvature of the field lines. This drift would be eastward for electrons and westward for protons; both drifts constitute a westward electric current. Besides this the spiral motion of the particles imparts diamagnetism to the trapped plasma. Figure 10 indicates how the equivalent current due to such spiral motion depends on the pressure gradient. In this picture imagine that there is a field extending uniformly outwards from the diagram towards you, and that there is a distribution of charges increasing in density from the top to the middle and then decreasing again down below. The spiral motion of the charges is here represented by a set of circular

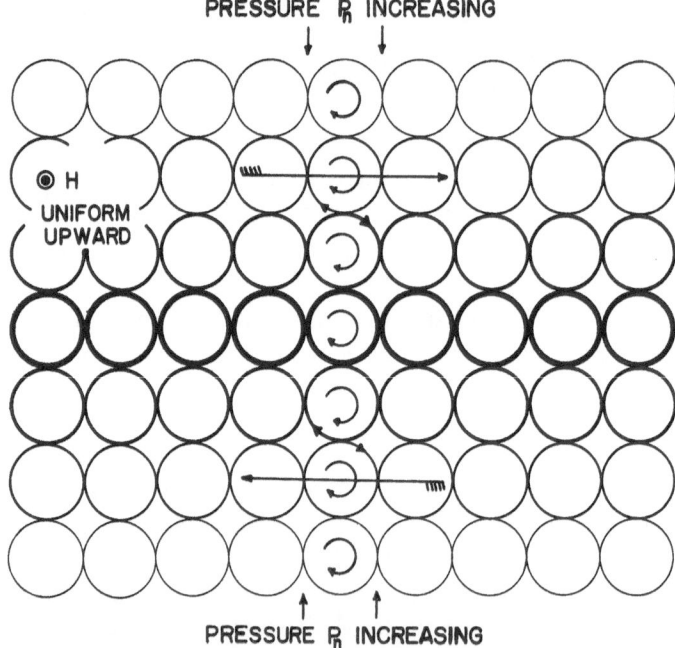

Fig. 10.   Representation of how equivalent current due to spiral motion of charged particles in a magnetic field depends on the pressure gradient.

paths of varying intensity, corresponding to the density of the particles, all of which are supposed to have the same sign in this picture (merely for simplicity of illustration) and to rotate in the same direction. The vertical part of the currents clearly cancels out. but the horizontal parts do not. Because of the greater strength of the current in say the second row compared to the first, there is a balance of current to the right in the upper part of the diagram, where the density is increasing downwards; in the lower

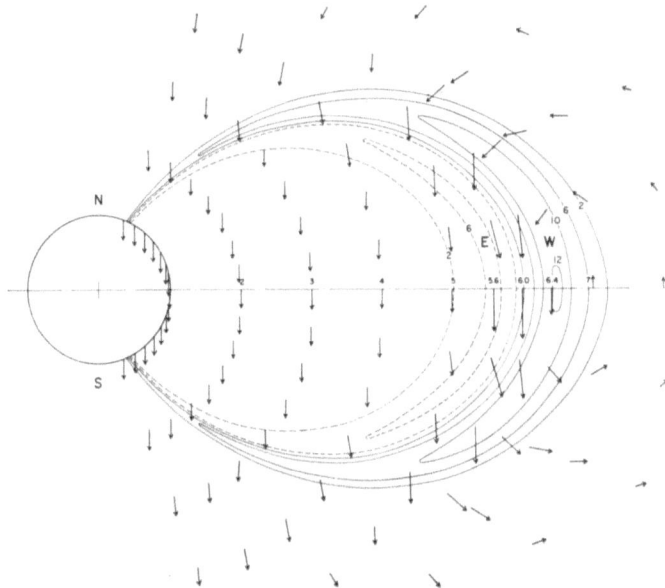

Fig. 11.   The DR field vectors, for the first approximation to the magnetic field of the model belt in a meridian plane; and isolines of the equivalent current intensity in the belt (solid lines indicate west-ward, dashed lines eastward current). The vector scale of force, and the unit in which the current intensity is expressed, are proportional to the energy density n_0E. (keV/cm^3) at the center line of the belt, at 6 earth radii from the earth's center (AKASOFU, CAIN and CHAPMAN, 1961).

part the current is to the left, where the density is decreasing downwards. This is an illustration of the fact that in the trapped plasma in the Van Allen belts there will be an equivalent current due to the diamagnetism associated with the spiral motions, in addition to the drift currents. The part of the equivalent current corresponding to the spiral motion of the particles will depend, not simply on the density and energy of the particles, but on the gradient of their pressure. So it comes out that the equivalent current will be eastwards on the inner side of the distribution if the density is increasing outwards, and westwards on the outside. Figure 11 shows the results of some calculations by AKASOFU, CAIN and myself (1961) on the electric current distribution corresponding to a model distribution of plasma located between 5 and 7 earth radii, with its density a maximum at 6 earth radii. We assumed a simple Gaussian distribution of density over this range, and also a pitch angle distribution. The result was a current pattern indicated by the isolines of current intensity shown in Figure 11. On the inner side the broken lines indicate eastward current, and on the outward side, westward current. But the net current is westward. Figure 11 also shows the field of such a current system for a chosen maximum density of the belt. This field is extremely uniform within about 3 earth radii; it is southward opposed to the main field at the earth's surface. In the inner part of the belt, near the center of its cross section, where an eastward current flows on the inner side and a westward current on the outer side, both add to the southward field, which is there intensified. Beyond the belt the field is northward; the ring current adds to the magnetic moment of the earth. Figure 12 shows the distortion of the dipole lines of force corresponding

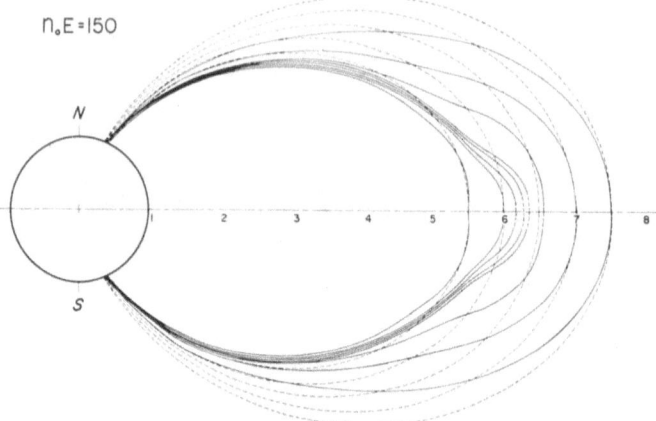

$n_0E = 150$

Fig. 12.   The lines of force of the first approximation to the combined field of the geomagnetic dipole and the belt. The dashed lines are lines of force of the dipole field (AKASOFU, CAIN and CHAPMAN, 1961).

to such a belt, for a certain density. Figure 13 shows in the lower part a graph (on a linear scale) of the magnetic field in the equatorial plane, along an equatorial radius, due to such a model distribution of trapped plasma. It indicates a diminution of the dipole field intensity by 50 gamma up to about 5 earth radii, and then an increased in diminution, to about 100 gamma, near the center-line of the belt, and then beyond the ring current the dipole intensity is increased. In the upper part, the broken line indicates the dipole field intensity (here on a logarithmic scale), and the full line shows the combination of that with the field of the ring current, showing a considerable dip in the field. This dip seems to resemble that found by Explorer VI, but I understand that the inferences of the data from the Explorer VI have been modified, and as thus changed, the correspondence is not so good. However we have seen this morning several slides suggesting that sometimes the field is below the earth's field in the region of the belts, as of course it is observed to be during the main phase of storms at the earth's surface. I think we have a fair general understanding of the ring current and its field, but exact correspondence requires us to know the energy spectrum and population of the belts down to all the energies that add significantly to the ring current. Now I pass on to some points about the aurora. Figure 14 is a

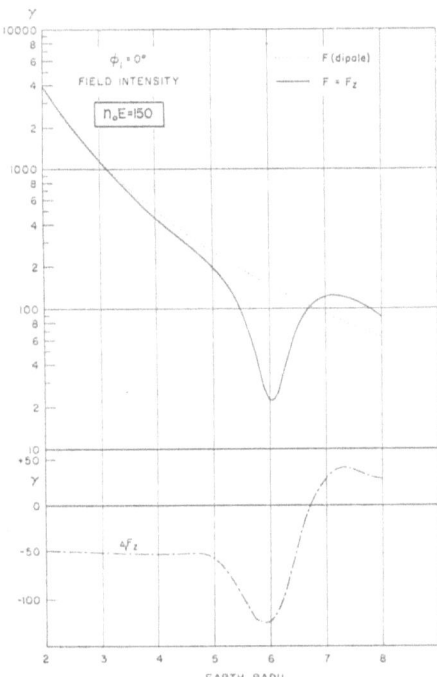

Fig. 13.   Lower curve: Magnetic field in the equatorial plane, along an equatorial radius, due to a model distribution of trapped plasma. Upper curve: Broken line indicates the dipole field intensity and the full line shows the combination of that with the field of the ring current.

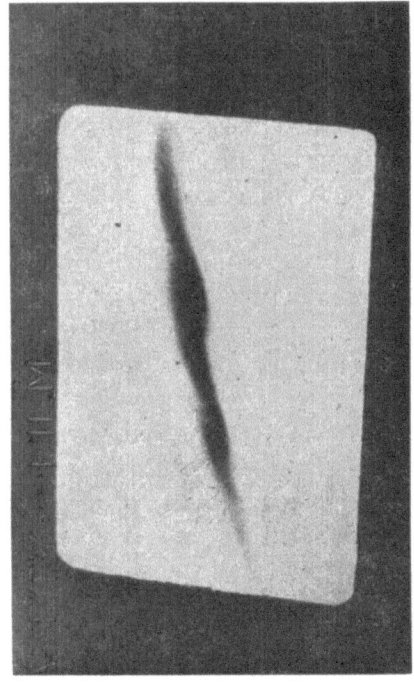

Fig. 14.   Auroral curtain.

picture showing an auroral curtain edge on. It is seen to be folded, and from the photograph it is also possible to measure the thickness of the luminous sheet; during quiet periods it is only about 3 km thick. During the active phase of auroras the sheet is about three hundred meters thick. This is one of the astonishing things about the aurora – its extreme thinness, compared with the extension in height, which is of the order of a hundred or a few hundred km, and the lateral extents horizontally, which is of the order of thousands of km. AKASOFU and CHAPMAN (1961) have attempted to associate this with the reduction of the earth's field by the ring current as indicated in Figure 11. We suggested that at times there might even be reversal of the field in the equatorial plane. Whether there be reversal or not, a great reduction of the field to a low value produces a pipelike region in the equatorial plane where the invariance of the magnetic moment of the particle ceases to be valid, and where the pitch angles can be scattered. In any distribution of plasma, there will be some particles with pitch angles so small that they can travel deep into the atmosphere along the field lines, and these can produce auroras. But they will very quickly be exhausted, and in order to have an aurora for some minutes, as we sometimes see, there must be a continuing supply of particles. It may be that variations in the field could scatter pitch angles, but we think that there must be some rather specific cause to produce the very thin sheet form of the aurora. We suggest that this is connected with the breakdown of the invariance of the magnetic moment along the region around the center line of the belt. This region lies mainly along the back of the earth, because on the front (sunward side) of the earth the currents in the surface of the hollow may still be increasing the field, and making it harder for the ring current to reduce the field to a low value. So we conceive that there is a region of low field intensity and of breakdown of the invariance of the magnetic moment, where particles are continuously produced with small pitch angles and able to travel along the earth's lines far into the auroral zone. We further suppose that sometimes, in great storms, there are irregularities in the distribution of the belts, as is sometimes shown by satellites to be the case. Because the diamagnetic part of the equivalent current depends on the pressure gradient, which may thus be irregular in the belt, we may have a region in which the earth's field is reduced, and in which there are a number of minima of the field. Each of these minima could be the source of a sheet of aurora, giving a multiple aurora. This is very common for the auroras seen in high latitudes; there may be a whole series of parallel arcs, up to five or more. The aurora has two phases, one of quiet arcs, which appear even in quiet magnetic periods at sunspot minimum (as found during the Second Polar Year). But associated with the onset of polar magnetic substorms, the aurora becomes activated and thinner, wavy and fluted, and great folds appear in it. These may move over the sky so that the sky is sometimes filled with a canopy of waving curtains, a most majestic and beautiful spectacle. Figure 15 shows a remarkable all-sky camera picture picked out by Dr. Akasofu from extensive examination of the all-sky camera material obtained during the IGY. It shows a fold of the aurora with fluting on it, occupying almost the whole field of view of the all-sky camera. As the aurora is generally located at about 100 kms above the ground, the field of view of a camera or a person on the ground has a radius of about 1100 kms. This fold in Figure 15 is of that order of magnitude. This makes it difficult to determine the whole distribution of the aurora on a larger scale, that is the whole auroral zone, even when there is a series of all sky camera stations whose fields overlap. It seems to me that a worthwhile project would be to try to photograph the aurora from above. If you go to an equal distance above the aurora from that on the ground below, you

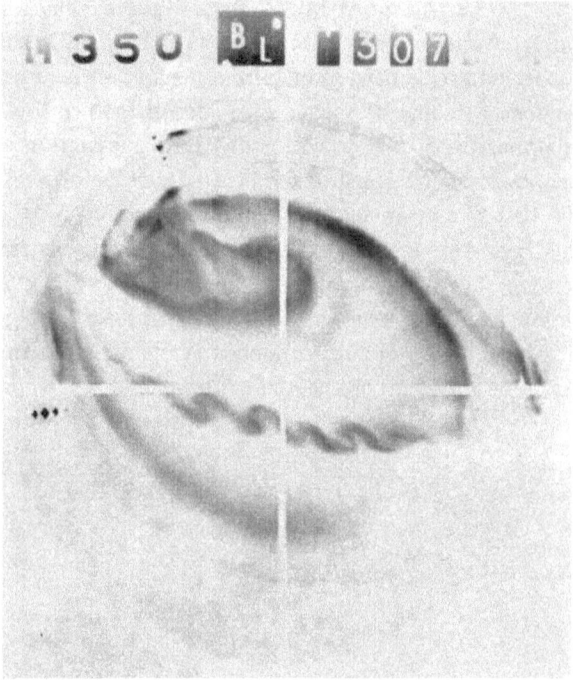

Fig. 15.   All-sky camera picture.

will not see it so well, because the edge of the field of view will be seen tangentially, whereas from the ground, even when the aurora is seen on the horizon, it is still somewhat inclined to the plane of the field of view. But if you go several hundred kilometers above the aurora you really can enlarge the field of view and see a larger portion of it, and so better estimate its main distribution. Dr. Akasofu in Alaska has recently been examining the large collection of all-sky camera films at the U.S. World Data Center for this IGY material. Thus he has been adding to the fine work that Dr. Davis did earlier by his examination of these films at the University of Alaska. Dr. Akasofu is studying what happens when the aurora breaks up. We suggest that at such a time, along this pipe-like region of low field intensity in the equatorial plane, there is an electric current flowing. It is very difficult for an electric current to flow anywhere else across the field. Such an electric current in free space is unstable; it will curve as typified on Figure 15, and wave about, but the form of the breakup of the aurora has some very definite characteristics. Most of the folds seem to be on the northward side of the (northern) aurora, indicating that the waves appear in the equatorial plane on the outer side of the normal location of the ring.

Examination of the magnetic and auroral data suggests that the theory of magnetic storms cannot be a simple straightforward theory. In July, 1959 there were three magnetic storms following three great solar flares, in each of which the first phase, the initial increase, was of similar magnitude, suggesting a similar intensity of the plasma flow. But the later development of these three storms was very different. They all had similar effects on the cosmic rays, but the middle storm had a much greater main phase than the other two. The third one had hardly any. Again, Figure 16 illustrates two magnetic storms by means of a record from a low latitude station, Honolulu (in the upper part), and below, a high latitude station, College, Alaska, near the auroral zone. In one storm shown on the left, of October 28, 1951, at Honolulu there was the first phase, an increase, followed by a considerable main phase, shown by the shading. And in that storm there were many intense substorms at College. In the storm shown on the right, with a similar first phase, suggesting a similar plasma intensity, there was very little main phase; and in that storm at College, there was very little polar substorm activity. There are many such examples, indicating that the development of the main phase and the appearance of polar substorms are closely related. Now the polar substorms are due to the entry of

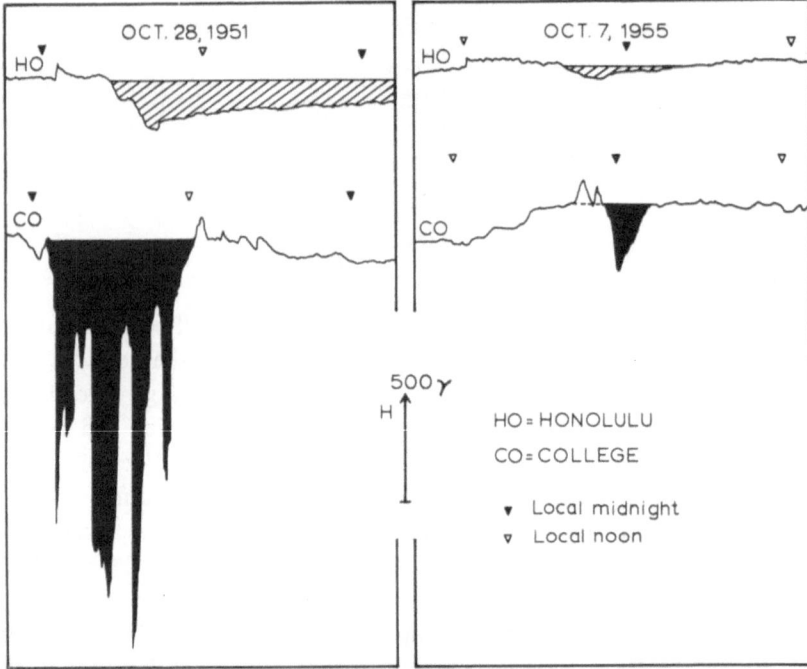

Fig. 16. The horizontal component magnetograms of a highly contrasted pair of storms from the Honolulu and College stations.

energetic particles into the atmosphere. It would seem that instabilities in the trapped plasma may accompany the building up of the ring current and of the van Allen belts, which we infer from the occurrence of the main phase. On the left, we again see two storms with many polar substorms and considerable main phases in low latitudes, and on the right, storms with a similar first phase but very little main phase and very few weak polar substorms. The greatest storms recover very quickly up to a certain point during the main phase, and then recovery goes on more slowly. This has led Akasofu and me to suggest that there are two kinds of ring current involved, an inner

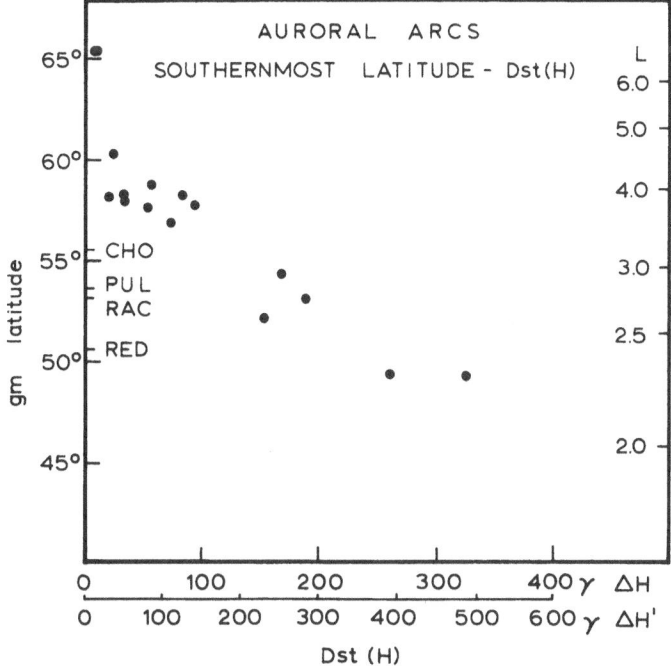

Fig. 17.   Relation between the intensity of main phase of storms (measured horizontally in gammas) and the lowest latitudes of aurora.

one which develops quickly and dies away quickly, and an outer one which develops and declines more slowly; it is the inner one that makes the storm a great one. Akasofu and I think that this corresponds to an intense belt at a relatively close distance to the earth, where for many reasons one must suppose the decay to be very rapid. Figure 17 illustrates the relation between the intensity of the main phase (measured horizontally in gammas) and the lowest latitudes of aurora (the ordinates). It indicates that quiet auroras are most frequent in latitudes of about 66°; as the storm intensity of the main phase increases up to say 350 gammas, the auroral latitude comes down to about 50° latitude; in some great storms during the IGY the minimum auroral latitude was decidedly lower than that. During the outstanding magnetic and auroral storm of February 11, 1958, a very low latitude aurora passed directly over Minneapolis, at a time between two polar substorms. When a polar substorm developed half an hour later, there was a very broad band of aurora indicating many auroral arcs. Akasofu and I think that this was due to a region of low field intensity of considerable radial extent during the occurrence of the polar substorms, while during quiet intervals between the substorms there was a deep dip in the field, at about 3.5 earth radii, corresponding to the low latitude of the quiet aurora shown on the top left. We have to combine and explain all the evidence we can obtain about the effects of the trapped plasma and of the plasma flow around the boundary of the cavity, from ground data (magnetic and auroral), and also the data in space which only satellites can produce. The two sets of data together are very much more valuable than one alone. Before the satellites we had to depend on ground data alone; now we have a harder task, in some ways, one that's only barely begun, of trying to satisfy the ground data and also to get consistency with the satellite data.

References

AKASOFU, S.-I., CAIN, J. C., and CHAPMAN, S.: 1961, 'The Magnetic Field of a Model Radiation Belt', *J. Geophys. Res.* **66**, 4013–4026.

AKASOFU, S.-I., and CHAPMAN, S.: 1961, 'The Ring Current, Geomagnetic Disturbance and the Van Allen Radiation Belts', *J. Geophys. Res.* **66**, 1321–1350.

AKASOFU, S.-I. and CHAPMAN, S.: 1963, 'The Development of the Main Phase of Geomagnetic Storms', *J. Geophys. Res.* **68**, 125–129.

CHAPMAN, S.: 1961, 'Sun Storms and the Earth: the Aurora Polaris and the Space Around the Earth', *Amer. Scient.* **49**, 249–284.

CHAPMAN, S.: 1963, 'Solar Plasma, Geomagnetism and Aurora' in *Geophysics: the Earth's Environment*, Gordon and Breach, New York.

CHAPMAN, S. and BARTELS, J.: 1940, *Geomagnetism*, Oxford University Press, London.

SILSBEE H. C. and VESTINE, E. H.: 1942, *Terr. Magn. Atm. Elec.* **47**, 206.

Discussion and Questions

E. Hones: (IDA): Does the time when the energetic particles precipitate in the atmosphere and produce aurora appear to correlate with the advent of low energy plasma into the magnetosphere? Could such advent be the mechanism whereby the field is locally reduced, permitting the higher energy, already trapped particles, to precipitate?

S. Chapman: I cannot quote any detailed evidence from the satellite measurements, but what you suggest might be the explanation. Akasofu's and my theory of aurora is a controversial matter and by no means established and accepted.

D. Beard (University of California): I'd like to mention that an additional effect besides that of the ring current, decreasing the magnetic field in the interior of the ring current to a certain low value, and increasing the magnetic field farther away from the surface of the earth. In addition to the ring current doing this, also the surface current of the magnetosphere will have the same effect. The surface current magnetic effect will also decrease the magnetic field very close to the earth, but it will increase the magnetic field, in a simple dipole field, at large altitudes.

S. Chapman: My belief is that it increases the magnetic field in the middle belt of the earth, but decreases it in the polar regions, where one cannot determine the various changes so accurately.

D. Beard: Perhaps I should have mentioned that the effect is very latitude dependent.

Unidentified: What about energy considerations? Isn't it true that there is not enough energy in the particles that have been observed in the belt to explain the field of intense aurora? I don't understand from your mechanism, when you perturb the pitch angle and dump them into the atmosphere, where you get enough electrons in order to supply the energy required to produce the aurora.

S. Chapman: An increase in the number of belt particles of lower energy than that sufficient to allow penetration to auroral levels could make an important contribution to the ring current. There may be many more particles having energies below 10 keV, and each of these will contribute to the ring current field. We need to have measurements of particles down to still lower energies before we can hope to get any real check on this.

IONOSPHERIC RESEARCH FROM SPACE VEHICLES

R. E. BOURDEAU

Space Sciences Division, Goddard Space Flight Center, Greenbelt, Mld.

1. Introduction

Previous to the advent of artificial satellites, our knowledge of the ionosphere was limited principally to the interpretation of data from the classical ground-based ionosonde and a few rocket measurements of ionizing radiation, electron density and ion composition, all applicable to altitudes below the F2 peak. The early rocket results of SEDDON and JACKSON (1958) which have been summarized by RATCLIFFE and WEEKES (1960), show that the lower ionosphere is characterized by an electron density which increases monotonically with altitude and thus that the term "layer" is an incorrect nomenclature for the D-, E-, and F-regions. In more recent years, a wealth of data has been gathered on the temporal, spatial and energy distribution of the charged particles for the ionosphere. This has been accomplished at previously relatively unexplored altitudes well above the F2 peak. Concurrently but unfortunately not simultaneously, measurements have been made of the time dependence of corpuscular, ultraviolet and X-radiation. The most serious observational gap is in the structure of the neutral atmosphere where the most significant contribution has been the computation of the neutral density parameter from satellite drag observations.

This paper compares the available interdisciplinary experimental data with theoretical models of the D-, E-, and F-regions as well as the upper ionosphere, eventually selecting those models which best fit the spaceflight observations. Because a significant amount of low-energy charged particle data were obtained by direct sampling techniques and because this methodology is relatively new to ionospheric research, the paper contains brief discussions of the validity of some of these results as they are presented. Although the theoretical models of all ionospheric regions have been enhanced significantly by increased spaceflights, refinement and changes in these models await the next two major steps – the launching of satellites and rocket probes which are truly geophysical in nature and the correlation of such interdisciplinary measurements with data resulting from recent breakthroughs in ground-based observational methods.

2. The D-Region Under Quiet Solar Conditions

The D-region, situated between approximately 50 and 85 kilometers, is the lowest ionospheric subdivision where a significant number of free electrons are found. This altitude interval is difficult to treat theoretically and experimentally because of the low charged particle concentrations, because the relatively high gas densities result in high

Chang & Huang (eds.), Proc. Plasma Space Sci. Symp. All rights reserved.

electron collision frequencies, and because of the high probability of negative ion formation.

Spaceflight radiation observations made during the absence of solar flares show that the three most probable ionizing agents are cosmic, Lyman-α (1215.7 Å) and X-radiations (2–8 Å). The most detailed theoretical analysis of D-region formation has been performed by NICOLET and AIKIN (1960) using an expansion of the following equation of ionization:

$$q = (\alpha_d + b\alpha_i) n_+ n_e, \tag{1}$$

where q is the equivalent electron production rate, α_d is the loss coefficient for recombination of positive molecular ions with electrons, α_i the loss coefficient for recombination of positive with negative ions, b is the ratio of negative ions to electrons, and n_e and n_+ are the electron and positive ion concentrations, respectively. The cosmic and X-radiations act on molecular nitrogen and oxygen, the major neutral constituents, whereas Lyman-α radiation acts on nitric oxide, a trace constituent. In their expansion of Equation (1), Nicolet and Aikin produced a summation of the separate effects of the three ionizing sources and concluded that:

(a) cosmic radiation is the most important ionizing agent at altitudes up to 70 km;

(b) in the 70–85 km region, assuming an NO^+ recombination coefficient of 3×10^{-9} cm^3 sec^{-1}, a nitric oxide abundance of only 10^{-10} of the total neutral concentration is required to make X-rays (2–8 Å) unimportant to the formation of the D-region under quiet solar conditions;

(c) negative ions are unimportant above 70 km;

(d) the ionization at altitudes between 85 and 100 km (base of the E-region) is the result of X-rays (30–100 Å) and ultraviolet radiation (Lyman-β and the Lyman continuum).

It is known principally from measurements made on the Greb satellite (KREPLIN, CHUBB and FRIEDMAN, 1962) and on the Orbiting Solar Observatory (LINDSAY, 1962) that the Lyman-α energy flux (3–6 erg cm^{-2} sec^{-1}) is relatively constant with solar condition. The X-ray energy flux, however, is extremely variable. Consequently, it is possible that Lyman-α and X-radiations alternate as the predominant ionizing agents of the normal upper D-region. POPPOFF and WHITTEN (1962) have challenged the Lyman-α hypothesis in attempting to account for the 70–85 km ionization entirely by X-radiation. However, they used energy flux values, which according to a recent review (FRIEDMAN, 1962), exceeds energies representative of a quiet sun even at the maximum of the solar cycle.

In this section it will be shown that to achieve agreement with spaceflight n_+ observations taken during quiet midsolar cycle conditions, it is necessary to invoke the Lyman-α hypothesis. Additionally it will be shown that to obtain consistency with ground-based measurements of electron density for similar conditions, existing D-region formation theories must be considered incomplete. Specifically, it appears that there is a tendency in most of the current models to underestimate the number of electrons lost in the various processes of negative ion formation.

It has been inferred in the previous discussion that care must be taken in comparing experimental results with theoretical models for the same solar condition. The extreme sensitivity of the X-ray energy flux to solar activity is demonstrated in Figure 1. It is

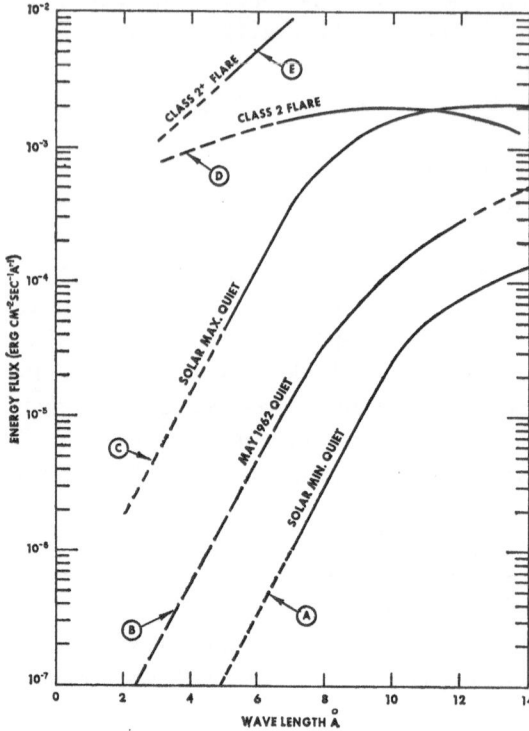

Fig. 1.   The variability of solar X-ray emissions. Curves *A*, *C* and *E* are from FRIEDMAN (1962)
Curves *B* and *D* are from POUNDS and WILLMORE (1962).

seen by a comparison of curves *A* and *C* (FRIEDMAN, 1962) that even in the absence of flares the intensity in the important wavelength region (2–8 Å) varies by two to three orders of magnitude during a complete solar cycle. The recent data (curve *B*) from the Ariel satellite (POUNDS and WILLMORE, 1962) correspond temporally to values used by NICOLET and AIKIN (1960) in computing their "quiet sun" model. The Ariel data also correspond most closely to the epoch of the solar cycle under which the few available D-region charged particle density profiles have been obtained.

In Figure 2 are presented two theoretical electron density profiles (AIKIN, 1962a) which are intended by their contrast to illustrate the enhancement of n_e that should occur in the 70–85 km region if the NO concentration is 10^{-10} of the total concentration and if the negative ion abundance is that computed with the assumption that O_2^- is the principal negative ion. The alternative theoretical profiles were computed by using X-ray energy fluxes approximately representative of curve B in Figure 1. The principal change from the original Nicolet-Aikin model is that the values of α_d for O_2^+ and NO^+ were increased by a factor of ten. The O_2^+ coefficient is increased in the light of more recent laboratory investigations (KASNER, ROGERS and BIONDI, 1961) whereas in a succeeding section it is shown that an increase of α_d for NO^+ from

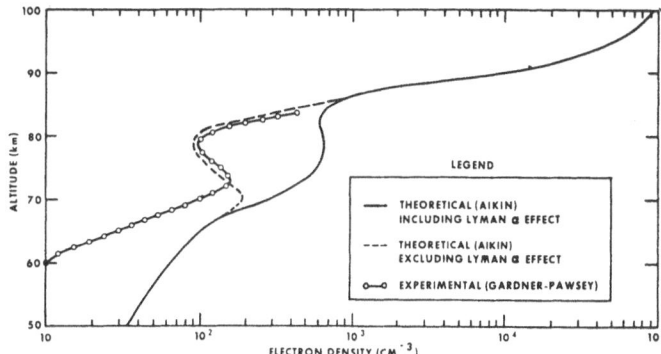

Fig. 2. Electron density profile of the quiet daytime D-region at mid-latitudes for the period between solar maximum and solar minimum.

3×10^{-9} to 3×10^{-8} cm^3 sec^{-1} is more consistent with an effective recombination co-efficient determined for the E-region from diurnal variation observations.

   To date, a reliable spaceflight method for measuring the low electron densities characteristic of the normal D-region has not yet been developed. USSR data from a rocket-borne impedance probe for altitudes as low as 75 km have been published (KRASNUSKIN and KOLESNIKOV, 1962) but too recently for critical review. The lag in the development of rocket-borne electron density experiments is due primarily to the extremely low electron to neutral gas density ratio and the associated high collision frequencies. Because of this lag, it is necessary to resort primarily to electron density profiles obtained by ground-based methods.

   In a recent review, RATCLIFFE and WEEKES (1960) conclude that D-region electron density profiles obtained by ground-based methods should be treated with caution. According to this review, the most reliable of these profiles was obtained by GARDNER and PAWSEY (1953) using observations of weak backscatter echoes. The experimental data were obtained at a time when the values of the X-ray energy flux most likely were intermediate between those represented by curves A and B in Figure 1. The experimental data are plotted in Figure 2 for comparison with the alternative theoretical models. The valley of ionization which shows a minimum at about 78 km is typical of most of the few electron density profiles reported by other observers using ground-based techniques. BELROSE and BURKE (1961), for example, report a valley with the minimum located at 70 km, also using a backscatter method. There is a large difference between both theoretical ionization estimates and the experimental data at altitudes below 70 km. It has been concluded by others (BELROSE and BURKE, 1961) that observational data by the backscatter method must be treated with caution in the lower D-region because of simplifying assumptions in the Appleton-Hartree formula used to derive the experimental result. If the experimental data below 70 km are assumed correct, the low electron densities could be explained either by a theoretical overestimate of the effective cosmic ray electron production rate or an underestimate of the processes leading to the formation of negative ions or a combination of the two.

It is seen by a comparison of the experimental data with the two alternative theoretical models in Figure 2 that all of the ionization in the 70–85 km region could be explained from the X-ray effect alone as proposed by Poppoff and Whitten without resort to the Lyman-α hypothesis of Nicolet and Aikin. However, a conclusion based on these few electron density data that the Lyman-α radiation has a negligible effect on the formation of the D-region would not be consistent with spaceflight observations of the positive ion characteristics, a discussion of which follows.

The first spaceflight measurements of D-region ion parameters (BOURDEAU, WHIPPLE and CLARK, 1959) were obtained by use of a rocket version of a Gerdien condenser similar to that flown extensively on aircraft and balloons by workers in atmospheric electricity. This experiment measures ion conductivity (λ_+). In the altitude region where the number of free electrons is negligibly small, λ_+ can be estimated theoretically from

$$\lambda_+ = ek_+ (q_+/\alpha_T)^{1/2}, \tag{2}$$

where e is the elementary charge, k_+ is the ionic mobility, q_+ is the rate of ion-pair production and α_T is Thomson's ion-ion volume recombination coefficient. The altitude dependence of k_+ and α_T are estimated from laboratory experiments. In the reported results, the altitude at which α_T can be extrapolated with confidence was overestimated. However, at altitudes up to 50 km where the use of α_T is appropriate, the experimental values of λ_+ are in good agreement with theoretical estimates based on Equation (2) in which q_+ is estimated from spaceflight observations of cosmic ray intensities. The experimental rocket values of λ_+ also are consistent with those obtained on balloons by other investigators. The importance of these agreements is that it lends some confidence to the theoretical conclusion (BOURDEAU et al., 1959; ICHIMIYA, TOKAYAMA and AONO, 1959) that possible errors in the rocket experimental data associated with shock waves (thermal ionization, adiabatic compression etc.) are small, even at the relatively high D-region gas pressures.

Positive ion densities now have been reported by SMITH (1961a) and AONO, HIRAO and MIYAZAKI (1961), each using a different method of computing n_+ from the observed current to an exposed electrode insulated from the rocket body. Their published values are compared in Figure 3 with a theoretical n_+ model (AIKIN, 1962a). Estimates of negative ion densities based principally on attachment and photodetachment processes are represented by the difference in the theoretical n_+ and n_e curves. Both sets of experimental data were obtained at a time when the values of the X-ray energy flux were most likely slightly larger than those represented by curve B in Figure 1.

It is seen from Figure 3 that the shape of the ion density profile experimentally obtained by SMITH is consistent with the shape of the theoretical n_+ curve and thus consistent with a hypothesis of three different ionization sources between 50 and 100 km, each predominating in a separate altitude region. However, it is seen also that the experimental values of n_+ are generally much higher than the theoretical values. The general excess of the experimental positive ion densities over the theoretical estimates may be interpreted as an altitude invariant error either in the experimental result or in one of the assumed theoretical parameters. It now will be shown that experimental uncertainties could account for the difference between the theoretical and experimental n_+ curves in Figure 3.

The results reported by SMITH (1961a) were obtained by the use of a standard asymmetric Langmuir probe in which the active electrode is the nose tip of the rocket.

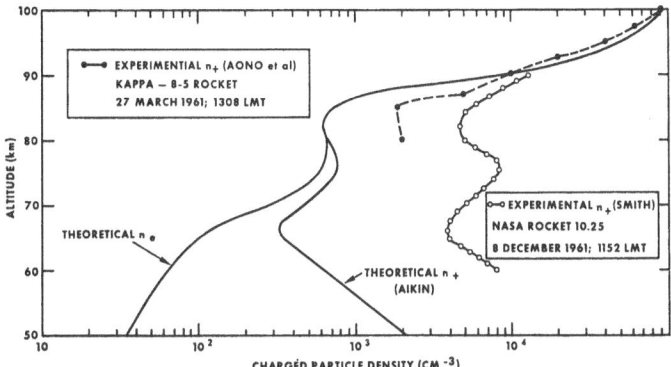

Fig. 3.   Ion density profile of the quiet daytime D-region at mid-latitudes for the period between solar maximum and solar minimum.

In the D-region this Langmuir probe functions as a conductivity meter, the principle of operation being analogous to that of the exploratory Gerdien condenser experiment (BOURDEAU et al., 1959). Positive ion current flowing from the ionosphere to the nose tip is measured as a function of a variable voltage applied between this electrode and the main rocket body. Values of λ_+ are computed by electrostatic theory from the slope of the volt-ampere curve when the active electrode is negative with respect to the ambient plasma. SMITH computed positive ion densities from the observed λ_+ according to the following simplified form of the pertinent electrostatic equation:

$$n_+ = (\lambda_+ m_+ v_+)/e^2,\tag{3}$$

where m_+ is the mean ionic mass and v_+ is the ion collision frequency. Laboratory investigations (LOEB, 1955) indicate that the simplifying assumptions leading to Equation (3) could result in uncertainties of a factor of two in the n_+ determination. SMITH used values for v_+ contained in the 1959 ARDC model atmosphere which are probably uncertain by another factor of two (AIKIN, private communication). Consequently, the values of n_+ computed by SMITH could be uncertain by a factor of four just from the geophysical interpretation of the measured conductivity alone. This uncertainty is greater at altitudes above 80 km because the methods used to obtain λ_+ from the observed positive ion current are no longer valid when the ionic mean free path becomes large compared to the dimensions of the active electrode (BOURDEAU et al., 1959).

Other possible errors in the measurement of D-region ion characteristics are associated with aerodynamic effects (discussed previously) and the photocurrent effect. The latter effect represents the predominant criticism of ionospheric direct measurements techniques. For this reason, it is necessary to present the following arguments for a neglible photocurrent effect on the experimental n_+ results shown in Figure 3.

Because solar radiation can induce electron emission from rocket surfaces, an artificial electron current will flow from this electrode into the surrounding medium. This unwanted photocurrent is of the same polarity as the desired positive ion current. Values for the photocurrent density (j_p) of 2.3×10^{-9} A/cm^{-2} at an altitude of 200 km have been measured on a rocket (HINTEREGGER, DAMON and HALL, 1959) and of 5×10^{-9} A/cm^{-2} in the upper ionosphere on the Explorer VIII satellite (BOURDEAU, DONLEY, SERBU and WHIPPLE, 1961). The approximate agreement of the j_p values measured at such radically different altitudes should not be surprising because the most responsible solar radiation, occurring at wavelengths between 2000 and 3000 Å (HINTEREGGER et al., 1959; ICHIMIYA et al., 1959), is not significantly absorbed even in the D-region.

The values of the perturbing photocurrent density listed above can be shown to be some two or three orders of magnitude larger than the ambient random ion current density flowing in a D-region undisturbed by the presence of a rocket. However, the actual effect is dependent on the ratio a_+/a_p where a_+ is the effective ion collection area and a_p is that part of the active electrode's surface area which is exposed to the sun. This ratio is extremely difficult to obtain theoretically for the D-region although orders of magnitude at large negative electrode potentials are indicated (ICHIMIYA et al., 1959). It is possible by this mechanism then, to overcompensate in magnitude for the photocurrent effect. Confidence in this conclusion can be acquired by examination of observed D-region volt-ampere curves. At negative electrode potentials, the current measured at the electrode is observed to be an order of magnitude larger than what would be expected from the product of j_p and a_p. Additionally even if the amplitude of the photocurrent was large, λ_+ is obtained from the slope of the volt-ampere curve at negative electrode potentials. This is a regime where the photocurrent is invariant with electrode potential. Hence it should not affect the measurement of the slope and thus the n_+ determination.

The n_+ data by AONO et al. (1961) shown in Figure 3 were obtained by measurement of the positive ion current to an effectively spherical electrode which was extended along the major rocket axis. The term "effectively spherical" is used because although the electrode actually consists of two crossed rings, they are so arranged that the analysis for ions is as though the electrode were a solid sphere. The virtue of the crossed rings is that they present a much smaller target to solar radiation and consequently the experiment is even less sensitive to the photocurrent effect than the device used by SMITH.

A form of the conventional Langmuir probe theory applied to ions is used by the Japanese in computing n_+ from the experimental volt-ampere curves. The general agreement of their published n_+ values with an independent simultaneous measurement of n_e in the E-region speaks well for the validity of their application of this theory at least for the simple case of no collisions in the plasma sheath surrounding the probe. However, their published data show slightly larger electron than ion concentrations in some portions of the E-region which are in turn indicative of second-order errors even for the simple case. The validity of Langmuir probe theory becomes more uncertain at D-region altitudes where multiple collisions within the sheath complicate the application of the conventional Langmuir probe theory. Thus, for different reasons associated with the interpretation of the observed volt-ampere curves the largest uncertainty in both sets of experimental data shown in Figure 3 can be expected between 80 and 90 km. It would not be surprising, therefore, if the same factor of 4 in the difference between the experimental data in the 80–85 km region were obtained even under identical solar conditions.

It generally is concluded by this author that the uncertainties in the D-region experimental n_+ values could be large enough to adjust to the theoretical n_+ estimates of AIKIN (1962a). On the other hand, it is unlikely that the experimental inadequacies are sufficient to adjust rocket observations of n_+ to the experimental electron density values shown in Figure 2, thus suggesting a much higher negative ion abundance than estimated theoretically by NICOLET and AIKIN (1960) and by WHITTEN and POPPOFF (1961). The high negative ion abundance is more in accordance with the theoretical model of MOLER (1960). The suggestion of high negative ion densities also is consistent with values computed from the original experimental conductivity data by WHIPPLE (1960) who suggests that dust particles believed responsible for noctilucent cloud formation at high latitudes could provide an important recombination surface for charged particles in the D-region. The simultaneous measurements of electron and positive ion densities made by AONO et al. (1961) show somehat larger n_+ values in the 95 km region again suggesting a large negative ion abundance at the lower D-region altitudes where their formation is even more likely.

The few data that are available support a tentative conclusion of ionization of the lower and upper D-regions by cosmic and Lyman-α radiation, respectively, during conditions of a quiet sun at the middle of the solar cycle. A firm conclusion awaits more accurate measurements of electron and ion densities obtained simultaneously with observations of the pertinent energy fluxes of the ionizing sources. Just as important is the need for improved knowledge of the pertinent reaction rates and of the number density of the trace neutral constituents.

3. The D-Region During Disturbed Solar Conditions

During periods of high solar activity, the D-region is characterized by enhanced ionization with associated electromagnetic wave attenuation strong enough to produce radio blackouts. There are many phenomena associated with solar flares which increase the normal D-region electron densities by up to two orders of magnitude. A comparison of the amount of enhanced ionization for different types of solar flare events is made with the Nicolet-Aikin quiet sun model (curve *A*) in Figure 4.

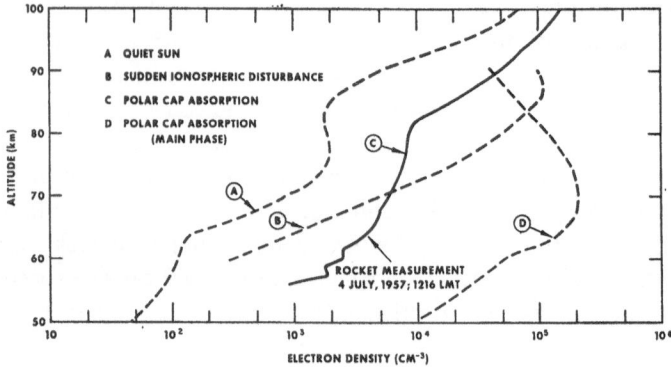

Fig. 4.   Comparison of the D-region under quiet and disturbed solar conditions.

Simultaneously with the appearance of the flare, radio absorption is observed in the D-region on the sunlit side of the earth for periods lasting up to approximately one hour. This particular phenomenon is called a Sudden Ionospheric Disturbance. CHUBB, FRIEDMAN and KREPLIN (1960) made rocket flights into such events and observed abnormally high X-ray fluxes at extremely low D-region altitudes. These observations have been used in Equation (1) by NICOLET and AIKIN (1960) to calculate an S.I.D. electron density profile (curve *B* of Figure 4). Here, the influences of cosmic and Lyman-α radiation become minor, the secondary layer in the 70–85 km region disappears, and the profile is characterized by overall enhancement and a monotonically increasing electron density. The S.I.D. profile does not include the contribution of radiation responsible for the formation of the E-region. Observations from the Greb satellite show that whenever the X-ray flux at wavelengths shorter than 8Å exceeds a critical value of 2×10^{-3} erg cm^{-2} sec^{-1}, radio fadeouts at short wavelengths and other sudden ionospheric disturbances occur (FRIEDMAN, 1962). It is seen from curves *D* and *E* in Figure 1 that this high flux value generally would be characteristic of flares stronger than Importance 2.

A second type of ionospheric storm predominates at night and is associated with active aurorae and magnetic disturbances. It generally is accepted that the ionizing agents are energetic particles comprising corpuscular emission from the sun, a belief founded on the observation that D-region absorption occurs some 21 hours after the appearance of a flare, an interval corresponding to the sun-earth transit time for these particles. During the storm, D-region electron densities increase to values high enough that echoes are observed on ionosondes. "Layers" appear at 90 km during weak geomagnetic activity and as low as 70 km for the more active events.

A third type of disturbance occurs only at auroral latitudes. These phenomena, called Polar Cap Absorption (PCA) events, are produced by energetic protons emitted from the sun during certain flares. Here echoes are observed from ionosondes at altitudes as low as 60 km. The phenomenon has been the subject of considerable

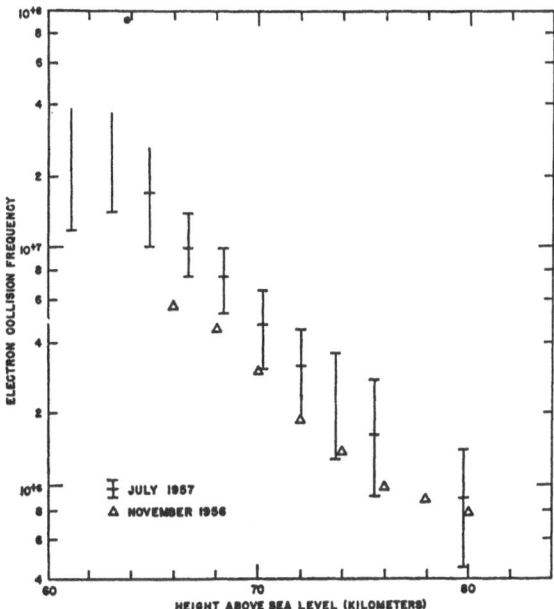

Fig. 5.   Rocket measurement (KANE, 1960) of the altitude dependence of electron collision frequency.

study during the past few years. Most recently, MAEHLUM and O'BRIEN (1962) have proposed a semi-quantitative time history of the D-region electron density profile during a PCA by inserting proton fluxes observed on the Injun satellite as a source function in the conventional equation of ionization. One of these profiles (curve D of Figure 4) coincides with the peak of 27.6 Mc cosmic radio noise absorption (17 db) observed at College, Alaska. Like the Nicolet-Aikin model for an S.I.D., curve D does not contain the contribution due to solar radiation. Inclusion of solar radiation should modify the result in the base of the E-region (above 80 km). JACKSON and KANE (1959) used a rocket-borne radio propagation experiment to measure a D-region electron density profile (curve C of Figure 4) during a less active phase of a PCA event (3 db absorption at 30 Mc). This profile compares favorably in shape but is generally less enhanced than that proposed by Maehlum and O'Brien for an interval when the observed radio absorption also was 3 db. If, as the evidence presented in the previous section indicates, there is a higher D-region negative ion abundance than is generally accepted, then the three dashed curves in Figure 4 could represent overestimates of the electron density.

The strong absorption of radio waves in the D-region is made possible by the high electron collision frequencies characteristic of this altitude interval. KANE (1959, 1960) has derived an electron collision frequency model from measurements of the difference in the absorption of ordinary and extraordinary radio propagation modes. His data which represent the best available rocket measurements of the altitude dependence of this important D-region parameter are presented in Figure 5. They were obtained on two separate rocket flights flown during auroral absorption events.

4. The Ion Content of the Lower Ionosphere

To a high degree of probability, the predominant ionizing sources of the E- and F-regions for quiet solar conditions at mid-latitudes are solar ultraviolet and X-ray radiations. It is possible to compute the rate at which different ion species are formed from a knowledge of the altitude dependence of the number of photons incident at each wavelength, of the densities of the individual neutral constituents and of the absorption and ionization cross sections of these constituents. There are available some altitude profiles of the photon fluxes as a function of wavelength but only for limited conditions. Early results from rocket-borne neutral gas spectrometers reported to date are extremely controversial. The validity of these results is questioned justifiably because recombination effects within the instruments distort the gas under study from its ambient condition. Thus most model atmospheres do not depend on experimental observations but rather are derived from the hydrostatic laws using an assumption for the critical altitude of diffusive separation. Our knowledge of the important cross sections also must be considered incomplete.

The most recent estimate of the altitude dependence of the rate at which various ions are produced was made by WATANABE and HINTEREGGER (1962) using solar radiation data obtained by Hinteregger under quiet solar conditions at midlatitudes. The rates for the ions formed in greatest abundance (N_2^+, O_2^+, O^+) are illustrated in the left-hand side of Figure 6. Watanabe and Hinteregger hasten to point out that these curves must be considered suggestive rather than quantitative principally because of the uncertainty in our knowledge of the neutral atmospheric composition. Some adjustment of these rates already are indicated from very recent neutral gas spectrometer results (SCHAEFFER, 1963). These results were obtained by use of a spectrometer which overcomes the inadequacies of the early devices by ionizing the gas under study before its entry into the instrument's analyzer section. The results show a larger ratio of atomic to molecular oxygen than that used to obtain the ion production rates shown in Figure 6.

Some indication of the solar radiation responsible for the production of various ionospheric regions can be made by comparing the inferred production rates of the individual ionic species with rocket-borne ion spectrometer results. Early rocket flights

(JOHNSON, MEADOWS and HOLMES, 1958) of a radio-frequency mass spectrometer made in the auroral zone showed that the diatomic ions, O_2^+ and NO^+, predominate below 200 km and that O^+ is the principal ion found above this altitude. These results have been confirmed by ISTOMIN (1960) and by TAYLOR and BRINTON (1961). The rocket data of Taylor and Brinton shown in the right hand side of Figure 6 are selected for comparison with the production rates because they were obtained for latitude and temporal conditions most closely representing those under which the photon fluxes used in estimating the photoionization rates were obtained. It is seen in the comparison that although the diatomic nitrogen ions are expected to be pro-

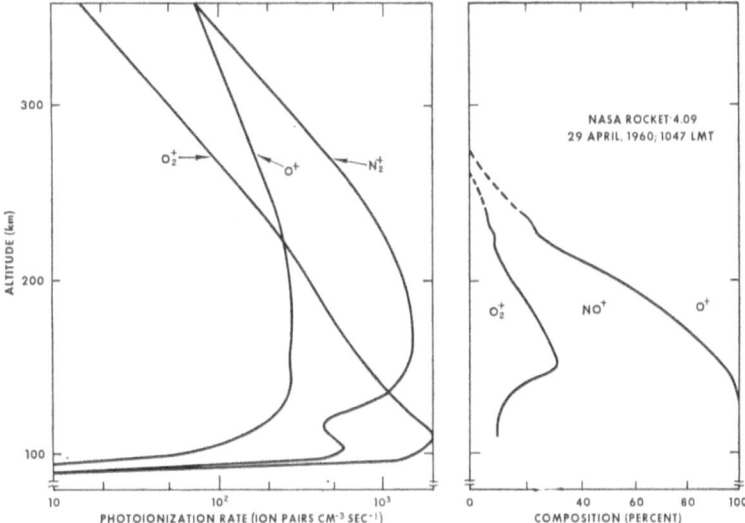

Fig. 6.   Comparison of the production rate and the actual abundance of ions in the lower ionosphere.

duced in great quantities, the spectrometer results show that they exist only as extremely minor charged constituents. The most likely reason is that N_2^+ ions dissociatively recombine ($N_2^+ + e \rightarrow N' + N''$) at a very rapid rate as demonstrated by FAIRE and CHAMPION (1959).

It was inferred in the previous section that there is general acceptance that X-radiation at wavelengths longer than 10 Å and ultraviolet radiation (Lyman-β and the Lyman continuum) are responsible for the base of the E-region (85–100 km). There remain two schools of thought for the formation of the remainder of the E-region: (a) general ionization of air by soft X-rays (VEGARD, 1958; HULBURT, 1958; BATES, 1956; FRIEDMAN, 1959) and (b) photoionization of O_2 by solar radiation in the wavelength region between 800 and 1026 Å (WULF and DEMING, 1938; NICOLET, 1945; WATANABE, MARMO and PRESSMAN, 1945). On the basis of more accurate solar radiation data, WATANABE and HINTEREGGER (1962) strongly support ionization of the 100–120 km region by direct production of O_2^+ from ultraviolet radiation at wavelengths between 911–1027 Å. Part of this conclusion is based on the agreement between experimentally observed E-region electron densities and those computed from the conventional ionization equation under quasi-equilibrium conditions:

$$n_e = (q/\alpha_d)^{1/2}, \tag{4}$$

where q is the photoionization rate for O_2^+ computed from the observed ultraviolet photon fluxes.

However, there are two sets of recent experimental evidence which indicate that Watanabe and Hinteregger overestimated the electron density produced in the 100–120 km region by ultraviolet radiation. Firstly, the recent rocket neutral gas spectrometer results indicate that the O_2 concentration and thus the O_2^+ production rate have been overestimated. Secondly, Watanabe and Hinteregger used a value for the O_2^+ recombinance coefficient taken from NICOLET and AIKIN (1960) of 3×10^{-8} cm^3 sec^{-1} which as discussed in the previous section appears from more recent laboratory work to be an order of magnitude too low.

A third set of experimental data which also tends to favor X-radiation as the more important agent for E-region formation involves the rocket ion spectrometer data shown in Figure 6. The diatomic ion NO^+ is observed to be predominant at E-region altitudes. Its existence can best be explained (AIKIN, 1962b) by ion-atom interchange involving the O^+ ion $(O^+ + N_2 \rightarrow NO^+ + N)$. The ultraviolet radiation (911–1027 Å) observed to penetrate below 120 km (HINTEREGGER and WATANABE, 1962) lies in a range above the ionization threshold for O^+. Consequently to be consistent with the rocket ion spectrometer observation that NO^+ is the predominant E-region ion, the ultraviolet hypothesis requires a reaction leading to NO^+ formation involving O_2^+, a reaction not considered by Hinteregger and Watanabe. Parenthetically, it is difficult to dispute the correctness of the observed ratio NO^+/O_2^- from an experimental point of view since the principal corrections which must be made to the measured ion spectrometer target currents involve the effects of vehicle motion and of electric fields due to the attracting vehicle potential. Both of these effects are expected to be nearly the same for ions of approximately the same mass. Hence, the measurement of their relative abundance should be reasonably accurate. As HINTEREGGER and WATANABE (1962) demonstrate, the ion spectrometer measurement of the increasing importance of O^+ with altitude leading to its predominance above 200 km is explained adequately by the simultaneously increasing role of extreme ultraviolet radiation in the wavelength region between 280 and 911 Å.

5. The Diurnal and Anomalous Behavior of the E-Region

The altitudes generally assigned to the E-region lie between approximately 85 and 140 km. The most accurate rocket method of daytime electron density determination for this altitude region remains the Seddon two-frequency radio-propagation experiment. Results from this experiment early demonstrated that the lower ionosphere is not characterized by discrete layers but rather that the electron density increases monotonically with altitude and thus that the F-region is a continuation of the upper E-region. Until recently, there have been no detailed observations of nighttime E-region ionization because neither this rocket-borne experiment nor ground-based ionosondes are sensitive enough. Now the introduction of low energy plasma probes into space-flight studies of the ionosphere permits our first insight into the diurnal characteristics of the E-region.

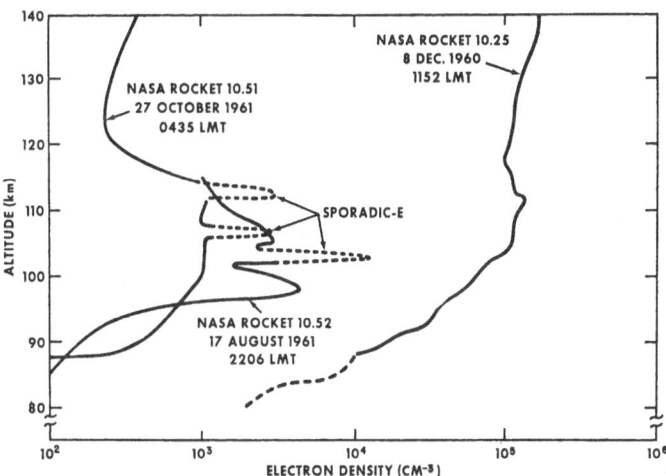

Fig. 7.   Rocket measurements of the diurnal and anomalous behavior of the E-region.

Electron density values reported from one daytime and two nighttime rocket launchings of an asymmetric Langmuir probe (SMITH, 1961a; SMITH, 1961b) are presented in Figure 7. It is seen that the electron density in the lower portion of the nighttime E-region is found to be 3×10^3 cm^{-3} at approximately 22h LMT and 1×10^3 cm^{-3} near sunrise, both about two orders of magnitude less than typical daytime values. SMITH (1961b) has used this observed diurnal variation to compute an effective E-region recombinance coefficient of 2×10^{-8} cm^3 sec^{-1}, which is perhaps mostly representative of the observed predominant ion, NO$^+$. In the view of this author, this value should be considered approximate, firstly because of possible second-order errors in computing absolute values of n_e from the observed volt-ampere curves and secondly because it is assumed that there is no nighttime ionization source. It is tempting from the observation of deep ionization valleys characteristic of the upper nighttime E-region profiles shown in Figure 7 to suggest that the ledges in the lower E-region result from observed scattering (DONAHUE, 1962) of ultraviolet radiation in the night sky.

The other significant data contained in Figure 7 are associated with the measurement of the ionization characteristics of the sporadic-E layer (E_s), which on an iono-sonde is characterized by echo reflection from a constant virtual height. The statistical occurrence of the E_s phenomenon from ionosonde data has been the subject of considerable study, yet the causes of sporadic-E layers remain topics for speculation. Early rocket observations of daytime E_s ionization were obtained by Seddon and Jackson (1958).

Two types of E_s ionization, differing in horizontal dimensions are contained in Figure 7. It is seen that the electron density observed in the sporadic-E layer (102 km) on the 17 August flight represents an enhancement of about a factor of four above the average lower E-region value. The thickness of the layer measured at half the peak electron density was 450 meters. The layer was detected at precisely the same altitude on both the ascent and descent portions of the rocket trajectory, indicating a horizontal dimension greater than 72 km. Similar characteristics were observed at an altitude of 112 km on the 27 October flight. It has been suggested (Whitehead, 1960) that E_s ionization with these characteristics results from re-distribution of electrons by wind shear in the presence of a geomagnetic field, rather than local changes in electron production or loss.

On the 27 October flight, another type of E_s ionization was observed. One such electron density peak is shown at an altitude of approximately 108 km in Figure 7. A larger number of such peaks than indicated in Figure 7 were observed with no correlation between the ascent and descent portions of the trajectory. The small horizontal dimensions lead Smith (1961b) to support current theories that some types of E_s ionization result from meteor trails.

It can be said that there is insufficient evidence to support conclusively any current theories to account for E_s ionization. Such evidence will come from planned correlative experiments performed simultaneously with charged particle density measurements. The rocket data are, however, serving to define the dimensions and ionization characteristics of sporadic-E layers. One of the most detailed altitude definition of the E_s ionization reported to this author was obtained by Boyd (private communication) from a daytime ion density profile on December 6, 1961. His observation of a depth of just under one kilometer and an ionization enhancement of about a factor of 3 is consistent with the results shown in Figure 7.

In addition to the measurement of electron density the asymmetric Langmuir probe is designed to provide data on the electron temperature, T_e. For the same rocket flight from which the daytime electron density profile illustrated in Figure 7 was obtained, Smith (1961a) reported electron temperature values of 1400 °K with little altitude variation between 100 km and the apogee altitude of 155 km. The disagreement of these T_e values with current hypotheses for this altitude region and with the results of other rocket experiments made under quiet daytime conditions justifiably has raised some questions as to the validity of the electron density results. There are two possible explanations for the differences. The first, geophysical in nature, is that this particular rocket flight took place within twenty-four hours of a geomagnetic disturbance. Current theoretical concepts of expected differences between the electron and neutral gas temperatures do not include the extreme variability of X-ray fluxes which perhaps could account for the high electron temperature. On the other hand it is just as likely, from this author's point of view, that the high electron temperatures result from simplifying assumptions used in deriving T_e from the observed volt-ampere curves. Specifically, n_e and T_e both are derived from the volt-ampere curve after unwanted photocurrent and positive ion currents are subtracted out. T_e is computed from the slope of the corrected curve at negative collector potentials and is sensitive to the accuracy of the correction. The electron density, on the other hand, is computed from the current observed at a distinct discontinuity in the curve where the collector is at plasma potential. At this point the electron current is more than an order of magnitude higher than the unwanted current and hence the derivation of n_e is insensitive to the assumed corrections. Thus the errors in deriving n_e are small enough to have little effect on the important conclusions from Figure 7 relative to (a) the magnitude of the diurnal variation of n_e in the lower E-region, (b) the observation of a deep ionization valley in the upper E-region at night, and (c) the major characteristics of sporadic-E ionization.

6. The Formation of the F2 Peak

The altitudes generally assigned to the F-region lie between 140 km and the altitude at which O^+ ceases to be the predominant ion. This region contains the altitude (F2 peak) of the maximum electron density found in the ionosphere. It was not until very recent years that significant experimental data on the characteristics of the upper F-region were obtained. Local charged particle density measurements now have been made on satellites using ion traps (KRASOVSKIJ, 1959; BOURDEAU, 1961) and rf impedance probes (BOURDEAU and BAUER, 1962; ULWICK and PFISTER, 1962; SAYERS, ROTHWELL and WAGER, 1962).

More meaningful to the physics of the F-region have been experimental measurements of the charged particle density distribution with altitude where latitudinal and temporal variations can be neglected. Such profiles have been obtained from ground-based experiments (VAN ZANDT and BOWLES, 1960), rocket-borne radio propagation experiments (BERNING, 1960; JACKSON and BAUER, 1961; KNECHT and RUSSELL, 1962) and ion traps (HANSON and MCKIBBIN, 1961; HANSON, 1962a). These investigators all have concluded from their observations of an electron density distribution with a practically constant logarithmic slope taken over a few hundred kilometers that the upper ionosphere can be represented as an isothermal medium in diffusive equilibrium.

The most favored theory of the formation of the F2 peak at mid-latitudes under quiet solar conditions can be illustrated by comparing these charged particle density profiles with the altitude dependence of electron production rate inferred by WATANABE and HINTEREGGER (1962). Such a comparison is made in Figure 8, using the electron density profile obtained by JACKSON and BAUER (1961) with a rocket-borne cw radio propagation experiment. It has been recognized that, as illustrated, the altitude of the F2 peak lies considerably above that of the maximum electron production rate. This is best explained from the following form of the continuity equation corresponding to quasi-equilibrium:

$$q = \beta n_e + \frac{d}{dz}(n_e W_d), \tag{5}$$

where β represents an attachment – like loss coefficient, z is the altitude and W_d serves to define changes in electron density by vertical motion or diffusion. In the region up to the F2 peak, βn_e predominates over the diffusion term and because it decreases more rapidly with altitude than the electron production rate, q, n_e increases with altitude up to its maximum value. Near the F2 peak, the diffusion term becomes larger than βn_e and the electron density begins to decrease. The diffusion mechanism

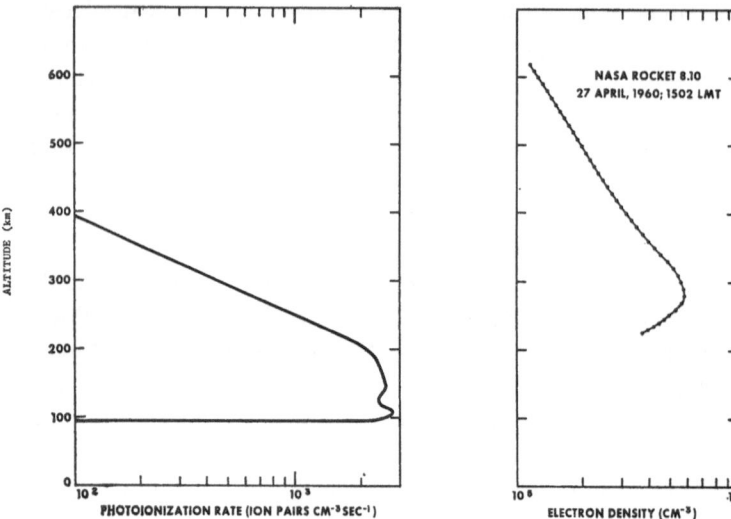

Fig. 8.   Comparison of the altitude dependence of electron production rate and electron density, illustrating the formation of the F2 peak.

is caused by gravitational forces acting upon the ions which, by coulomb attraction, cause the electrons to diffuse downward. At altitudes well above the F2 peak where q and βn_e in Equation (5) can be considered negligible, the electron density distribution is given by the diffusive equilibrium equation:

$$\frac{d(\ln n_e)}{dz} = \frac{m_+g}{k(T_e + T_i)}, \tag{6}$$

where g is the acceleration of gravity, k is the Boltzmann constant, and T_e and T_i are the electron and ion temperatures. This most favored model of the formation of the F2 peak perhaps requires some modification near the geomagnetic equator where because of the existence of a horizontal magnetic field, diffusion cannot readily produce vertical movements.

Many empirical attempts have been made to solve the continuity equation, Equation (5), by making assumptions about the nature of the vertical movement, the production term, q, and the loss coefficient, β, and comparing the calculated results with experimental ones. Principally because of the numerous variables involved, these models vary considerably – from attempts to explain the F2 peak without diffusion

(SAGALYN and SMIDDY, 1963) to agreement with the most-favored theory discussed above. As an example of the latter case, CHANDRA (1963) finds that the attachment-loss and diffusion terms are of equal importance near the F2 peak by a computation involving n_e and q from Figure 8 and a loss coefficient behavior given by RATCLIFFE, SCHMERLING, SETTY and THOMAS (1956).

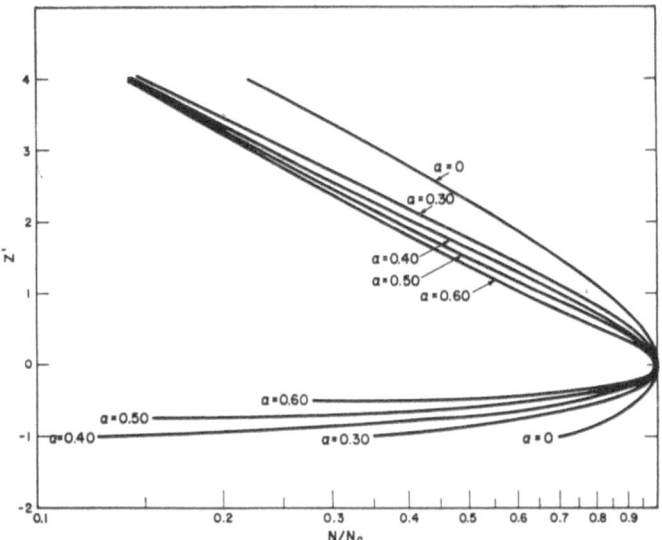

Fig. 9. Empirical model (CHANDRA, 1963) of the shape of the F2 peak.

Other empirical models have been developed to describe the electron density distribution of the F-region from a knowledge of the electron density and an arbitrarily-defined scale height both measured at the F2 peak. These models vary in complexity from the use of a simple CHAPMAN distribution with a constant scale height (WRIGHT, 1960) to a variable scale height (YONEZAWA and TAKAHASHI, 1960) and finally a variable scale height gradient (CHANDRA, 1963). The latter model, applicable to an isothermal region where O^+ dominates, provides good empirical agreement with all available mid-latitude profiles attempted at height ranges between 100 km below the peak to about 700 km. CHANDRA's empirical model is illustrated in Figure 9 where the abcissa (N/N_o) is the ratio of the electron density at a given altitude to its value at the F2 peak, and the ordinate z' is given by

$$z' = (h-h_m)/H_o, \tag{7}$$

where $(h-h_m)$ is the height measured from the F2 peak and H_o is the scale height corresponding to atomic oxygen. The parameter α is a measure of the departure from a simple CHAPMAN distribution and is given by

$$\alpha = (H_o-H_m)/H_o, \tag{8}$$

where H_m is the "scale height" at the F2 peak. CHANDRA finds that the "scale height" values computed from available electron density profiles correspond closely to neutral gas scale heights taken from the 1961 COSPAR International Reference Atmosphere.

7. The Helium Ion Layer and the Protonosphere

For a given value at a reference altitude, the upper ionosphere electron density which is given by Equation (6) will be governed by the mean ionic mass and the sum of the electron and ion temperatures. Charged-particle density profiles obtained at mid-latitudes under quiet solar conditions predominantly show that $T_e + T_i$ is practically constant at altitudes above 450 km. Consequently, changes in scale height of the electron-ion gas in the upper ionosphere predominantly reflect the transition from one ion species to a lighter one. It will be shown in this section that early concepts of an upper ionosphere characterized only by O^+ and H^+ ions and by a reasonably constant transition altitude now should be discarded in favor of a three-constituent medium having an intervening helium ion layer with a thickness which varies considerably with the atmospheric temperature.

Drag observations on the Echo satellite led NICOLET (1961) to first suggest that helium should be considered an important constituent of the upper ionosphere. The first experimental evidences for the existence of helium ions in the upper ionosphere were obtained independently by an indirect method (HANSON, 1962a) and by direct measurement (BOURDEAU, DONLEY, WHIPPLE and BAUER, 1962). HANSON, working from an ion density profile obtained by HALE (1961) concluded that the thickness of the helium ion layer is of the order of 2000 km, extending from 1200 to about 3400 km, and that the measured scale height for helium ions corresponds to an atmospheric temperature of 1600 °K. From the observed helium ion content (10^{12} ions/cm²) and an assumed rate of ionization, HANSON computed an equilibrium time for helium ions of about three days. He thus postulated that there should be no large diurnal variations

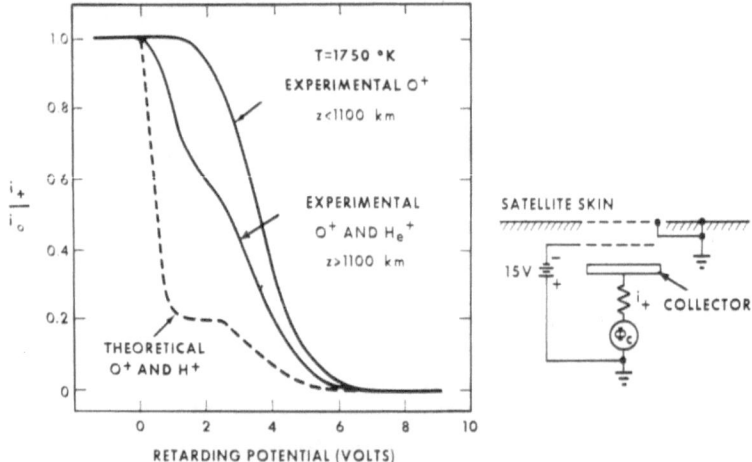

Fig. 10.   Direct detection of helium ions from the Explorer VIII retarding potential experiment.

in the helium ion layer. This conclusion has not been corroborated by other experimental observations and by the theoretical work of BAUER (1962, 1963).

The Explorer VIII results (BOURDEAU et al., 1962) relative to direct detection of helium ions are summarized in Figure 10. This retarding potential experiment differs only in geometry from the device used by USSR investigators (KRASOVSKIJ, 1959) on Sputnik III. The sensor consists of three electrodes arranged in planar concentric geometry. The inner grid is biased negatively so that the collector is responsive only to the flow of positive ions from the ionosphere. Because the satellite velocity largely exceeds the thermal velocity of the ions, the latter have a kinetic energy relative to the spacecraft in proportion to their mass. In accordance with an expression given by WHIPPLE (1959) the resulting volt-ampere curve, illustrated in Figure 10, will be characterized by an inflection point for a mixture of O^+ and He^+ and by a plateau for a mixture of O^+ and H^+. The abscissa of the volt-ampere curve is the total retarding potential given by the sum of the satellite and applied collector (ϕ_c) potentials while the ordinate is the ratio of the ion current at a given retarding potential to its value at zero potential. Planar ion traps have a good history for positive ion density determination but generally yield too high an ion temperature (HANSON and McKIBBIN, 1961; BOURDEAU et al., 1961), possibly due to an enhancement in the normal component of the particle energy caused by non-planar sheath geometry (HINTEREGGER, private communication). This factor is not considered in Whipple's theoretical relationship. However, because of the relative insensitivity of the shape of the volt-ampere curve to ion temperature, the ability to determine ion composition is not altered significantly. For daytime conditions, the Explorer VIII observations during the active life of the satellite (November–December, 1960) were that O^+ predominated below 1100 km and that He^+ was the principal ion from 1100 km to the maximum altitude (1600 km) of the measurements. Unfavorable vehicle orientation prevented valid measurements for nighttime conditions and at altitudes above 1600 km.

Subsequent to the reporting of the first experimental evidence for the importance of upper atmosphere helium ions, BAUER (1962) theoretically showed that the thickness of the layer would change with atmospheric temperature even for the simple case when the relative concentrations are invariant at a reference altitude. Assuming that the ionic species are distributed according to diffusive equilibrium (DUNGEY, 1955; MANGE, 1960), the geopotential altitude (h'_{ij}) at which two ionic species have equal concentrations has been given by BAUER (1962) as

$$h'_{ij} - h'_o = H_{ij} \ln n_{ij}, \tag{9}$$

where h'_o is the geopotential altitude at which the concentrations of these ions are in the ratio n_{ij}, and where

$$H_{ij} = kT/(m_i - m_j)g_o, \tag{10}$$

T being the atmospheric temperature and g_o the acceleration of gravity at the earth's surface.

BAUER (1963) has proposed a more detailed theoretical model for the thickness of the helium ion layer as a function of atmospheric temperature by considering the case where the relative concentrations of the three principal constituents vary at the reference altitude. He points out that the relative concentrations at a given altitude should vary with temperature in the same proportion as the ratio of the corresponding neutral constituents. Using the temperature dependent escape rate given for hydrogen (BATES and PATTERSON, 1961) and for both hydrogen and helium (NICOLET and KOCKARTS, 1962) and Equations (9) and (10), he computed two transition altitudes as a function of atmospheric temperature, a lower transition altitudes where O^+ and He^+ ions have equal concentrations and an upper transition altitude where He^+ and H^+

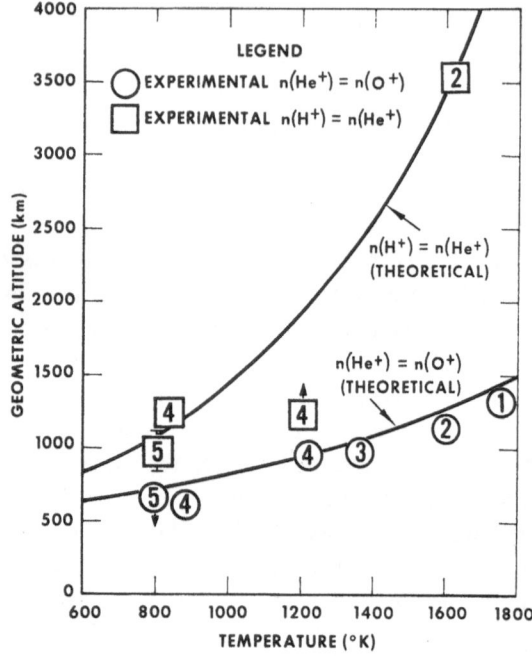

Fig. 11.   Comparison of experimental measurement with theoretically-obtained model (BAUER, 1963) of ion transition altitude as a function of atmospheric temperature.

are of the same density. These curves are shown to be in reasonably good agreement with experimental observations in Figure 11. The circles in Figure 11 refer to observed altitudes at which $n(He^+) = n(O^+)$ while the squares represent measured altitudes at which $n(He^+) = n(H^+)$. The corresponding temperatures were obtained on the same spacecraft and represent either direct measurements of T_e or T_i or values of $(T_e + T_i)/2$ computed from the observed electron-ion scale height. The abscissa, then, corresponds to the neutral gas temperature only if it is assumed that the electron, ion and neutral gas temperatures all are in equilibrium at the altitudes indicated.

Experimental points 1 and 2 are taken from BOURDEAU et al. (1962) and HANSON (1962a), respectively. Experimental points 4 represent the first preliminary results from the Ariel satellite (WILLMORE, BOYD and BOWEN, 1962). The Ariel investigators found that under daytime conditions helium ions predominated at altitudes between

950 km and to at least 1200 km, the satellite apogee. For a nighttime condition, it was found that the base of the helium ion layer was below 650 km and that hydrogen ions dominated at 1200 km.

The Ariel experiment is based on the same principle as that of the Sputnik III and Explorer VIII retarding potential experiments and yet, in the opinion of this author, represents a major advance in the development of ion spectrometers for upper iono-spheric studies. Because of the spherical geometry used, it should not be as sensitive to the direction of the ion trajectory and thus provides valid measurements of ion temperature in contrast to the planar ion traps. Additionally, the Ariel investigators set up the experiment so that the second derivative of the volt-ampere curve is tele-metered directly. This reduces the effects of slowly-varying vehicle potentials and other ion sheath parameters. These two basic differences make for a higher ion com-position resolution than is possible with the Sputnik III and Explorer VIII ion traps.

The Ariel satellite observations are extremely important in explaining the failure (ULWICK and PFISTER, 1962) to detect He^+ ions from scale-height changes in nighttime electron density profiles. As Bauer correctly points out, the thickness of the helium ion layer is smaller than a helium ion height at low temperatures characteristic of a nighttime profile. As a result, the detection by indirect methods requires an extremely accurate charged-particle density profile. The detection of this layer is expected to become even more difficult as we approach the year of minimum solar activity.

Experimental points 3 and 5 in Figure 11 were obtained by the indirect method from electron and ion density profiles obtained on Scout rockets. These data are presented in the succeeding section to illustrate the diurnal variation of electron density in the extreme upper ionosphere.

8. The Diurnal Variation of Electron Density in the Upper Ionosphere

The diurnal variation of electron density in the upper ionosphere is a complicated function of the electron density and altitude of the F2 peak, the atmospheric temperature, and the extremely variable heights of the two ion transition altitudes. This variability can be illustrated by comparison of two charged particle density profiles taken near the diurnal maximum and minimum during this portion of the solar cycle. That representative of diurnal maximum conditions is illustrated in Figure 12. The circled points are electron density values obtained by BAUER and JACKSON (1962) from FARADAY rotation observations at 72 Mc. The left-hand ordinate is geopotential altitude which takes into account the altitude dependence of the acceleration of gravity while the right-hand ordinate is the true or geometric altitude. HANSON (1962a) and BAUER (1962) have developed expressions based on an expansion of Equation (6) for the electron density distribution in a multi-constituent isothermal medium in diffusive equilibrium. These expressions suppose a given ratio of the con-centrations of the pertinent ions at a reference altitude. In Figure 12, two such theoretical curves (BAUER, 1962) are shown to illustrate the dependence of the electron density distribution on the ionic constituents. It can be seen that the circled points are more consistent with a theoretical curve for a mixture of O^+ and He^+ (solid line) than a mixture of O^+ and He^+ (dashed curve).

The charged particle density profile obtained near diurnal minimum conditions is illustrated in Figure 13. Both ion density (DONLEY, 1963) and electron density were measured. In the latter case, data were obtained only to a geopotential altitude of

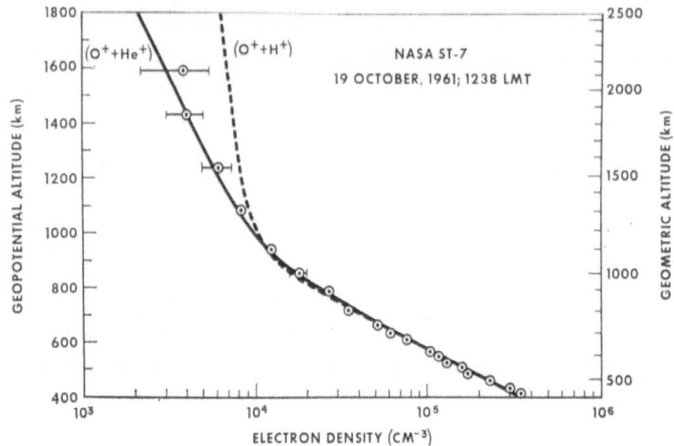

Fig. 12.   Indirect detection of helium ions from electron density profile obtained on a Scout rocket.

700 km. The good agreement of the experimental n_+ values with the electron density obtained by rocket measurement and with the value of n_e measured by an ionosonde at the F2 peak speaks well for the validity of the data.

It can be shown from the theoretical curve provided by Bauer (curve A in Figure 13) that the experimental ion density profile corresponds most closely to a three-

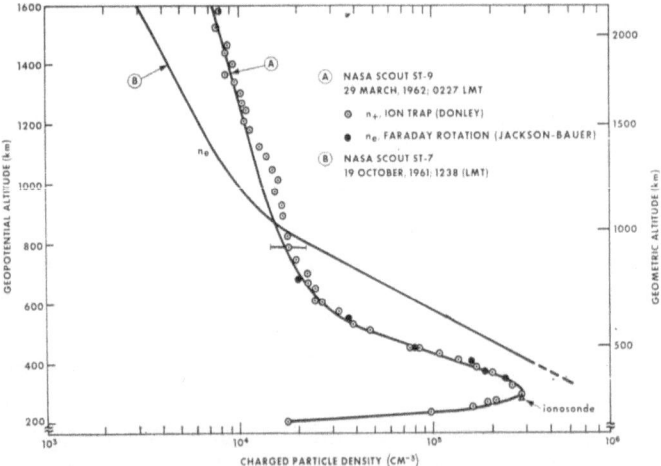

Fig. 13.   Scout rocket measurements of the diurnal variation of the upper ionosphere.

constituent mixture rather than O^+ and H^+. The closest fit of the experimental points to the theoretical curve in the ion transition altitude leads to an estimate of less than 300 km for the thickness of the helium ion layer. The discontinuity in the experimental n_+ values at geopotential altitudes between 900 and 1100 km probably can be explained from the observation that the spacecraft went from a dark to a sunlit condition

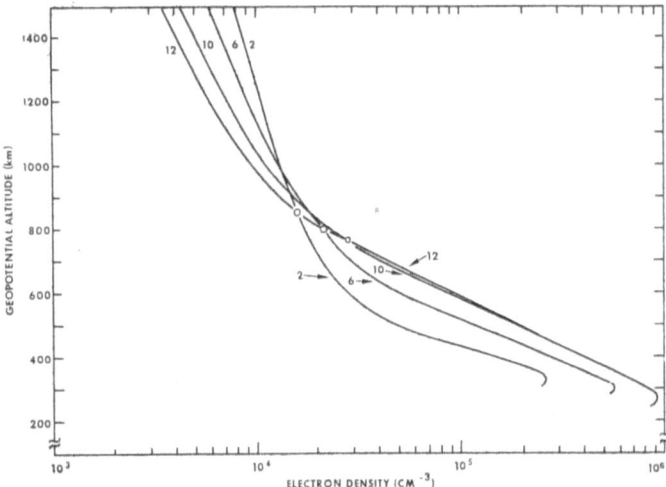

Fig. 14. Qualitative model of the detailed diurnal variation of the upper ionosphere.

in this region. The discontinuity probably represents a transient departure from temperature equilibrium.

Curve B in Figure 13 is the electron density profile replotted from Figure 12 to illustrate an observed diurnal variation of charged particle density in the upper ionosphere. The observation that the nighttime electron densities actually exceed those for daytime conditions at geometric altitudes greater than 1000 km should not be surprising since the lower ion transition altitude and the large scale height for protons overcompensate for the higher daytime atmospheric temperature and the higher daytime electron density at the F2 peak.

SAYERS, ROTHWELL and WAGER (1962) claim to have discovered a new ionization ledge with peak densities located at geometric altitudes varying from 1200 km at midnight down to 600 km at midday. The apparent negative scale heights gradients which were reported carry an implication of an ionization source in this altitude region. Discontinuities in charged particle density profiles perhaps can be expected at sunrise and at the equatorial anomaly, yet the discovery of this ledge is claimed throughout the day and for all latitudes traversed by the Ariel satellite. SAYERS et al. constructed their profiles from rf probe measurements of electron density at the satellite position. The nature of the Ariel orbit during the time the data were obtained was such that the altitude profiles are not time independent but rather represent observations taken over a change of several hours in local mean time. An alternative explanation, then, is that these ledges are only apparent in that they do not represent altitude discontinuities but rather changes in ionization brought about by the difference in local time at various points in the Ariel orbit. This alternative explanation is illustrated qualitatively in Figure 14. The experimental curves in Figure 13 already illustrate the diurnal variation of the electron density in the upper ionosphere. The four curves shown in Figure 14 are inferred profiles for the hours indicated. They were constructed from a typical variation of the electron density at the F2 peak as observed for May (the month of the Ariel observations) by ionosondes, an atmospheric temperature variation taken from reference atmospheres, and the ion transition model of Bauer illustrated in Figure 11. It can be shown for a satellite which requires several hours of local time to traverse the 400–1200 km altitude region, that an apparent ionization ledge would be observed when the satellite passes through the higher altitude region at night and conversely when it passes through the lower altitude region during midday. Parenthetically, an interesting feature to note from Figure 14 is the approximate constancy with diurnal time of the electron density at a geopotential altitude of about 800 km (approximately 1000 km geometric).

9. The Altitude Dependence of Charged Particle Temperature

It is of considerable importance to compare electron (T_e), ion (T_i) and neutral gas (T_g) temperatures as a function of altitude because their interdependence perhaps is the most sensitive index of complex reactions, the understanding of which is the ultimate goal of upper atmosphere physicists. For the time being, it will be assumed that T_i and T_g are equal.

Because direct and indirect measurements of charged particle temperatures have been made under radically different conditions and because of the limitations of the kinetic gas temperature models, various investigators have provided conflicting answers to the important question of temperature equilibrium between electrons and heavy constituents under daytime conditions. It generally is accepted that temperature equilibrium predominates at all altitudes during the night under quiet solar conditions. In this section it will be shown for a restricted set of latitude and atmospheric conditions that during midday temperature equilibrium exists below 150 and above 450 km, in accordance with recent review (BAUER and BOURDEAU, 1962; BOURDEAU and BAUER, 1962). It will be shown, on the other hand, that there is a substantial difference between spaceflight results and those from some ground-based radar incoherent backscatter experiments.

Before proceeding, it is important to define the term "temperature equilibrium". Actually because in the ionization process the electrons are created with high initial energies, their temperature will always be somewhat higher and only can approach T_g asymptotically in time depending on the efficiency of the energy transfer mechanism. Temperature equilibrium in this paper is defined as a difference between T_e and T_g which is smaller than the uncertainties in reference atmospheres and in spaceflight experimental methods of measuring charged particle temperatures. It is estimated, perhaps optimistically, that these uncertainties together are about ten percent of the absolute value of T_g.

Our knowledge of the behavior of the neutral gas temperature is based principally on atmospheric densities computed from satellite drag observations and an assumed atmospheric composition. Values of T_g so derived must be considered approximate

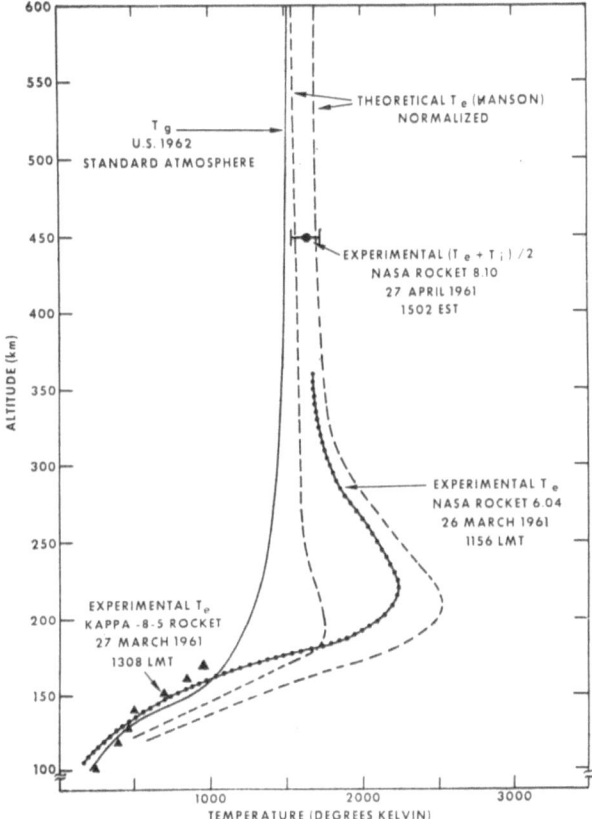

Fig. 15.   Comparison of the altitude dependence of neutral gas and experimental and theoretical electron temperatures at mid-latitudes under quiet solar conditions.

principally because of the observational gap in atmospheric composition. Plotted as a solid line in Figure 15 is the altitude dependence of the neutral gas temperature published by the U.S. Committee on Extension to the Standard Atmosphere (SISSENWINE, 1962). The COESA reference atmosphere was selected because it depicts "typical mid-latitude year-round conditions averaged for daylight hours and for the range of solar activity that occurs between sunspot minimum and sunspot maximum", conditions which most closely approximate those under which the charged particle temperature measurements have been made.

The earliest theoretical study of the temperature difference between T_e and T_g was advanced by HANSON and JOHNSON (1961). Because of improved cross-sectional and solar radiation data and because of the omission of an important process (electron impact excitation of vibrational levels of molecular nitrogen), this work now must be considered superseded by more recent theoretical models (HANSON, 1962b; DAL-GARNO, MCELROY and MOFFETT, 1962). These models, which consider ionization by solar radiation only, involve several expansions (depending upon the energy transfer mechanisms) of the equation:

$$(T_e - T_g)/\tau = \frac{2q'E}{3n_e k} \tag{11}$$

where q' is the rate at which photoelectrons of energy E are released and τ is the time which it takes for these electrons to transfer their excess energy in various processes. Considerations are given to loss of this excess energy by (a) inelastic ionizing collisions with neutral particles, (b) inelastic exciting collisions with neutral particles where the energy is stored either in metastable or vibrational states, (c) energy transfer to ambient ions and (d) elastic collisions with ambient electrons. In all but the latter case, much of the excess energy of the newly-created photoelectrons is removed before they selectively heat the ambient electron gas.

HANSON (1962b) computed four models for the case where the daytime temperature in the isothermal region is 1200 °K. These four models are based on two different altitude-dependent values for the electron production rate and two different assumptions for the excess photoelectron energy. The two extremes of these four curves are plotted in Figure 15. It is seen that the maximum predicted departure from temperature equilibrium proposed by Hanson corresponds closely to the altitude of maximum electron production illustrated in Figure 8. Possible small errors have been introduced in Figure 15 by assuming that the value of $T_e - T_g$ actually computed by Hanson are invariant when T_g in the isothermal region is raised from 1200 °K as used in his reference atmosphere and the 1500 °K value used here.

At present, an experimental determination of the complete altitude dependence of $T_e - T_g$ can be accomplished only empirically by comparing Langmuir probe measurements of electron temperature obtained on vertical sounding rockets with a reference atmosphere. Langmuir probes have required considerable development in order to overcome problems associated with the disturbance introduced into the medium by a conducting body. Extremely high electron temperatures were reported from observations of large potentials on the Sputnik III spacecraft (KRASOVSKIJ, 1959). However, it is dangerous to infer electron temperatures by this indirect method because of the probability of a vehicle potential which is artificially controlled by differing surface work functions, for example, than by ambient ionospheric parameters.

It wasn't until 1961 that electron temperatures close to accepted kinetic gas values were first reported for the E-region by Japanese investigators (AONO et al., 1961) and for the upper ionosphere by use of the Explorer VIII Satellite (SERBU, BOURDEAU and DONLEY, 1961). However, these data by themselves do not provide a complete altitude profile of T_e. The Japanese data, which represent an average from two separate devices flown simultaneously, were obtained over the altitude interval between 100 and 170 km and are plotted in Figure 15. A separate rocket (NASA 6.04) containing an ejectable symmetric bipolar probe was launched at approximately the same latitude and time. These results (BRACE, 1962) are plotted as the dotted curve in Figure 15. Parenthetically, they represent a significant modification by Brace of the data which was first reported in a preliminary form (SPENCER, BRACE and CARIGNAN, 1962). The error spread of the overall curve has been estimated by BRACE to be ± 10 percent. Also plotted in Figure 15 is the atmospheric temperature computed on the assumption of temperature equilibrium by JACKSON and BAUER (1961) from their experimental electron density profile. This value, applicable to the region above 450 km, was selected because it was taken under identical solar conditions (indexed by observed solar decimeter flux observations) as the two electron temperature profiles. Considering the uncertainty in the reference atmosphere and the experimental error flag, it can be concluded *for these particular solar conditions* that temperature equilibrium exists below 150 km and above 450 km. Also there is reasonable agreement between the experimental results and the theoretical work of Hanson.

DALGARNO *et al.* (1962) have proposed four theoretical electron temperature models for a given noontime value of T_g. The end result in the form of the altitude dependence of $T_e - T_g$ for $T_g = 2000\,^\circ$K is presented in Figure 16. The alternative curves are based on some reduction of excess photoelectron energy by (a) ionizing collisions alone, (b) including possible energy transfer through exciting collisions leading to the metastable state, (c) excluding (b) but including possible energy transfer through exciting collisions leading to the vibrational state and (d) a combination of all energy transfer mechanisms. Also plotted in Figure 16 are experimental values for $T_e - T_g$ taken from Figure 15. It is tempting from the agreement of the experimental data with curves *a* and *b* in Figure 16 to suggest that excess photoelectron energy in the lower altitude regions is removed principally by a combination of ionizing collisions and exciting collisions leading to the metastable state.

HANSON (1962b) has advanced a theoretical possibility based on low thermal contact between the ions and neutral particles that the ion temperature at very high altitudes could depart from the neutral gas temperature and approach the electron temperature. He estimates that such a transition could possibly occur in the 600–1000 km region. It is extremely difficult to corroborate this possibility from experimental observations because the differences between T_e, T_i and T_g are small at these altitudes. The daytime electron density profile illustrated in Figure 12 shows that the sum of the electron and ion temperatures taken in both the He$^+$ and O$^+$ regions are approximately the same. The constancy of this sum with altitude also can be inferred in published data (KING, 1963) from the Alouette Topside Sounder Satellite. This is evidence either that the differences in the electron, ion and neutral gas temperatures are vanishingly small above 450 km or that the ion temperature transition proposed by Hanson occurs above 1000 km, if at all.

The agreement of Langmuir probe results with the theoretical work of Hanson and DALGARNO *et al.* (1962) suggests that, at midday, temperature equilibrium exists below 150 km and above 450 km whenever ionization is the result of solar radiation alone. *However, because of the small number of observations, generalization of this conclusion may be premature.* SPENCER, BRACE and CARIGNAN (1962) have made rocket flights at auroral latitudes and into a disturbed mid-latitude ionosphere. For these conditions they report E and F2 region electron temperatures which are systematically higher than the quiet mid-latitude results. The difference is perhaps rep-

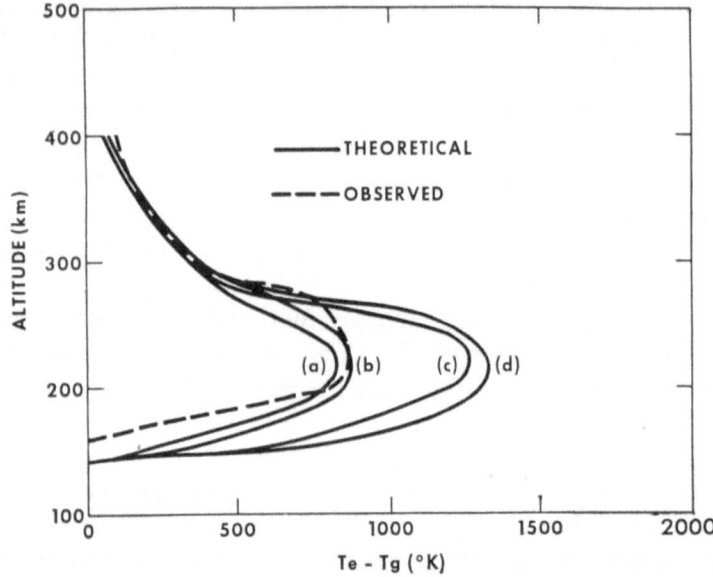

Fig. 16. Theoretical (DALGARNO, 1962) and experimental (BRACE, 1962) models of the altitude dependence of the difference between electron and neutral gas temperatures at mid-latitudes under quiet solar conditions.

resentative of the influence of corpuscular radiation. There is also a small possibility that the electron temperatures reported for disturbed conditions by SPENCER *et al.* (1962) are not representative of the most abundant electrons. Their symmetric Langmuir probe samples only those electrons which can overcome a retarding potential of a few tenths of a volt. Since a Maxwellian electron energy distribution has been established for quiet solar conditions, the data reported by Brace and plotted in Figure 15 and 16 should be representative of the entire electron population. However, the existence of a Maxwellian energy distribution under disturbed ionospheric conditions has not been verified.

The spaceflight observations in a quiet daytime ionosphere are consistent with the conclusions (BOWLES, OCHS and GREEN, 1962) from ground-based radar incoherent backscatter experiments that the electron and ion temperatures are in equilibrium in the upper ionosphere except at sunrise and during disturbed conditions. EVANS (1962), on the other hand, also using backscatter experiments reports different results. The principal differences for which no explanation can be given here are: (a) Evans reports temperature equilibrium at 200 km, an altitude at which both theory and the spaceflight observations show a significant departure from equilibrium; (b) the ratio T_e/T_i is 1.6 and constant above 300 km as reported by EVANS whereas both theory and the spaceflight observations show this ratio decreasing with altitude above 250 km; (c) EVANS reports that the sum of T_e and T_i increases with altitude up to 700 km whereas it is concluded from many rocket measurements of charged particle density profiles that this sum is constant with altitude in the upper ionosphere. The results obtained by EVANS also are inconsistent with the diurnal variation of charged particle temperatures at altitudes above 450 km, which is discussed in the succeeding section.

10. Diurnal Variation of Upper Ionosphere Temperatures

In this section, a comparison will be made of the diurnal variations of charged particle and neutral gas temperatures for the isothermal altitude region above 450 km. At these altitudes our knowledge of T_g generally is derived from satellite drag observations and an assumed atmospheric composition. The drag observations show that density variations are correlated with solar activity. Although not the source of upper atmosphere heating, solar decimeter radiation which is observable at the earth's surface is an indicator of this interrelationship. For a given level of solar activity JACCHIA (1962), HARRIS and PRIESTER (1962) and PAETZOLD (1962) conclude:

(1) The neutral gas temperature has its maximum value in mid-afternoon;

(2) This maximum would occur at sunset if absorption of solar ultraviolet radiation was the only source of atmosphere heating;

(3) Consequently, there is a second source of heating associated with the solar wind;

(4) These conclusions apply to the atmosphere at low latitudes.

In the preceding section it was shown for a specific set of observations taken at mid-latitudes that the electron and neutral gas temperatures have approximately the same value above 450 km, which is consistent with theoretical computations based on solar ultraviolet radiation as the sole ionizing source. If the low latitude ionization generally can be explained entirely by solar ultraviolet radiation and if, as claimed, the diurnal variation characteristic of the neutral atmosphere requires a second heat source, then the additional heating mechanism should be one which is least likely to selectively heat either the light or heavy atmospheric constituents. The heating by hydromagnetic waves proposed by DESSLER (1959) probably meets these criteria. It is not possible from the charged particle temperature data presented in this section to infer a diurnal neutral gas temperature variation precisely enough to confirm or deny the hypothesis of a second heat source which is effective at mid-latitudes.

There is some variance in reference atmospheres in the absolute magnitude of the temperature at diurnal maximum and minimum. The model attributed to PRIESTER by JASTROW (1961) which is plotted in Figure 17 has an average daytime value consistent with the COESA standard atmosphere. It corresponds to a level of solar activity for which the observed solar decimeter (10.7 cm) radiation flux is 150 $W/m^2/cps$. Other models (HARRIS and PRIESTER, 1962; JACCHIA, 1962) generally show somewhat lower temperatures at diurnal maximum. Also plotted in Figure 17 are experimental values all taken at mid-latitudes of (*a*) the neutral gas temperature measured directly by BLAMONT (1961, 1962) using rockets containing sodium vapor release experiments, (*b*) direct measurements of electron temperature from the Explorer VIII Langmuir probe (SERBU *et al.*, 1961) and (*c*) values of T_g computed from rocket measurements of electron scale height on the assumption of temperature equilibrium, for which $T_g = \frac{1}{2}(T_e + T_i)$. These data were obtained at various times and consequently have

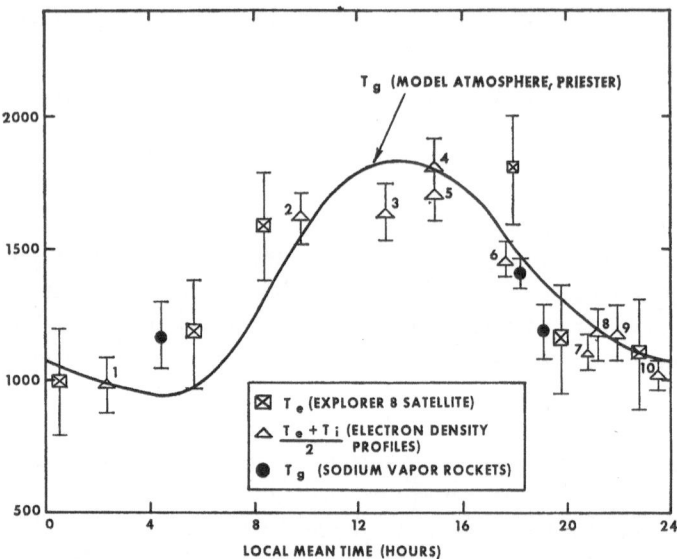

Fig. 17.   Diurnal variation of upper atmosphere temperature for altitudes above 450 km at mid-latitudes, solar decimeter flux = 150×10^{-22} W/m²/cps.

been normalized by this author to the same solar activity (10.7 cm flux = 150 W/m²/ cps) using the empirical relation given by HARRIS and PRIESTER (1962). The temperatures derived from electron or ion profiles are taken (1) from DONLEY (1963), (2), (8) and (9) from BERNING (private communication), (3) from BAUER and JACKSON (1962), (4) from JACKSON and BAUER (1961), (5) from HANSON (1962), (6) and (10) from KNECHT and RUSSELL (1962), and (7) from HANSON and MCKIBBIN (1961).

It is tempting from the general agreement of the neutral gas temperatures with the Explorer VIII electron temperatures and with the values of $\frac{1}{2}(T_e + T_i)$ derived from charged particle density profiles, all illustrated in Figure 17, to infer support for the conclusion reached in the previous section that temperature equilibrium exists above 450 km for most of the day under quiet ionospheric conditions at mid-latitudes. However, the empirical method of comparison used in Figure 17 can only be an approximate test of equilibrium, because except for BLAMONT's data, the neutral gas temperature is an inferred parameter and because the preciseness of normalizing data taken at different times to the same level of solar activity has not been established.

The Explorer VIII electron temperature device is a gridded trap which experimentally eliminates unwanted photocurrent effects. The error flags shown in Figure 17 result from a limited volt-ampere curve resolution imposed by the telemetry system. There possibly is an additional small error associated with the electrical transparency characteristics of the grids used to remove the photocurrent effect (SERBU et al., 1961). The absence of an on-board tape recorder prevented detailed analyses of the diurnal electron temperature variation and the detection of latitude effects.

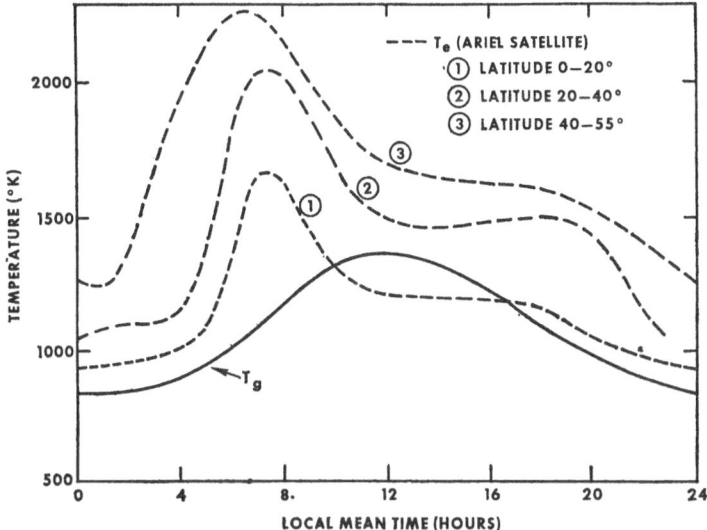

Fig. 18.   Possible latitude dependence of the diurnal variation of electron temperature for altitudes above 450 km, solar decimeter flux = 100 × 10⁻²² W/m²/cps.

A different type of Langmuir probe now has been flown on the Ariel satellite. Basically an asymmetric probe with the sensor located on a boom away from the main spacecraft, the experiment tends to suppress such unwanted effects as varying vehicle potentials and photocurrent and magnetic field modulation of the electron current by measuring the electron temperature in an extremely short time interval. The availability of an on-board tape recorder has permitted the investigators to report the most detailed diurnal electron temperature variation yet obtained and additionally to separate the data according to latitude. Their preliminary results as reported (WILLMORE, BOYD and BOWEN, 1962) are shown in Figure 18. The three curves, each representing a different latitude region were obtained at a time when the 10.7 cm solar radiation flux averaged 100 W/m²/cps. The Priester model of the diurnal variation of the neutral gas temperature in Figure 17 has been normalized to this lower level of solar activity and replotted in Figure 18 for comparison.

The Ariel data were analyzed on the assumptions that the temperature above 450 km is isothermal and that there is a symmetrical distribution about the equator. WILLMORE et al. (1962) point out that the salient features are a general rise in temperature and a less pronounced diurnal variation at high latitudes and that both of these observations are consistent with significant atmospheric heating by particles being dumped at high latitudes.

Another salient feature of the Ariel data is the high electron temperature observed in the sunrise period. This is not inconsistent with the experimental results in Figure 17 because of the observational gap in the latter data during this period. The sunrise effect in the Ariel data is evidence for a departure from temperature equilibrium. The diurnal shapes of the curves suggest that temperature equilibrium is re-established toward midday. The observation of a departure from temperature equilibrium at sunrise is consistent with the conclusions from one set of ground-based backscatter results (BOWLES, 1962) but not with EVANS (1962) who reports maximum departure from equilibrium at noon. A higher ratio T_e/T_i at sunrise than at noon is more consistent with what would be expected from a cursory extension to the theoretical observation (HANSON, 1962) that T_e/T_i is directly proportional to q/n_e^2 above the F2 peak. The electron production rate, q, is in turn proportional to atmospheric density. It can be shown from a comparison of the atmospheric density variation derived from satellite drag with the electron density profiles illustrated in Figure 14 that the ratio q/n_e^2 and thus that T_e/T_i should be higher at dawn than at noon.

It would be premature to draw firm conclusions regarding additional ionizing and heating sources from the diurnal comparisons of upper ionosphere electron and neutral gas temperatures shown in Figures 17 and 18. The experimental charged particle observations in Figure 17, for example, were obtained over a period of more than one year. The accuracy of the process of normalizing these data to the same level of solar activity quite likely introduces enough error that the diurnal shape of the indirectly-derived neutral gas temperature cannot be confirmed. Consequently, the hypothesis of a second heat source at low latitudes also cannot be confirmed.

The midday values of T_g proposed by PRIESTER and which is applicable to low latitudes fall in between the two lower latitude electron temperature curves measured on the Ariel satellite. No conclusion can be drawn from the different diurnal shapes of T_e and T_g in Figure 18, principally because two refinements in the analysis of the Ariel data are required. Firstly, there is no evidence that the Ariel results have been normalized for the day-to-day fluctuations in solar activity which can amount to a ten percent correction. This normalization could alter both the diurnal shape of the individual electron temperature curves and the amplitude of the latitude effect. Secondly, it would be expected again from a cursory theoretical analysis that the electron temperature in the upper ionosphere during the sunrise period would be altitude dependent. Since the present analysis of the Ariel data assumes no altitude effect, this also could alter the diurnal shape of the individual electron temperature curves. With reference to the diurnal variation of T_g, the Ariel data suggest a latitude effect which up until now has not been seriously considered in neutral gas models. This latitude effect in turn suggests that other heat and ionization sources in addition to direct solar radiation need to be considered in a description of even the normal auroral ionosphere.

11. Summary

Spaceflight observations made in recent years are most consistent with the following general conclusions:

(1) The principal ionizing agents for the D-region under quiet solar conditions at an epoch corresponding to the middle of the solar cycle are cosmic radiation for the region below 70 km and Lyman-α radiation at altitudes between 70 and 85 km. The negative ion abundance at all D-region altitudes has been underestimated theoretically.

(2) X-radiation (2–8 Å) and energetic protons, which are respectively responsible for Sudden Ionospheric Disturbances and Polar Cap Absorption Events, lead to a D-region electron density enhancement of about two orders of magnitude.

(3) The principal ions produced in the lower ionosphere are N_2^+, O_2^+ and O^+. The ions which exist in greatest abundance are NO^+, O_2^+ and O^+. These ionization characteristics result from the combined influence of solar ultraviolet and X-radiation in the E-region with ultraviolet radiation becoming the dominant source in the F-region.

(4) The average E-region electron density decreases by about two orders of magnitude at night. Some types of sporadic-E ionization have a depth of less than one kilometer and a horizontal dimension greater than 70 km. Other types have a smaller horizontal dimension. There is a first order increase of the electron density in a sporadic E-layer.

(5) The formation of the F2 peak is best explained by the combined altitude dependence of the electron production rate, electron loss by an attachment-like process and diffusion.

(6) The constant electron-ion scale height which is observed for altitude intervals of a few hundred kilometers above the F2 peak is evidence that the upper ionosphere is isothermal and in diffusive equilibrium.

(7) The upper ionosphere is characterized by the presence of three ionic constituents (O^+, He^+, H^+) each predominating in a different altitude region. The thickness of the helium ion layer has a large diurnal variability. The control which these ionic constituents exercise over the electron density distribution is such that the electron density is generally higher at midday below 1000 km, shows a small diurnal variation near 1000 km, and is generally higher at altitudes between 1000 and 2000 km at night.

(8) In an undisturbed mid-latitude ionosphere, it is generally accepted that temperature equilibrium can be expected at night. Significant departures from equilibrium can be expected at sunrise and during disturbed conditions. Toward midday under quiet mid-latitude conditions significant departures from equilibrium would be expected in the lower F-region, but small differences between T_e, T_i and T_g are observed below 150 and above 450 km.

(9) Except for the sunrise anomaly the diurnal maximum and minimum values of electron temperature at mid-latitudes approximate the diurnal extremes of the neutral gas temperature in the upper ionosphere. An observed latitude dependence of electron

temperature suggest that corpuscular radiation is an important heat source at high latitudes.

12. Acknowledgements

I wish to acknowledge the helpful discussions with A. C. AIKIN, S. J. BAUER, L. H. BRACE, S. CHANDRA, L. G. SMITH and E. C. WHIPPLE. I also wish to thank L. H. BRACE, J. L. DONLEY, J. E. JACKSON, S. J. BAUER, E. SCHAEFFER and especially the investigators associated with the Ariel satellite for making their results available in advance of publication.

References

AIKIN, A. C.: 1962a, *International Symposium on Equatorial Aeronomy*, Lima, Peru, September 25.
AIKIN, A. C.: 1962b, *Goddard Space Flight Center Report* X-215-62-132.
AONO, Y., HIRAO, K., and MIYAZAKI, S.: 1961, *J. Rad. Res. Lab.*, Japan **8**, 453.
BATES, D. R.: 1956, in: *Solar Eclipses and the Ionosphere*, Beynon and Brown, editors, Pergamon Press, p. 184.
BATES, D. R. and PATTERSON, T. N. L.: 1961, *Plan. Space Sci.* **5**, 257.
BAUER, S. J. and JACKSON, J. E.: 1962, *J. Geophys. Res.* **67**, 1675.
BAUER, S. J.: 1962, *J. Atm. Sci.* **19**, 276.
BAUER, S. J.: 1963, *Nature* **197**, 36.
BAUER, S. J. and BOURDEAU, R. E.: 1962, *J. Atm. Sci.* **19**, 218.
BELROSE, J. S. and BURKE, M. J.: 1961, *URSI meeting*, Washington, D.C., May 4.
BERNING, W. W.: 1960, *J. Geophys. Res.* **65**, 2589.
BLAMONT, J., 1961, in: *Space Research II* (ed. by H. C. VAN DE HULST *et al.*), North-Holland Publ. Comp., p. 974.
BOURDEAU, R. E., WHIPPLE, E. C., and CLARK, J. F.: 1959, *J. Geophys. Res.* **64**, 1363.
BOURDEAU, R. E.: 1961, in: *Space Research II* (ed. by H. C. VAN DE HULST *et al.*) North-Holland Publ. Comp., p. 554.
BOURDEAU, R. E., DONLEY, J. L., SERBU, G. P., and WHIPPLE, E. C. 1961, *Journ. Astronaut. Sci.* **8**, 65.
BOURDEAU, R. E. and BAUER, S. J.: 1962, *Third Intern. Space Sciences Symposium*, Washington, D.C., May.
BOURDEAU, R. E., DONLEY, J. L., WHIPPLE, E. C., and BAUER, S. J.: 1962, *J. Geophys. Res.* **67**, 467.
BOWLES, K. L., OCHS, G. R., and GREEN, J. L.: 1962, *J. Research, National Bureau of Standards* **66**, 395.
BRACE, L. H.: 1962, *University of Michigan Scientific Report*. JS 3, 03599-11-F.
CHANDRA, S.: 1963, *J. Geophys. Res.* **68**, 1937.
CHUBB, T. A., FRIEDMAN, H., and KREPLIN, R. W.: 1960 *J. Geophys. Res.* **65**, 1831.
DALGARNO, A., MCELROY, M. B., and MOFFETT, R. J.: 1962, *Geophys. Corp. of America, Tech. Report* 62-11-N.
DESSLER, A. J.: 1959, *J. Geophys. Res.* **64**, 397.
DONAHUE, T. M.: 1962, *Space Sci. Rev.* **1**, 135.
DONLEY, J. L.: 1963, *J. Geophys. Res.* **68**, 2058.
DUNGEY, J. W.: 1955, *The Physics of the Ionosphere*, London, Physical Society, p. 229.
EVANS, J. V.: 1962, *J. Geophys. Res.* **67**, 4914.
FAIRE, A. C. and CHAMPION, K. S.: 1959, *Phys. Rev.* **113**, 1.
FRIEDMAN, H.: 1959, *Proc. IRE* **47**, 272.
FRIEDMAN, H.: 1962, *Astronautics* **14**, August.
GARDNER, F. F. and PAWSEY, J. L.: 1953, *J. Atm. Terr. Phys.* **3**, 231.
HALE, L. C.: 1961, *J. Geophys. Res.* **66**, 1554.
HANSON, W. B. and JOHNSON, F. S.: 1961, *Les Congrès et Colloques de L'Université de Liège* **20**, 390.
HANSON, W. B. and MCKIBBIN, D. D.: 1961, *J. Geophys. Res.* **66**, 1667.
HANSON, W. B.: 1962a, *J. Geophys. Res.* **67**, 183.
HANSON, W. B.: 1962b, *Third Intern. Space Sciences Symp.*, Washington, D.C., May.
HARRIS, I. and PRIESTER, W.: 1962, *J. Geophys. Res.* **67**, 4585.

HINTEREGGER, H. E., DAMON, K. R., and HALL, L. H.: 1959, *J. Geophys. Res.* **64**, 961.
HINTEREGGER, H. E. and WATANABE, K.: 1962, *J. Geophys. Res.* **67**, 3373.
HULBURT, E. C.: 1938, *Phys. Rev.* **53**, 344.
ICHIMIJA, T., TAKAYAMA, K., and AONO, Y.: 1959, *J. Radio Res. Lab.*, Japan **13**, 155.
ISTOMIN, V. G.: 1960, *Artificial Earth Satellites* **2**.
JACCHIA, L.: 1962, *Third Intern. Space Sciences Symp.*, Washington, D.C., May.
JACKSON, J. E., and KANE, J. A.: 1959, *J. Geophys. Res.* **64**, 1074.
JACKSON, J. E., and BAUER, S. J.: 1961, *J. Geophys. Res.* **66**, 3055.
JASTROW, R.: 1961, *25th Wright Brothers Lecture*, Washington, D.C., December.
JOHNSON, C. Y., MEADOWS, E. B., and HOLMES, J. C.: 1958, *J. Geophys. Res.* **63**, 443.
KANE, J. A.: 1959, *J. Geophys. Res.* **64**, 133.
KANE, J. A.: 1960, *AGARD meeting*, Athens, Greece, March.
KASNER, W. H., ROGERS, W. A., and BIONDI, M. A.: 1961, *Phys. Rev. Letters* **7**, 321.
KING, J., 1963, *Nature* **197**, 639.
KNECHT, R. W. and RUSSELL, S.: 1962, *J. Geophys. Res.* **67**, 1178.
KRASOVSKIJ, V. I.: 1959, *Proc. IRE N.Y.* **41**, 289.
KRASNUSKIN, P. E. and KOLESNIKOV, N. L.: 1962, *Dokl. Akad. Nauk. SSSR* **146**, 596.
KREPLIN, R. W., CHUBB, T. A., and FRIEDMAN, H.: 1962, *J. Geophys. Res.* **67**, 2231.
LINDSAY, J. L.: 1962, *Third International Symposium on Space Phenomena*, Detroit, Michigan, October.
LOEB, L. B.: 1955, *Basic Processes of Gaseous Electronics*, University of California Press, p. 42.
MAEHLUM, B. and O'BRIEN, B. J.: 1962, *J. Geophys. Res.* **67**, 3281.
MANGE, P.: 1961, *J. Geophys. Res.* **66**, 2263.
MOLER, W. F.: 1960, *J. Geophys. Res.* **65**, 1459.
NICOLET, M.: 1945, *Mem. Inst. Roy. Meteorol.*, Belgium **19**, 124.
NICOLET, M. and AIKIN, A. C.: 1960, *J. Geophys. Res.* **65**, 1469.
NICOLET, M.: 1961, *J. Geophys. Res.* **66**, 2263.
NICOLET, M. and KOCKARTS, G.: 1962, *Third Intern. Space Sciences Symposium*, Washington, D.C.,
    May.
PAETZOLD, H. K.: 1962, *J. Geophys. Res.* **67**, 2741.
POPPOFF, I. G. and WHITTEN, R. C.: 1962, *J. Geophys. Res.* **67**, 2986.
POUNDS, K. A. and WILLMORE, A. P.: 1962, *International Conference on the Ionosphere*, London, July.
RATCLIFFE, J. A., SCHMERLING, E. R., and SETTY, C. S.: 1956, *Phil. Trans. Roy. Soc.* **248**, 621.
RATCLIFFE, J. A. and WEEKES, K.: 1960, *Physics of the Upper Atmosphere* (ed. by J. A. RATCLIFFE)
    Academic Press.
SAYERS, J., ROTHWELL, P., and WAGER, J. H.: 1962, *Nature* **4847**, 1143.
SAGALYN, R. C. and SMIDDY, M.: 1963, *J. Geophys. Res.* **68**, 199.
SCHAEFFER, E. J.: 1963, *J. Geophys. Res.* **68**, 1175.
SERBU, G. P., BOURDEAU, R. E., and DONLEY, J. L.: 1961, *J. Geophys. Res.* **66**, 4313.
SEDDON, J. C. and JACKSON, J. E.: 1958, *Ann. Geophys.* **14**, 456.
SISSENWINE, N.: 1962, *Astronautics*, August, p. 52.
SMITH, L. G.: 1961a, *AGU meeting*, Washington, D.C., April.
SMITH, L. G.: 1961b, *AGU meeting*, Los Angeles, California, December 27.
SPENCER, N. W., BRACE, L. H., and CARIGNAN, G. R.: 1962, *J. Geophys. Res.* **67**, 157.
TAYLOR, H. A. and BRINTON, H. C.: 1961, *J. Geophys. Res.* **66**, 2587.
ULWICK, J. C. and PFISTER, W.: 1962, *Third Intern. Space Science Symposium*, Washington, D.C., May.
VAN ZANDT, T. E. and BOWLES, K. L.: 1960, *J. Geophys. Res.* **65**, 2627.
VEGARD, L.: 1938, *Geophys. Publ.* **5**.
WATANABE, K., MARMO, F. F., and PRESSMAN, J.: 1955, *J. Geophys. Res.* **60**.
WATANABE, K. and HINTEREGGER, H. E.: 1962, *J. Geophys. Res.* **67**, 999.
WHIPPLE, E. C.: 1959, *Proc. IRE.* **47**, 2023.
WHIPPLE, E. C.: 1960, *Proc. Intern. Astronaut. Congress*, Stockholm, p. 99.
WHITEHEAD, J. D.: 1960, *Nature* **188**, 567.
WHITTEN, R. C. and POPPOFF, I. G.: 1961, *J. Geophys. Res.* **67**, 2986.
WILLMORE, A. C., BOYD, R. L. F., and BOWEN, S. J.: 1962, *Conference on the Ionosphere*, London, July.
WRIGHT, J. W.: 1960, *J. Geophys. Res.* **65**, 335.
WULF, O. R. and DEMING, L. S.: 1938, *Terr. Magn. and Atmos. Elec.* **43**, 283.
YONEZAWA, T. and TAKAHASHI: 1960, *J. Radio Research Lab.*, Japan **7**, 335.

Discussion and Questions

E. Hones: Can you elaborate on the latitude dependence of spread-F and the correlation between this phenomena and energetic particle fluxes?

R. E. Bourdeau: Yes. At mid latitudes it seems that spread-F is more prevalent at night than in the daytime. But observations from Alouette at higher latitudes indicate that these spread-F conditions occur.

E. Hones: Well, what I meant was that the variation around the earth at which the spread starts to occur as you go north.

R. E. Bourdeau: The data from Alouette were obtained along the 75th Meridian, predominantly.

Capt. Stevens (Air Force): I'm interested in the effects that a meteor entering the atmosphere has on the ionosphere. Can you tell me if a meteor causes ionization and if so at what altitude?

R. E. Bourdeau: Yes, there are some theories that say meteors are possibly the causes of some type of sporadic-E ionization. If they do cause sporadic-E, then they would give you more than one of the layers I have shown you in a given profile, spaced a few kilometers apart and to answer your question about the altitude it would be somewhere in the 100 and 115 kilometer region if it does occur.

OBSERVATIONAL MANIFESTATIONS OF THE INTERACTION OF THE LUNAR SURFACE WITH INTERPLANETARY SPACE*

ZDENĚK KOPAL

Department of Astronomy, The University, Manchester

In the past three days of this symposium, we heard much about different probes sent out by human hand to explore the contents and various properties of the interplanetary space. Before the conclusion of our proceedings it should be of interest to point out that – apart from the comets discussed earlier by Professor Biermann – we have, veritably at our doorstep (astronomically speaking) another permanent probe, ever present and exposing no small target to all constituents of the interplanetary space which it intercepts on its peripheral journey through space: namely, our moon. To be sure, the moon can scarcely be regarded as yet as an instrumented probe – though the day when this may be true is not too far in the future – but at least in one sense it deserves this name even today, for its surface can be regarded as a natural wavelength converter; and a discussion of this particular feature of lunar physics will constitute the principal topic of my talk this afternoon.

It is indeed interesting to consider the full range of phenomena produced on the lunar surface – unprotected by any atmosphere to speak of – by its interaction with the full contents of the interplanetary space. In this survey we shall forego completely the consequences of the impact of asteroids, meteorites, or comets – not because they are not interesting (and sometimes spectacular), but because they are outside the scope of this symposium. The constant downpour of micrometeorites seems, in particular, responsible for the rough microstructure of the lunar surface giving rise to the observed photometric characteristics of our satellite – in particular, for the strong back-scattering of sunlight manifested by the light changes of any element of the surface with the phase, or the absence of limb-darkening of lunar disk at optical frequencies.

The question can, however, be asked: does scattered sunlight (plus thermal radiation due to that portion of sunlight which is absorbed and re-emitted at generally longer wavelengths) represent the sum total of the observable moonlight? The answer is now known to be in the *negative*; and my main task should be to explain to you why, as well as to demonstrate the extent to which this is true.

It is well known that, apart from visible sunlight, the Earth-Moon system receives also continuous bombardment by high-energy electro-magnetic and corpuscular radiation from the sun. One phenomenon which may confidently be expected as a result, is the photoionization of some of the lunar surface material, consisting as it does of

* Work supported by Contract AF61 (052)–380 between the Department of Astronomy, University of Manchester, and the Cambridge Research Laboratories, Office of Aerospace Research, U.S. Air Force, through their European Office in Brussels.

predominantly lighter elements of the periodic table (Si, O, Mg, etc.) with weakly
bound electrons. Such an effect is indeed giving rise to serious considerations as to
whether or not the surface of the Moon may possess a net electric charge. Corpuscular
bombardment is again bound to give rise to X-ray emission due to bremsstrahlung.
The bremsstrahlung of the lunar surface at wavelengths between 10–100 Å cannot,
of course, be observed from ground-based terrestrial facilities, and has so far been
detected only during the relatively brief times while rocket borne X-ray spectrographs
operated above the atmosphere. As a result, the resolving power remains still rather
low; and this line of research, which could potentially enable us to undertake a
quantitative chemical analysis of the lunar surface (by elements) at a distance, constitutes still largely a task for the future.

Suppose, however, that the recombination of atoms ionized by energetic particles
or quanta of solar origin does not occur by a single transition, but a cascade process
giving rise to luminescent emission at longer wavelengths which can penetrate the
terrestrial atmosphere and thus become observable on ground. Indications that
luminescent emission may constitute indeed an appreciable contribution to total
moonlight have indeed been piling up for some decades. To begin with the most
elementary one – it has been noted long ago (Danjon, Rougier, Link) that the apparent
brightness of the Moon (when due regard is paid to the phase as well as to the variation of its distance from the Earth) is not quite the same from month to month,
but fluctuates somewhat; and these fluctuations appeared to be correlated in general,
with the cycle of solar activity (though their amplitude seemed larger than that
manifested by the solar constant). Another indication in the same direction has come

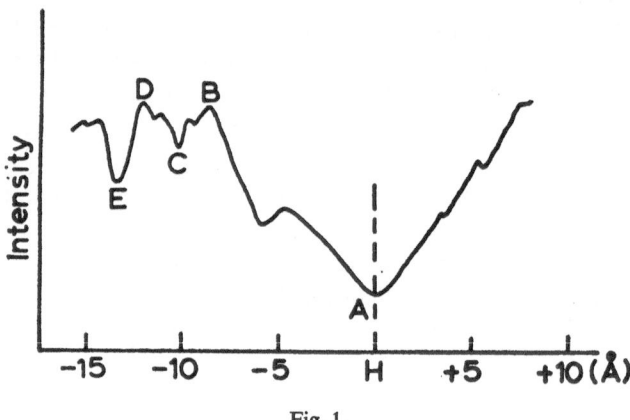

Fig. 1.

from photometric studies of lunar eclipses: LINK (1946 and 1962) was the first to point
out that the observed diminution of light, during eclipse, was not proportional to the
eclipsed part of the sun as visible from the moon, but smaller – indicating that significant effect was produced by illumination of the lunar surface by the solar corona.

Both these photometric methods are, however, mainly qualitative, and cannot tell us anything about the spectral composition of the lunar luminescence. In order to obtain some information, we must obviously resort to spectrographic methods; and this can be done along the following lines. Consider the profile of any absorption line of the solar spectrum – such as that of the H-line of Ca II as shown on the accompanying figure. The scattering of sunlight on the lunar surface is (in the visible part of the spectrum) largely independent of the frequency: therefore, if moonlight consisted only of scattered sunlight, the line profiles in the spectrum of moonlight should be exact replicas of those in the solar spectrum. If, on the other hand, the lunar surface contained an additional source of light, the lunar line profiles should become correspondingly shallower than those in the solar spectrum; and if $R_{m,s}$ denote the ratio of intensity of any particular point of the line profile to that of the adjacent continuum in the spectrum of the Moon and of the Sun, respectively, the ratio

$$\rho = \frac{R_m - R_s}{1 - R_m} \tag{1}$$

then denotes the fractional luminosity of the luminescent radiation superimposed on the scattered component of moonlight (and expressed again in terms of the intensity of adjacent continuum taken as the unit).

This "method of line depths" for studying the luminescence of the lunar surface was first proposed likewise by LINK (1951), and employed first by KOZYREV (1956) and, subsequently by DUBOIS (1957, 1959) for interpretation of actual observations. Both Kozyrev and Dubois found indeed the indications of lunar luminescence to be present to the extent of several percent of the intensity of neighboring continuum; its intensity varying with the position of the slit on the Moon, with the wavelength, as well as with the time. The work by Kozyrev on the luminescence of lunar surface around the crater Aristarchus foreshadowed, indeed, most of the essential features of the phenomenon as it is known today. However, because both Kozyrev, as well as Dubois, employed the techniques of photographic photometry which are really marginal from the point of view of accuracy, their results failed to attract the attention they deserved; and were also not followed up till with the inception of the Manchester program of lunar studies early in 1960.*

At that time it has become obvious that in order to carry out significant measurements of lunar luminescence which may amount to a few percent of the incident

---

* We may mention in passing that there exists a third; independent, approach by which the existence of lunar luminescence could be established by the observations: namely by measuring the dependence of the degree of polarization of moonlight on the wavelength. The basic idea goes back to the fact that, whereas that part of moonlight which represents scattered sunlight becomes distinctly polarized by this process (and the degree as well as plane of polarization vary with the phase), the luminescent emission is non-polarized. The polarization of scattered moonlight varies, moreover, but slowly with the wavelength; and, therefore, any irregularities on the curve of polarization versus wavelength should indicate (no less distinctly than the filling up of the absorption lines) the presence of the luminescent bands at the respective frequencies. The results obtainable by this method could usefully supplement the measurements by the line-depths method; and are the only ones that could be obtained in between the absorption lines of the lunar spectrum.

flux by the line-depth method, the resolving power of the spectrograph will have to be increased considerably over that employed by Kozyrev or Dubois; and photographic techniques replaced by photoelectric registration. With the financial support of the Cambridge Research Laboratories, Office of Aerospace Research of U.S. Air Force, a new model of a photoelectric spectrograph suitable for this work has been designed and built at the Department of Astronomy, University of Manchester, by Dr. J. Ring in collaboration with Dr. J. Grainger. The details of this instrument have been described before on several occasions (cf. GRAINGER and RING, 1962abc) so that few details need to be added in this place. As the dispersive medium, the instrument utilizes a Bausch and Lomb diffraction grating, operating in an optical system at a relative aperture of $f/5$. It has been specifically designed for attachment at the Newtonian focus of the 50-inch reflector of the University of Padua Observatory at Asiago, Italy, where all observations with this instrument have so far been carried out by kind permission of its Director, Professor L. Rosino. The dispersion attained by this spectrograph is 5 Å/mm, with spectral resolving power down to 0.1 Å; registration is photoelectric (using EMI 6256B photomultiplier), with separate channel employed to compensate the seeing.

The observations were started in 1961, and have been continued ever since to the full extent of available telescope time. The region of the spectrum selected for examination was the neighborhood of the H-line of Ca II, partly because of Kozyrev's previous work in this domain, but mainly because of the great half-width (~ 9Å) of this absorption line and its low absolute intensity, making the detection of luminescence easier. The actual observations have been carried out by Dr. Grainger and Messrs. Gregory, Hindle, Jefferson, Roberts (present or past research students in my department), and are continuing at the present time. While all the work which we have been able to accomplish so far has not more than scatched the surface of many problems arising in this connection (mainly because of the limitations of available observing time). The preliminary results have been indeed found to be thought-provoking.

First, the existence of the luminescence of the lunar surface, suggested by the previous work by Kozyrev and Dubois, has been confirmed beyond any doubt; and its amount found to vary from place to place on the Moon, as well as with the time. Near the core of the H-line of Ca II (λ 3970 Å) strongest luminescence – amounting to 10% of the intensity of adjacent continuum – was found to be emitted by the lunar ground covered by the Tycho bright ray passing through the crater Bessel in Mare Serenitatis (the size of the slit corresponded, very approximately, to a rectangular area of 1×10 miles on the surface of the Moon). A second area exhibiting almost equally intense luminescence – $8 \pm 1.5\%$ of the adjacent continuum – was located by Grainger in the neighborhood of the crater Plato. The plains of the Mare Serenitatis itself appear to luminesce noticeably less ($2 \pm 0.7\%$); and so does the creater Copernicus. On the other hand, the crater Aristarchus luminesces more ($3 \pm 1\%$), and this amount appears to fluctuate with the time.

After this brief survey of the essential facts of observation, (fuller details of which will be published elsewhere) let us consider briefly, the *energetic aspects* of the observed phenomena, in the hope of narrowing down the search for their possible cause. The observations indicate that the flux of the lunar fluorescent radiation appears to amount to a few percent of that of the neighboring solar continuum. What is the corresponding amount in absolute units? As on the Earth (outside its atmosphere), each square centimeter of the lunar surface receives from the sun the energy flux equivalent to 1.97 cal/min or 1.37×10^6 ergs/sec. If we assume, reasonably enough, that the sun radiates approximately like a black body of temperature T, the amount of energy L received through a wavelength window $\Delta\lambda$ at a particular wavelength λ should then be given by

$$L = 1.37 \times 10^6 \frac{J_\lambda \Delta\lambda}{\int_0^\infty J_\lambda \, d\lambda} \frac{\text{ergs}}{\text{cm}^2 \, \text{sec}} \tag{2}$$

where J_λ denotes the Planck function

$$J_\lambda = \frac{2hc^2}{\lambda^5} \left\{ e^{c_2/\lambda T} - 1 \right\}^{-1} \tag{3}$$

h being the Planck constant; c, the velocity of light; and $c_2 = hc/k$ (k, the Boltzmann constant).

Inserting (3) in (2) and integrating J over all wavelengths we readily find that

$$L = 1.37 \times 10^6 \times 15 \left(\frac{hc}{\pi k \lambda T} \right)^4 \frac{\Delta\lambda}{\lambda} \left\{ e^{c_2/\lambda T} - 1 \right\}^{-1} \frac{\text{ergs}}{\text{cm}^2 \, \text{sec}} \tag{4}$$

If we adopt now $\Delta\lambda = 1 \text{ Å} = 10^{-8}$ cm to be situated in the neighborhood of $\lambda = 3970 \text{ Å}$ (i.e., the wavelength of the H-line of Ca II) we find that, for

$$T = 5500°, L = 139 \text{ ergs/cm}^2 \text{ sec per Å};$$
$$T = 5700°, L = 158 \text{ ergs/cm}^2 \text{ sec per Å}.$$

The mean albedo of the apparent lunar disk in the blue is very close to 0.073. Accordingly, only about 7% of the flux L computed above is scattered from the lunar surface – i.e., only 10–11% ergs/sec Å reach us from each cm² in the form of "moonlight" at this wavelength. The observations indicate that the lunar luminescent flux amounts to a few percent – say, conservatively, 3% – of the scattered moonlight. This would correspond to about one-third of one erg of luminescent radiation per Å per sec (around 3970 Å) from each-square centimeter of the surface of the moon.

At present, we have as yet no definite information on the width of the observed luminescent band; but it is probably several tens of Ångströms wide (since the variation in luminescence over the 15 Å or so of the observed scans appears to be small). Let us, for the sake of argument, assume a bandwidth of 40–50 Å, which is fairly typical of known terrestial luminophors. The efficiency with which the stimulating energy is converted into luminescence is likewise unknown; but guided again by the properties of known substances let us consider it to be between 1–10%. If so a stimulating flux of the order of 100–1000 ergs fall, per second, on each square centimeter of the lunar surface, from the sun or interplanetary space.

What could be the source of this energy? It is much easier at present to say what it is not; and much safer to try to arrive at the correct answer by gradual elimination of alternative possibilities. Thus it is possible to assert that, contrary to what we believed earlier, the observed lunar luminescence is *not* caused by the impact of solar wind; for what we have heard earlier in this symposium about the properties of the latter leaves no room for doubt that the energy of its flux (as recorded, for instance, by Mariner 2 last year for several months) is by at least two orders of magnitude too small to account for the observed luminescence.

The solar electromagnetic radiation in the short-wave domain remains a possibility. It should be stressed that the flux of its most intense bright lines (such as the hydrogen $L\alpha$, or the He II line at $\lambda\,304$ Å) is again totally inadequate to produce any noticeable luminescence of the lunar surface. However, the measured flux of solar X-rays at the time of the *disturbed* sun would be ample to account for the observed phenomena. A suspicion that the lunar phenomena may be invoked by the electromagnetic rather than corpuscular radiation from the sun is further strengthened by the fact that the intensity of luminescent radiation appears to follow roughly that of insolation. The observed variation of the intensity of lunar luminescence with the time suggests furthermore, that the exciting frequencies should indeed be sought at the extreme end of solar short-wave spectrum; for, at longer wavelengths, the solar energy output is known to be remarkably invariant. The search for a possible correlation between the solar X-ray activity and lunar luminescence would seem very desirable; but, unfortunately, it cannot be carried out from the same observing station (as the sun and the moon could not be surveyed at the same time; and the transit-time of the disturbance would in both cases be equal).

As a third alternative, we cannot dismiss from our mind the possibility that at least a part of the observed luminescence may be excited by the radioactivity of the lunar surface, which may constitute an accumulated effect of long-lived isotopes of the spallation products of primary cosmic rays impinging on the moon since time immemorial (and, in particular, during the outbursts of nearby supernovae). Such part of luminescence as may be induced by this agent should, of course, persist throughout the long lunar night; and should then be easily detectable by observation through interference filters with passbands centered at the respective bands of luminescent emission. Neither spectrographic observations, nor photographic observations through sufficiently narrow-passband filters, of the dark hemisphere of the moon have so far been made to search for possible indications of lunar luminescence; and as long as this remains so, the central point of our inquiry concerning the proper source of the energy causing the observed luminescence of the lunar surface must still await a more definite answer.

And with this, not entirely satisfactory, statement we must take leave of our subject at the present time. What the Manchester astronomers have so far discovered – or rather, confirmed by methods which no longer admit of doubt about the significance of the results – is the fact that the surface of the moon emits in daytime luminescent radiation showing indications of band structure, the intensity of which may amount up to 9% of the neighboring continuum; that this intensity varies from spot to spot on the moon indicating the presence of different materials covering the ground; and that it also varies with the time (though the scarcity of the data does not permit us to correlate this fluctuation as yet with any specific aspect of solar activity); but what the cause that stimulates this radiation may actually be we do not as yet pretend to know.

References

DUBOIS, J.: 1957, *Journal of Physics* **18**, 13S.

DUBOIS, J.: 1959, *Czech. Acad. Sci.* **69**, Pt. 6.

GRAINGER, J. F. and RING, J.: 1962a, *Physics and Astronomy of the Moon* (ed. by Z. KOPAL), Academic Press, London and New York, Chapter 10, pp. 385–404.

GRAINGER, J. F. and RING, J.: 1962b, *I.A.U. Symposium on the Moon,* No. 14 (ed. by Z. KOPAL and Z. K. MIKHAILOV), Academic Press, London and New York pp. 445–452.

GRAINGER, J. F. and RING, J.: 1962c, *Mon. Not. Roy. Astron. Soc.* **125**, 93.

KOZYREV, N. A.: 1956, *Izv. Krimsk. astrofiz. obs.* **16**, 148.

LINK, F.: 1946, *Comptes rendus* **223**, 978.

LINK, F.: 1951, *Bull. Astron. Inst. Czechoslovakia* **2**, 131.

LINK, F.: 1962, *Physics and Astronomy of the Moon* (ed. by Z. KOPAL), London and New York, Ch. 6.

Note added in proof (January 1965). Since the delivery of this paper, new observational data accumulated in the latter part of 1963 (cf., in particular, KOPAL and RACKHAM, *Icarus* **2** (1963) 481; *Nature* **201** (1964) 239; GREENACRE, *Sky and Telescope* **26** (1963) 316; **27** (1964) 3) have greatly increased our knowledge of this subject. In particular, it is probable now that lunar luminescence constitutes a fairly frequent phenomenon, caused (or, at least triggered) by corpuscular radiation associated with specific disturbances of the solar surface.

Discussion and Questions

N. Roman (NASA): What did you use for a comparison source for your spectra, that is your solar spectra?

Z. Kopal: We used sunlight both scattered from clouds and the sky.

N. Roman: Were the moon observations made in the daytime or did you assume that everything remained constant from day to night?

Z. Kopal: Yes. The tracings of the comparison spectra were taken in daytime; but the moon observations were made only at night.

ROUND TABLE DISCUSSION ON FRIDAY

Participants: S. Chapman (chairman), H. Alfvén, R. Bourdeau, L. Cahill, L. Davis, J. Heppner, Z. Kopal, R. Lehnert, N. Nakada, N. Ness, E. Schmerling.

D. Beard (University of California): Without trying to appear too controversial, this comment is made in reference to both Dr. Ness's talk and Prof. Alfvén's comments earlier. If there were an interplanetary magnetic field and it were compressed against the magnetosphere, whether or not the boundaries would appear as Dr. Cahill's discontinuity would depend upon the direction of the compressed field, which would have the same intensity as the geomagnetic field at the edge of the magnetosphere. It would have the same intensity but might have the same direction or an opposite one and in the case that it were in the same direction, no discontinuity would appear. Also, if you carried the compressed field around to the dark side of the earth it would tend to flare out and it could use the kind of thing that I think Dr. Ness observed.

S. Chapman: Do you have any comments?

N. Ness: No. I thought about this before the lecture; and I felt that if there were an external magnetic field in the same direction and with the same magnitude as the earth's field, we had no way of detecting it on the dark side. The only way that one could detect the boundary then, or a discontinuity if you like, is by a trapped particle discontinuity.

S. Chapman: Maybe I'll put my foot in my mouth here but I have an extremely naive interpretation of a solar flare. (Blackboard) If we have the sun emitting particles having an associated magnetic field which is, of course, carried along with the particles, then out here there is a magnetic field which has the same configuration as the earth's field and there is some distance where, when averaged, the earth's field is about the same order of magnitude as the solar field which is being dragged by particles; and this means that out here there will be a region which is somewhat turbulent. Consequently, the earth's boundary is not quite as sharp, that is, one would not expect it to be quite so sharp.

I'd like to ask Dr. Ness a question. On the last slide you showed today you increased the magnetic field by a constant amount and came up with a curve that agreed quite well with experimental results; I missed the explanation of the curve.

N. Ness: The fit was not that excellent very close to the earth but the fit was accomplished empirically, actually, by forming the differences vectorially between the measured and computed field and observed that these differences seemed to approach a constant value as the probe departed from the earth. We then took this vector and added it to the field, extrapolated all the way back along the entire trajectory. There is no theoretical basis why this particular geometry leads to a cavity field having these characteristics; but it is, however, because of the nature of the magnetic field measured by detectors on the spacecraft, indicative of what sort of cavity one must derive theoretically.

H. Alfvén: This is a repetition of some parts of my lecture which may be useful. There were two different models which can be based on the experiments we have done. There were two cases – namely, parallel field by which I mean the field was parallel to the dipole field in the equatorial plane and the other case antiparallel. The model for the case of parallel fields was something like this. (Blackboard) Plasma from interplanetary space can penetrate rather far toward the dipole center but stops where the dipole field equals the interplanetary field. So if we looked out here and we had a parallel field we would not see any discontinuity. However in other regions of this model there would be a discontinuity. In the case of the anti-parallel field however, we have a configuration which looks like this. (Blackboard) Here the field goes up, here down, and here is the Cahill discontinuity where the magnetic field flops 180 degrees. However in the direction out here, on the dark side there is no discontinuity, as you go out here as was reported by Dr. Ness, you see a steady decline down to the interplanetary level not noticing any significant discontinuity. These models are open magnetosphere models as opposed to the closed models we have heard about and it is characteristic of these open models that in some directions you have discontinuities and in other directions a rather continuous change over from the magnetosphere to interplanetary space. Of course there may be turbulence which causes rapid fluctuation but no changes of the Cahill discontinuity type.

N. Ness: I'd like to make a few comments about this problem. We have in detail, from the Explorer X, results on what we refer to here as the boundary which were not clearly presented. We have data coverage over an interval of approximately two days outside the trapped boundary or where interplanetary space and the trapped boundary were exchanging places. The strength of the magnetic field had a 24-hour character which seems to indicate very strongly that it is associated with the geomagnetic field and would not be representative of interplanetary conditions. That is to say, the magnetic field when in region *A* decreases very much for 24 hours and in the next 24 hours it jumps back up to a higher value. Then it decreases again in the next 24 hours which to my mind – and I think to most of those who looked at the data in detail – indicates that the magnetic field orientation resides in a geomagnetic origin and clearly not in any mixing of the interplanetary field with the geomagnetic field. Perhaps Dr. Heppner would like to comment further on this point.

J. Heppner: I was just going to add that I don't think that the laboratory plasma conditions fit. One additional factor: on the front side here in your experiment, you're dealing with a highly magnetized plasma in which the ratio of energy densities is essentially unity; where as, it was observed in the case of Explorer X outside the boundary, to be about 10. In other words the plasma energy density is a factor of 10 greater than that of the field energy. I think you have quite a different case in the laboratory experiment. I think the same thing is indicated by Mariner results that the plasma energy density is much greater than the field energy density in interplanetary space.

H. Alfvén: The kinetic energy of the plasma is larger than the magnetic field density, but we are going to change the parameter within a very wide range.

J. Heppner: Can you change that ratio in your experiment?

H. Alfvén: Oh yes, we have already made experiments where the discharge has changed by a factor of one million.

E. Hones (IDA): I wonder if Dr. Cahill could say a little more about the direction of the field that he's found in Explorer XIV on the night side of the earth. Presumably there in the region where the field becomes essentially constant, you are looking at the pulled out, at least in the model where you think of the earth's field as greatly distorted, you are looking at pulled out high latitude lines. Is the direction proper for this or do you look at the northern hemisphere high latitude lines? Or are all your data in the southern hemisphere?

L. Cahill: Well, the inclination is 30 degrees and most of the orbit is in the southern hemisphere however the inbound path and in most of these that I've shown, the satellite is very close to the geographic equator and hence not far from the geomagnetic equator and so these are essentially equatorial passes that we show here. There's another factor here that I believe probably bears on this. That is that it's just shortly after the winter solstice that the earth is tipped away from the sun; and regarding the models of the tail of the magnetosphere that have been made, for example, by Spreiter and Briggs, one would think that perhaps the tail of magnetosphere is strongly effected by the solar wind. Now we are well below the ecliptic plane in this case and I think that while the field lines are consistent with the stretching out of the high latitude field lines as you say, I haven't been able to determine the direction of the spin axis with sufficient accuracy to tell you exactly if there is a correspondence with the Explorer X results. You see we are on the opposite side of the midnight meridian, if you like, to what Explorer X was. Isn't that right?

E. Hones: Yes.

L. Cahill: Yes, we are then on the opposite side. But both of these orbits are at least comparably below the geomagnetic equatorial plane. The Explorer X results are somewhat lower than ours are but we are at roughly the same location. Thus we may have a chance to compare these when we get a better analysis of the direction, after the spin axis has been determined more accurately.

Unidentified: I'd like to ask Dr. Cahill, could you give a comment on the existence of ring current? Or is it impossible to make a clear cut statement.

L. Cahill: Well, I can say where a ring current isn't. I can't very well say where it is. I believe that we are very sure on the basis of the Explorer XII results on the sunlight hemisphere of the earth that there is no ring current between something like 4 earth radii up to apogee. Nowhere in this region is the field depressed sufficiently in a local sense that you would expect a ring current. Now just at the limit of the Explorer XII magnetometer which is near 1000 gammas we did see effect which you would attribute to ring currents. This was during the storm of September 30, 1961. We discussed these before and what we see is that at altitudes greater than 3 earth radii, the field rises after the storm of September 30 and slowly comes back to the empirical value again. We would assume we are out of the region of ring currents at this altitude.

However, I was quite interested to check the next storm which occurred a few days later in October. Hoping we would find the same pattern, that the field would rise and then decay again. I am sorry to say that we did not find this same simple pattern in October so we would like a third storm to see what happens in that case. The first storm was promising and the second not so, that is about all I can say at this time. We just do not go in far enough to see the depression in the field that you would expect to see if you continued in further.

J. Dungey: I just wanted to follow up on the previous discussion. As I understand it, the field direction is very crudely pointing away from the earth. Do you ever see it pointing toward the earth?

L. Cahill: The exterior field?

J. Dungey: Yes, well when you are far out with Explorer XIV.

L. Cahill: If you mean on the tail, no. No we don't.

J. O'Hara: A question for Dr. Bourdeau. In your slides you had several references to both a quiet and disturbed sun. I was wondering what criteria you used to establish either a quiet or a disturbed sun.

R. Bourdeau: It is not a black or white condition because there are so many gray areas here. But in general the world wide magnetic index is an indication that small flares can produce disturbances which are barely distinguishable. Then you can go from a condition like that over to a Class I or a Class 2 flare which cause sudden ionospheric disturbances and then over to events more severe than that such as those which cause polar cap absorption events or to world wide magnetic storms. But depending on the specific ionospheric region one is examining there are various levels of disturbances. There are some indications that even non-visible flares that are detectable by the increase in X-ray radiation on satellites are still significant enough to enhance the D-region but not strong enough to disrupt radio communication.

L. Cahill: Someone went through two years of K indices and picked two of the quietest nights he could on the basis of the K index. And he found on one of these two nights in the records that the entire E-region was lifted up several hundred kilometers and came down again in the span of half an hour. So this doesn't seem to be too satisfactory a method for determining what is quiet or what is a disturbed condition.

R. Bourdeau: In addition to that there is the possibility of particles precipitating from the radiation belt for one reason or another and here the indices we use to differentiate between the quiet and disturbed condition is whether or not the characteristics of the charged particles are explainable on the basis of solar radiation. If they are not then there is another source of ionization present.

J. O'Hara: I was thinking that there would be some term properly applied to quiet and disturbed sun, whereas you were referring to a steady state and unsteady state of the sun.

R. Bourdeau: The definition I try to stick to is where you can explain the characteristics of the charged particles on the basis of either solar X-rays or solar UV radiation.

E. Hones: I just want to clear up one more point in my own mind on this question with Dr. Cahill. Were you ever in a position where you would expect to see a stretched out field line heading back toward the earth? You said a minute ago in response to Dr. Dungey that you never observed a line heading toward earth. Have you ever made any observations where you would expect to find this condition in the stretch out model of the earth's field?

L. Cahill: Well, this disturbed me somewhat and I thought it peculiar that we always observed lines pointing outward. However, it does seem to me that since we are south of the ecliptic plane and considering the fact that the direction of the stretching out must be dominated by the direction of the solar wind perhaps I can rationalize this so that I would always expect to see it stretched outwards. Now you must remember there data are only five or six passes although it is observed to be stretched outwards on both inward and outbound passes. Now I think as time goes on when we move around to the other side of the tail later on in February that we will see some coming back the other way. Of course this will be at higher latitudes and higher inclinations to the ecliptic plane.

E. Hones: What we are talking about here is the direction of the field vector.

L. Cahill: That's right I have never seen it stretched toward the earth.

E. Hones: Pointing to the earth.

L. Cahill: Yes.

N. Ness: I'd like to ask Dr. Cahill a question concerning the results of Explorer XIV close in. How sensitive is the apparent uniform depression of the magnetic field to slight changes in trajectory.

L. Cahill: Well, it will of course be quite sensitive to trajectory. I think if you look at the graphs you would have to slide back the empirical curve about 1000 kilometers or so to make it coincide with the measured field on the date of greatest depression, that is the 4th of January.

N. Ness: Do you recall the value?

L. Cahill: No.

N. Ness: Well, I think it was something like 50 gammas approximate depression and was quite uniform between negative depression on the dark side and positive depression on the sunward side. Thus the characteristics of the cavity field near the earth are changing very rapidly.

L. Cahill: It certainly is. Of course you recall that I said that I did not have a tremendous amount of confidence in the relation between the theoretical and observed field here. It is also right that usually within this period it is depressed; however, you will remember on January 9 it was not depressed. It was quite close to the theoretical value. Now maybe this was fluctuation in the trajectory or it might be a real depression that slowly creeps up to the theoretical value.

N. Ness: I wonder if Leo Davis might have some comments about the possibility of trajectory accuracy on say Explorer XIV affecting the detection of particles.

L. Davis: I have no data on the accuracy of trajectories. There is some on Explorer XII. The predicted crossing into the earth's shadow and the data we have shown discrepancy of the order of 2 minutes.

L. Biermann: With respect to the expected geometry of the magnetic field in the geomagnetic cavity, first, there is the systematic difference due to the motion of the earth which means that plasma comes from a direction 5 or 6 degrees away from the direction it left the sun. Second, we have the rotation of the earth which carries along with it the geomagnetic field which somehow might distort the geometry of the field lines.

L. Cahill: As I take it you are asking if we have taken into consideration the offset of the direction of the solar plasma due to the rotation of the earth. I don't think that at this time we are far enough along to take these requirements into account. Certainly I think there are effects due to this, for instance the wobble in the angle alpha which is not the same each time around. Though these effects are visible we haven't done too much about them yet.

E. Hones: In connection with the question Dr. Biermann just asked, Dr. Ness in his talk indicated that the 24-hour variation in the magnetic field was more or less what you would expect due to the earth's rotation. So presumably you have made some sort of analysis of the data which shows you that you are getting the effect inside the cavity that one would associate with a wobbling dipole. Is that correct?

N. Ness: I said that the data experienced a 24 hour periodicity which strongly suggested that the field we were seeing was strongly associated with the geomagnetic field. I also stated that we tried, by turning this truncated cylinder approximation around, to take into account the aberration of the solar wind by the velocity of the earth in its orbit around the sun. We had no success. But we felt this was due to the fact that the cylinder was a very poor approximation.

Unidentified: I'd like to ask a general question. At smaller distances say 5 or 6 earth radii, do magnetic field measurements indicate that the earth's magnetic field is rotating with the earth? Is this assumed or what?

N. Ness: In the theoretical calculations we assume that the earth's field co-rotates with the earth. In our data you must remember that the satellite went through this region in a period of 3 or 4 hours. Dr. Cahill's data doesn't extend to this region. It stops at 3 earth radii. We have no indications from the data, which show only small difference from the theoretical values which can be attributed to experimental error.

Unidentified: In the transition region where you are going from a zone where the magnetic field rotates with the earth as it does at small distances to a region where it doesn't, namely in the solar wind, you must expect electric fields.

N. Ness: I certainly expect a lot of things. And I think I'll refer this to a statement made concerning solar flares that in this region things are very complex.

L. Davis: Just one remark here. Are you thinking that you might have a discontinuity here if you don't have co-rotation. I don't think that it is necessary. Your computed fields become invalid if there is not co-rotation. Were you implying this was so?

Unidentified: No.

Unidentified: I would like to ask Dr. Kopal a question. It should be noted that ionospheric densities are very sensitive to changes in X-ray and ultra-violet light from the sun and I wonder if you have been able to make any correlation along these lines?

Z. Kopal: The energy in the spectrum of the sun at these wavelengths is much too small to account for these phenomena. Below 100 Ångströms or so these certainly do occur but unfortunately the energy here is not large enough. Now if you go to longer wavelengths the energy is high enough.

S. Bowyer (Catholic University of America): I'd like to ask Mr. Davis a question. Did you find a double peak in your data or are you outside the region of second maximum in protons?

L. Davis: If you're referring to the second maxima that McIlwain's data clearly shows, well he is working with about 30 MeV protons which are 30 MeV greater than ours. He sees a secondary peak around 2 earth radii. We, of course, see only one peak depending on what energy particles we are talking about. It could be anywhere below 3 to 3.5 earth radii.

S. Bowyer: You observe only one peak?

L. Davis: Yes.

S. Bowyer: How far do you go in?

L. Davis: Two earth radii. Well, we have measured below 2 earth radii but the intensities are dropping already on higher energy protons when we reach altitudes of 2 earth radii.

S. Bowyer: You are sure there can be no maximum below this?

L. Davis: We can set an upper limit but it would be extremely high.

M. Wagdi (Catholic University of Washington): I would like to ask Prof. Alfvén a question. Is there any possibility of a shock wave appearing at the geomagnetic boundary due to the solar wind and if so what is the critical velocity for such an occurrence.

H. Alfvén: I think the answer is yes and I think Dr. Chang is more competent to answer your question than I.

C. Chang: If the Mach number or perhaps we should say the Alfvén Mach number is of the order of 2 or larger you should have a shock wave forming.

J. Dungey: I'd like to try something on Bob Bordeau about helium. If I understand this correctly, the processes for helium such as photoionization are slow relative to those for oxygen. If this is correct then ionization by energetic electrons is relatively more important for helium than for oxygen. If so then the thickness of the helium ion layer could be an indication of the amount of ionization by electrons. And we had the privilege of a visit from Bob Bourdeau in London recently to see the British satellite Ariel results upon which were found some rather strange latitude variations in helium.

R. Bourdeau: The curves that I have shown you that relate the altitude with the thickness of the helium layer are based on a theory of the hydrostatic law for the density of neutral constituents. It varies with the atmospheric temperature, in other words, the escape rates. That is the behavior one would expect if solar UV radiation produced the ionization of the helium. Now if there is indeed a correlation of the thickness of the helium belt with geomagnetic latitudes then what you suggest, that energetic particles are ionizing the helium, is a very definite possibility. I don't know whether the Ariel investigators have proceeded with their results far enough to state that there is a geomagnetic variation, but the last time I heard they thought there might be a change in the thickness of the belt with latitude.

S. Chapman: I think that it is true to say that particle ionization would really quite inadequate when you consider the high altitudes. In the helium belt you are in a region where the ion production rate is of the order of 10 per cc so you don't need much particle ionization to make an appreciable effect.

R. Bourdeau: I'd like to add one more thing. There are some recent observations which quite strongly suggest that the night time electron temperature often exceeds the neutral gas temperature by as much as 50 percent. Now here we have the sun turned off; so we have an ionization source here, and if one looks at the energetic particles data, one concludes precipitating particles might be this source.

S. Chapman: I'd like to call upon Prof. Alfvén to make a few remarks.

Prof. H. Alfvén's Closing Remarks

Ladies and gentlemen, we are approaching the end of an extremely interesting symposium on plasma space science. We have a great number of extremely interesting new facts here and one must remember the similarity between the present space age and the age 4 or 5 hundred years ago when the great geographical discoveries were made. At that time also, people got an enormous number of new facts which they tried to arrange with the help of some parameters and the parameters they used were latitude and longitude. Today when we study plasma space science we ask what parameters should we use. Well, the motion of charged particles, or more properly the acceleration of charged particles, in a plasma depends on two parameters, the magnetic field and the electric field. The magnetic field is relatively easy to measure, the electric field is very difficult and has not yet been measured. As a matter of fact this is a great similarity with the age 4 or 5 hundred years ago. Because the latitude was easy to measure but it was very difficult to measure the longitude. It was so difficult that quite a few people left for India and ended up in America. The situation was so bad that I think if I remember history correctly, the English Parliament deposited a large sum as a prize to be given to the man who first could determine the longitude. And this was very important because this made geography an exact science. What should we do today? I propose that we do the same and deposit a sum of one cent in the hands of Prof. Chang to be given to the first man who can clear up the second parameter, the electric field. And when the Chang prize is awarded we will all know that our science has entered the new area of much higher exactness and we will also remember with the best thanks this symposium which Prof. Chang has so generously invited us to and which he has run with so much energy and confidence. Thank you very much Prof. Chang.

APPENDIX

THE NATIONAL AERONAUTICS AND SPACE ADMINISTRATION SPACE SCIENCE PROGRAM - PROGRESS AND POTENTIAL

JOHN E. NAUGLE

Director of Geophysics and Astronomy Programs,
Office of Space Sciences,
National Aeronautics and Space Administration, Washington, D.C.

Introduction

Ten of the speakers of this Symposium on Plasma Space Science are basing their talks on experiments performed on spacecraft flown as a part of the NASA Space Science Program. There are nearly as many additional talks by physicists on theories which can only be checked by experiments flown on spacecraft. This is an indication of the progress made in the NASA Space Science Program since it began in 1958. By the end of this week you will have been given a picture of the significant results which have come from the program. You will also have seen some of the potentials for major scientific discoveries in space science.

Tonight I want to review the NASA Space Science Program. I always have trouble when I try to decide how to talk about the scientific program at NASA, and I talk about it in different ways depending upon the listeners. It can be described administratively by the lines along which the authority, responsibility and money flow. This is the way I usually discuss it when I address a group of industrialists.

For administrative reasons we have assigned primary responsibility for each earth satellite to one group at Headquarters and to one field center. We have assigned primary responsibility for each spacecraft which leaves the earth and goes to a planet to a separate group at Headquarters and to one field center. It is logical for the experience and technology that are required for planetary flights to be grouped at one place, and this is the Jet Propulsion Laboratory; that for scientific satellites in the physical sciences is at the Goddard Space Flight Center.

It is equally logical to describe the program as a coherent set of experiments to increase our knowledge and understanding of particular scientific phenomena. This is the way we usually discuss the program with scientists. It makes good sense to have plasma probes and magnetometers on a Mariner II going to Venus and on an Explorer XIV satellite going back and forth through the earth's magnetosphere, because from the two we can begin to understand the solar wind and its effect on the interplanetary magnetic field and the magnetosphere. It makes good sense to observe the activity on the sun at the same time with X-rays and ultraviolet light. We may also need data from magnetometers and solar observatories on the ground in order to complete our understanding of the phenomena involved. We plan our program and establish the objectives by looking at the scientific phenomena and asking, "What are the fundamental questions and how can we get the answers?"

Chang & Huang (eds.), Proc. Plasma Space Sci. Symp. All rights reserved.

After looking over the topics of this symposium, I decided that it would be better at this time to discuss the mechanics of the program, how we make our decisions, what spacecraft we are using and even how much they cost. You have had two days of discussions of the scientific results of the program and you have two more to go. Tonight I will try to give you a change of pace. I will describe the spacecraft which are now being used and the ones which are being developed. I will show you the flight schedules. This will show you the size and kind of program that is underway. I will discuss some of the problems and worries which we have in the program. Since this is a national program, supported by you and me, and involving a large number of institutions and people, I think it is proper that you should share with us some of these concerns.

I have another reason for discussing the program in this way. The Space Program is the subject of a great national debate. Some very knowledgeable scientists have likened the program to a kind of scientific WPA project. Other equally knowledgeable people have compared the program to the exploratory voyages of Columbus. Some have said that the money we are spending would be more wisely spent for the study of cancer or for the relief of depressed areas. Others have said that the very survival of this nation depends upon our having a vigorous Space Program. Some people see the exploration of space as a challenge to the creative talent of the nation. Others see it as a disruption of the orderly development of science and technology. I want you to understand the scope of the scientific program so that you can form your own opinion.

I am at NASA because I find that the exploration of space is a challenge. It is an interesting and exciting challenge. Each year at this time a large number of creative young people must decide where and how to apply their talents. They desperately want to do something exciting and something for which history will remember them. All too often in the past the only outlet which national needs and national goals offered to satisfy this drive was war. Back in 1948 when I graduated from college one of the major opportunities for scientists was with the Atomic Energy Commission to work on atom bombs. Later in 1953, at the time I got my Ph.D., the major opportunities were to work on ICBM's to carry these bombs. At that time we were competing with Russia in terms of destructive power. Today the new graduates face still a different challenge. They have the opportunity to participate in the exploration of space. We are now competing with the Russians in a race to explore space. I realize that I may be naive but I think that this is a more logical race in which to compete than the race to increase our mutual destructive power. I think that positive benefits may result from a race in the exploration of space; whereas competition in destructive power benefits all competitors only in a negative sense.

Clearly the outcome of this debate is of major concern to us. Each of you will participate in it to various degrees. You should certainly, as a result of this symposium, be able to participate in that debate on a more knowledgeable basis.

Tonight I will first tell you about the NASA Office of Space Sciences; what it is, how it operates, how much resources are available to it and how it decides what to do with them. You will see that we plan our program around the objectives of various

scientific disciplines. I will discuss briefly the experiments which have been performed –
the progress that has been made. I will discuss the future program as it is planned –
the experiments which are underway – the potential of the program. I will close with
a brief discussion of the problems of the program.

Organization and Functions of the Office of Space Sciences

When Mr. Webb became Administrator of NASA he established at NASA Headquarters four program offices, Office of Manned Space Flight, Office of Applications,
Office of Advanced Research and Technology and the Office of Space Sciences. The
directors of these offices are responsible for planning the programs in their areas
of responsibility, for estimating the resources required, for presenting their program
to Congress to obtain these resources, and for directing the execution of their programs. Dr. Newell, as Director of the Office of Space Sciences, is responsible for
determining what scientific experiments and investigations should be performed in

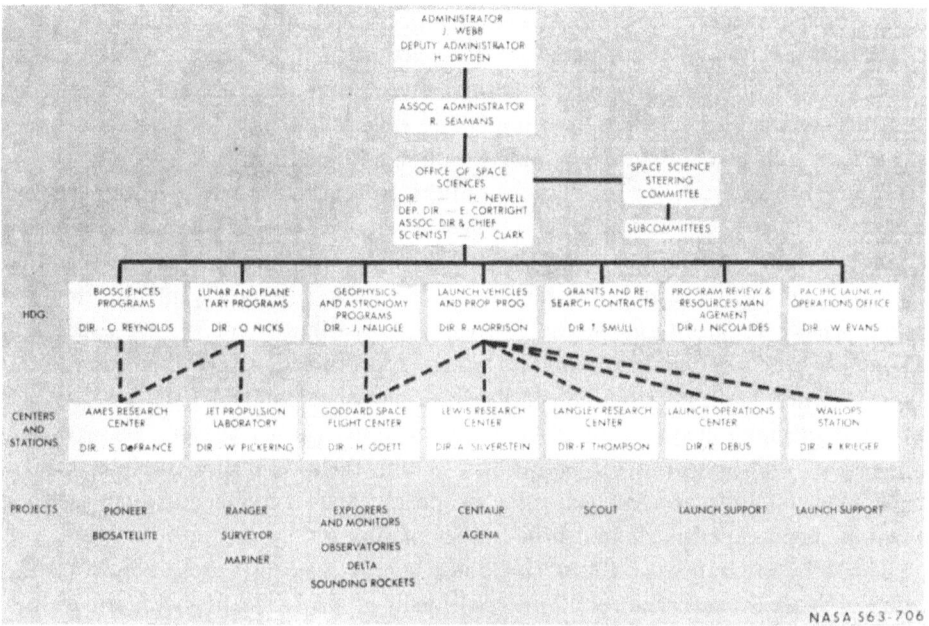

Fig. 1.   Space sciences program. Organization.

space; what kinds of spacecraft are required to carry these experiments; what kind
and how many launches are required to boost the spacecraft into orbit; what new
launch vehicle must be developed, if required; what are the costs of the program;
and he is responsible for defending the program before Congress. He has asked
Congress for about 837 million dollar for the program this year. About 4000 scientists,
engineers and technicians at the nine NASA field centers, the Jet Propulsion Laboratory and some 40 universities will be involved in its execution. An additional 10 000

people in industrial organizations are engaged in the design, development and manufacture of the spacecraft and launch vehicles used in the Program.

Figure 1 shows how we are organized to do this job. Early in the life of NASA the decision was made to divide the scientific exploration of space into two major divisions, a scientific satellite program to explore the earth's atmosphere, ionosphere, magnetosphere and the region of interplanetary space near the earth, and a Lunar and Planetary Program to study the moon, the planets and interplanetary space. In short, spacecraft which remain close to the earth were the responsibility of one part of the Office of Space Sciences and the responsibility of a particular NASA field center, the Goddard Space Flight Center; spacecraft which escape from the earth were the responsibility of another portion of the Office of Space Sciences and a separate center, the Jet Propulsion Laboratory. Later a third scientific segment of the Office of Space Sciences, the Bioscience Division, was established. A fourth such segment, the Manned Space Science Division, has recently been established. This means that Dr. Newell has not only the responsibility for the Space Science Program using unmanned spacecraft but also has the responsibility for the selection of the scientific experiments to be flown on manned spacecraft and for the scientific training of the astronauts, and participates in the selection of the scientists which will fly on the later Apollo mission. Along with these divisions in the Office of Space Sciences there are three additional divisions, Grants and Research Contracts, Launch Vehicles and Propulsion, Program Review and Resources Management.

In order to advise him on the conduct of the program and to involve the scientific community in the formulation of the program, Dr. Newell has established a Space Sciences Steering Committee composed primarily of the directors of the scientific divisions and their deputies. We involve the scientific community in planning the Space Science Program and in the selection of experiments for NASA spacecraft through seven discipline subcommittees which report directly to the Chairman of the Space Sciences Steering Committee. These seven subcommittees have as members scientists of high scientific caliber from NASA centers, other government agencies, universities, and non-profit organizations. Within their respective discipline areas, these subcommittees review scientific proposals and recommend space science missions, flight experiments and broad areas of supporting research.

The Office of Grants and Research Contracts provides contracting support for the Office of Space Sciences and for all other elements of NASA dealing with universities and non-profit organizations. In addition, this office provides program direction for the sustaining university program which I will discuss later.

The Program Review and Resources Management Division provides staff assistance to the Director of the Office of Space Sciences.

The responsibility for execution of a flight project, once it has been approved, is given to one of the NASA centers or the Jet Propulsion Laboratory. In general, the Goddard Space Flight Center is responsible for the flight projects supporting the Geophysics and Astronomy Program. The Sounding Rocket Program, the scientific satellites and the communication and meteorology satellites, are all managed by the

Goddard Space Flight Center. The Lunar and Planetary flight projets are primarily managed by the Jet Propulsion Laboratory. The Ames Research Center is responsible for the flight projects of the Biosciences Program and for two of the Lunar and Planetary projects. The Langley Research Center is responsible for one of the scientific satellite projects.

What is a program office? What are its functions? Why do we need one? A Headquarters program office has three major functions. The first is to formulate a coherent program which can be presented before Congress in order to obtain the funds to carry out the program. The second major function is to provide the technical direction of the program, and the third major function is to perform a continuing evaluation of the progress of the program. We need a program in space science because of the magnitude of the program. Vehicles need to be developed. We need to know what kind of vehicles are required. In some cases it may take four or five years to develop a new vehicle for the programs. One has to decide, for instance, when we should land a capsule on Mars. Ultimately, of course, it is you and I, the people, who decide through our elected representatives in Congress what the magnitude of the program should be, but in order for Congress to make that decision someone must have some idea what experiments are to be performed in order to decide how large a spacecraft is needed. We need to know what information is to be transmitted back. This gives us the size of the telemetry system. This ultimately tells us what size of a rocket we must develop. After we have made that decision then we have to estimate how much it will cost and take this whole program before Congress and obtain the money to do the job. This is one reason for having a program office. Another reason is that we try to make use of the creative people with creative ideas not only in all NASA field centers but also in all the universities throughout the country. Unfortunately there are insufficient funds for us to be able to give each university that wants to be in the program its own vehicle and its own spacecraft. Therefore, we must put together groups of experiments from many sources, from universities, from our field centers, and in some cases from industry. Some mechanism must be found for selecting the best experiments, for integrating these into a common spacecraft, and for providing funds and resources to all these various groups.

Formulation of the Program

How do we formulate the program? Do we design a spacecraft and then go out and look for experiments to fill it, or do we see what experiments we want and then design the spacecraft to carry these experiments? The program is formulated by scientific disciplines. We have a number of scientists at Headquarters who are responsible for the various scientific disciplines. We call such a person a "Scientific Program Head". Typical areas are energetic particles, ionospheric physics, and atmospheric structure. A program head should know the workers in his field; he should know the problems in his area, and the merits of the various methods of attempting to solve these problems. He, himself should have some stature in his field. There is a considerable amount of authority and responsibility centered in these people. These are the people who make the decisions with regard to which research proposals shall be funded and at what level. These are the people who are responsible for judging whether or not a particular project at a university or a NASA center is progressing satisfactorily. We like to fill these positions with mature, able, creative scientists. Unfortunately, mature, able, creative scientists are not particularly interested in administering the programs of other scientists. It is extremely difficult to find the proper people to fill these positions. There are many people who can give a lengthy dissertation on the shortcomings of NASA scientists, but the number of such people who are willing to forego their own research and come to Washington to help run the program is negligible.

In order to obtain competent people for these jobs, we are, if necessary, hiring people for a one year tour of duty. We try to use people who have served on a NASA Subcommittee for a year prior to coming to work full time. This enables them to learn what the program is and how we operate, and therefore they are able to make meaningful recommendations on program content and management almost as soon as they arrive.

So far this arrangement has worked reasonably well. We have had some very competent people to help us, and those who have helped say that this has been a very valuable experience for them and well worth a year's interruption of their own research. However, we cannot operate a program of this size with temporary people. There must be a number of permanent full-time scientists who are around long enough so that they have to face the consequences of the decisions which they make and who can train and work with the temporary people. I have digressed from my discussion on how the program is formulated in order to discuss the people who are responsible for its formulation. Now I would like to return to the discussion of program planning.

I have said that these people are responsible for the formulation of the programs in their area of competence. How do they do this? Do they sit at a desk and dream up an ideal program and then send out directives? The answer of course is no. This might be the way a totalitarian society would conduct its program but I doubt it. The ideas for experiments, the concepts to be tested, are born in the minds of individuals. It is the role of the Headquarters program scientists to assemble these ideas for individual programs or projects into a coherent program and to estimate the resources required to carry it out. The proposals and suggestions come primarily from: (1) individual scientists at universities and field centers; (2) scientific advisory groups such as the Space Science Board and the seven NASA subcommittees; (3) technical requirements for environmental data; (4) technological developments; and (5) from the head of the program. NASA involves directly some fifty outside scientists as advisors through the seven discipline subcommittees which each meet about four times per year to review the NASA program, and to advise Dr. Newell on the experiments to be selected and the research to be conducted.

The proposals and suggestions which go to make up a program may vary from a program to launch several satellites for a particular set of measurements to a suggestion for a measurement which may require a program head to interest someone else in doing the experiment. In every case the program head must interest a NASA center in undertaking the management of the project and convince his management at NASA Headquarters that it is a worthwhile project in order to get the funds and resources necessary to formulate and carry out a successful program. We find that we must have the confidence and cooperation of three major groups, the scientific community, the NASA centers who execute the program, and NASA management who must obtain the funds and resources from Congress and who have the final authority and responsibility for the use of these funds and resources. In planning a program we are responsible not only for obtaining the maximum scientific results from the currently available resources but we must also provide for a healthy scientific effort five to ten years from now. We must have a program which insures a continuing influx of young creative scientists. We must also recognize at this particular time that there will be a need for a vast coordinated program of solar physics at the time of the next solar maximum in about 1967 or 1968, and we must cajole and connive with the scientific community, Congress and industry to insure that the proper launch vehicles and spacecraft are developed, that new instruments are perfected, and that funds and resources are available.

Each year all of the proposed ongoing programs are reviewed, and a decision is made as to what the total program should be. Some projects may have to be deferred because the launch vehicles are not ready. We have had many such delays in the planetary program due to the difficulties which we have had with Centaur. Many others will have to be deferred because the budget within which we have to work does not provide for all the projects which are ready from a purely scientific and technological viewpoint. After further reviews within NASA and presentations before Congress, the final program is approved by Congress and the necessary funds and resources are allocated to accomplish the approved programs. This brings me to the second major responsibility of a program office.

Technical Direction of Approved Program

How is the program carried out? Who makes the day to day technical decisions? There are three major groups which carry out the program, universities, NASA field centers, and industry.

A. UNIVERSITIES

Universities participate in the program in two ways. They carry out a program of advanced research. With these funds data from previous experiments are analysed. Laboratory and theoretical studies are carried out. New experiments are conceived and developed to a breadboard stage. Students participate and receive training in space sciences. These programs are funded under contracts and grants.

Universities participate also by providing experiments for spacecraft. The design, development and construction of a breadboard model of an experiment is usually carried out under a grant or contract from Headquarters. If the experiment is good the experimenter will want to fly it on a spacecraft. To do so he submits a proposal to NASA. When an experiment is selected for flight on a spacecraft a number of prototype and flight units must be constructed, calibrated and tested. Since these must be constructed to meet the schedule of the spacecraft involved, the contracting and technical monitoring are often handled by a NASA center responsible for the spacecraft. Some universities prefer to manufacture the hardware themselves. Others may elect to subcontract this work to industry. In either case the principal investigator is responsible for the successful operation of his experiment. If the mission is successful, the data are given to him and he analyses and publishes them. He has exclusive use of the data for a specified mutually agreed-upon time after he receives them. It is his responsibility to analyse, interpret and publish the results of the experiment.

B. NASA FIELD CENTERS

These centers are responsible for the technical management of the various flight projects in the Space Science Program. A project may be as simple as a single sounding rocket flight or as complex as development of an Orbiting Astronomical Observatory or landing a capsule on Mars. A center is responsible for the design and manufacture of spacecraft or for the preparation of the specifications for the selection of an outside contractor to build the spacecraft.

Scientists at a center often contribute experiments for spacecraft. They submit proposals for review and for selection by the Space Sciences Steering Committee just as scientists from universities do. These scientists also contribute very substantially to the success of the overall program. They work with project engineers and spacecraft designers to insure that the spacecraft is designed to meet the objectives of the scientific experiments. Project scientists serve as the communication link between the project engineers and the experiment scientists, whether in a university or in the Center. By working closely with the project people they are able to understand and to interpret to the other scientists the necessary constraints which must be placed on experiments to insure the overall success of the mission. By being scientists and working closely with experimenters on a spacecraft they are able to explain to the project people the constraints which a particular experiment must place on a space-craft.

We at NASA feel that a strong university scientific program is necessary to insure a successful program. We feel just as strongly that a strong scientific program within NASA is necessary to insure a successful and continuing program. We find that the majority of the scientific experiments performed in space originate in universities. A substantial minority of experiments are developed by scientists in the NASA field centers. The remaining experiments originate in other government laboratories or in industrial organizations.

C. INDUSTRY

The third major contributor to the program is industry. Industrial organizations participate in a program in many ways – from supplying electronic parts to the design, manufacturing, testing and integration into the spacecraft of an experiment. In all cases this work is done under the direction of a project manager at one of the NASA field centers.

The Evaluation of Program Progress

What other criteria can we use to judge whether a project is progressing or not? In the case of spacecraft, there is really only one criterion, and that is a harsh one, success. In a scientific program we are not developing vehicles. We are looking for scientific data; therefore, although we may learn from our failure, there is no logical mechanism by which a failure can be turned into a success. This is the criterion by which the project people operate. In the case of the scientific experiment, the same criterion must be applied. This means that experiments for spacecraft must be built and tested to exacting specifications. Here friction and misunderstanding may arise between the scientist and the project people. The scientist may feel that some of the tests are unnecessary. However, if a project manager does not insist on the proper test, and the experiment fails, the manager has done no favors for the experimenter. We expect the university program, in addition to providing experiments which will yield scientific results, also to provide training for graduate students. Therefore one of the more significant criteria by which we evaluate the progress of the university program is the number of good students which graduate each year.

Selection of Experiments

Some of you may not be familiar with the method by which we select experiments. I would like to discuss this for you. Shortly after we have obtained approval to initiate a new mission, we notify all the scientists who have demonstrated competence or have expressed interest in the program, that payload space is available on a particular spacecraft. We send them a description of the spacecraft, of the orbit in which it will fly, of its payload capability and give the expected launch date. A deadline set for submission of proposals is given in the letter. We request each scientist interested in placing experiments to send 35 copies of his proposal to the Office of Grants and Research Contracts. These proposals are sent to the appropriate program office. If it is a lunar spacecraft it will be sent to the Lunar and Planetary Programs. If it is for a scientific satellite it will be sent to the Geophysics and Astronomy Programs. The program office is then responsible for the review and evaluation of the experiments. Copies of the proposals are sent to the appropriate discipline subcommittees where they are reviewed and evaluated, and placed in categories according to their desirability for flight on this particular mission. Copies of the proposals are also sent to the project manager and project scientist at the field center responsible for the flight project, and they review the proposals for compatibility with the spacecraft. Those experimenters placed in the top category by the subcommittees are called to an experimenters meeting at Headquarters. At this meeting they describe and defend their proposal before the other experimenters, the Space Sciences Steering Committee, and people from the appropriate program offices. The responsible program office recommends a payload for the spacecraft after this hearing. The Space Sciences Steering Committee reviews the payload procedures which have been followed, and the recommendations of the various subcommittees and the program office; and then makes a final recommendation to the Director of the Office of Space Sciences. The Director of Space Sciences, Dr. Newell, has final responsibility for the selection of the payload which is to be flown. After a payload has been selected the Director notifies the center responsible for the spacecraft which experiments have been selected and authorizes the center to contact the experimenters and negotiate contracts to provide funds for the flight hardware. At the same time the principal investigator for each of the experiments is notified of his selection and informed that he will be contacted by the center. The center is responsible for integrating experiments into the spacecraft and for assuring that the spacecraft meets the mission objectives. The principal investigator is responsible for the design, manufacture, calibration and testing of his own particular experiment.

Resources

So much for how we operate. Now let us look at what the program costs. Figure 2 shows the amount of money which has been spent in the various elements of the program since its inception in 1959. It shows that the program has grown steadily. This year we have asked Congress for $837 million. As shown on the chart this is divided among the six programs as follows:

Program	*(Millions)*
Lunar and Planetary	$331.3
Geophysics and Astronomy	$232.6
Biosciences	$ 41.3
Launch Vehicle Development	$149.6
Sustaining University Program	$ 56.4
Construction of Facilities	$ 25.5
	$836.7

There has been considerable concern expressed in the press that the resources required for the Manned Space Flight Program would jeopardize the Space Science Program. You can see by the growth of the Space Science Program that this has not happened. Mr. Webb and the entire top management of NASA understand the fundamental importance of the scientific exploration of space.

The funds listed for Launch Vehicle Development do not include the costs of the vehicles used for flight projects which are included in the individual projects. The Launch Vehicle Development funds are used primarily for the development of Centaur which we need so badly for our Lunar and Planetary Program.

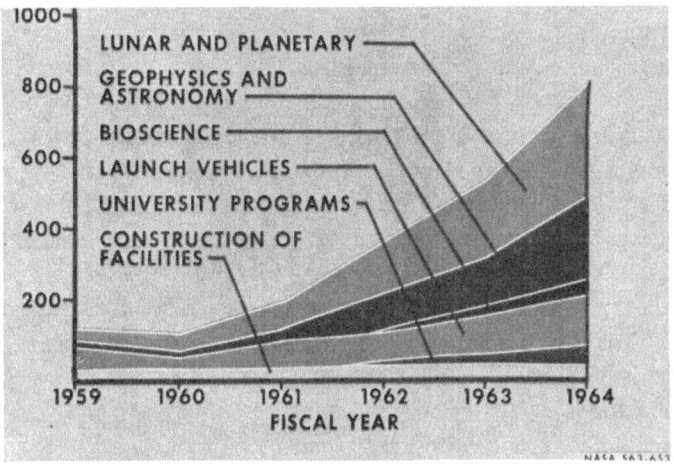

Fig. 2.   Space sciences program funding in millions of dollars.

Launch Vehicle Program

The Director of the Office of Space Sciences, in addition to being responsible for the formulation of the scientific program, is also responsible for the launch vehicles which are used in the program. Dr. Richard B. Morrison, Director of Launch Vehicle Program is responsible for the procurement and development of the vehicles to be used in the program. Figure 3 shows the major vehicles that we use, together

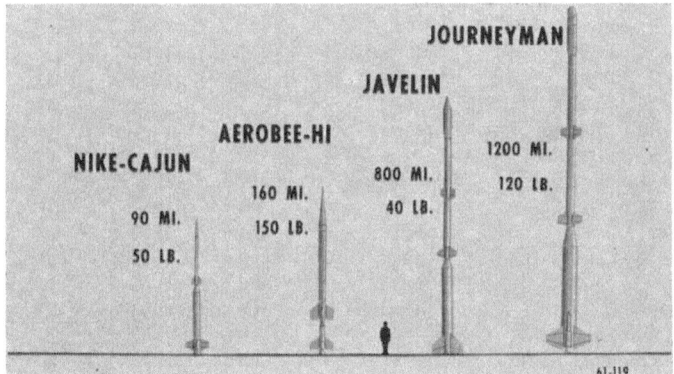

Fig. 3.   Light and medium launch vehicles. Average cost of vehicles.

with their cost. The Scout is a four-stage solid propellant rocket. It can place about 220 pounds in a 1670 km circular orbit. The Scout is almost completely developed. However, it will require another four or five shots before it has the same high probability of success in each firing that has been achieved with the Delta. The Delta is a three-stage vehicle, two liquid first stages and a solid third stage. It can place about 500 pounds in a 1850 km circular orbit or carry about 120 pounds to escape from the earth. Thor-Agena is a two-stage vehicle with two liquid stages. It can place about 1000 pounds in a 1200 km circular orbit. The Atlas-Agena is a bigger two-stage vehicle using the same second stage as a Thor-Agena. It can place a 3600 pound Orbiting Astronomical Observatory in a 2500 km circular orbit or place a 1000 pound payload in an eccentric orbit with an apogee of about 100000 km.

We are still developing the Centaur. As you know, the first test was unsuccessful. A second test is scheduled in the fall of 1963. We firmly believe that we are well on the way to developing a successful Centaur. At present we are expecting the Centaur to become operational in late 1964 or early 1965.

I think that there is no group of scientists who are more dependent on supporting technology for their results than the space scientists. If a launch vehicle malfunctions or a spacecraft fails, not only are several million dollars and two or three years of work by the experimenters wasted, but another year or two may elapse before we can schedule another spacecraft as replacement.

One of the reasons that so many scientific results are available for this symposium has been the remarkable success of the Delta Rocket. Table I shows the shots which

TABLE I

DELTA PROGRESS

Delta No.	Experiment	Popular name	Launch date	Remarks
1	Passive communications	Echo	May 13, 1960	Unsuccessful
2	Passive communications	Echo	Aug. 12, 1960	Successful
3	Weather – infrared	Tiros II	Nov. 23, 1960	Successful
4	Magnetic field	Explorer X	Mar. 25, 1961	Successful
5	Weather – infrared	Tiros III	July 12, 1961	Successful
6	Energetic particles and electromagnetic radiation	Explorer XII	Aug. 15, 1961	Successful
7	Weather – infrared	Tiros IV	Feb. 8, 1962	Successful
8	Solar Observatory	OSO I	Mar. 7, 1962	Successful
9	Ionosphere	Ariel I	Apr. 26, 1962	Successful
10	Weather	Tiros V	June 19, 1962	Successful
11	Active communications	Telstar I	July 10, 1962	Successful
12	Weather	Tiros VI	Sept. 18, 1962	Successful
13	Energetic particles	Explorer XIV	Oct. 2, 1962	Successful
14	Energetic particles	Explorer XV	Oct. 27, 1962	Successful
15	Active communications	Relay I	Dec. 13, 1962	Successful
16	Active communications	Syncom I	Feb. 14, 1963	(improved delta)

have been fired with Delta. You will be discussing results from Explorers X, XII, OSO-I, Ariel and Explorers XIV and XV launched with Deltas.

We have improved the performance of Delta since its initial use. Figure 4 shows the performance of the old and the new versions. This improved Delta is very important for our future program. We will be able to place a 130 pound Interplanetary Moni-

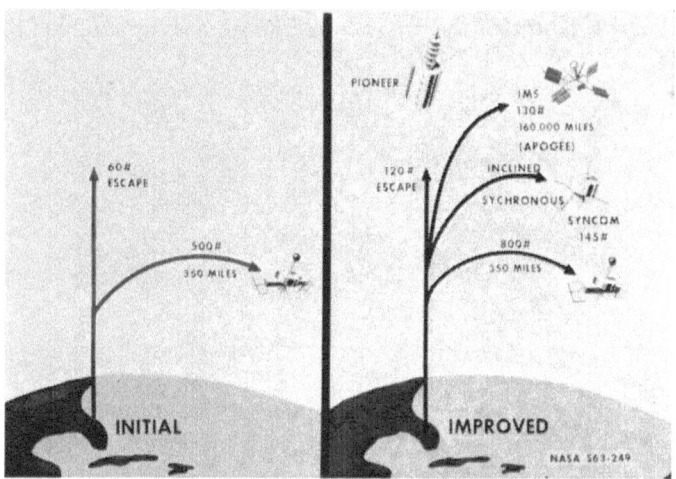

Fig. 4.   Delta improvement program.

toring Probe in an eccentric orbit extending out well beyond the magnetosphere. We could send a Pioneer weighing somewhat more than 100 lbs. to escape and place a 145 pound Syncom in a Geostationary orbit.

Sustaining University Program

When the decision was made to expand and accelerate the national space effort, it was immediately apparent that neither the magnitude nor the usual mechanisms of the sponsored research program was adequate to make the best use of university scientists or to insure that there would be an adequate supply of scientists and engineers for the future.

In order to enlarge university participation in the program and at the same time to strengthen the universities, the Sustaining University Program was established in the Office of Grants and Research Contracts. To augment and complement sponsored research and in-house activities in support of NASA's mission, grants are made to the universities to increase the future supply of scientists and engineers required in space-related science and technology; to help universities provide facilities urgently needed for space research; and to enable universities to increase their role in support of NASA's program through encouragement of creative multidisciplinary investigations, development of new capabilities, and consolidation of space-oriented activities.

For Fiscal Year 1964, we have asked Congress for 55 million dollars for this program. Of this amount, 25 million dollars is planned for training grants, 18 million dollars for new laboratory facilities, and 12 million dollars for support of research (Figure 5).

NASA began its graduate training program in 1962 with grants to ten universities for three-year support of 100 graduate students at a total cost of about 2 million dollars. In 1963, we increased the program to 786 trainees at a total of 88 universities

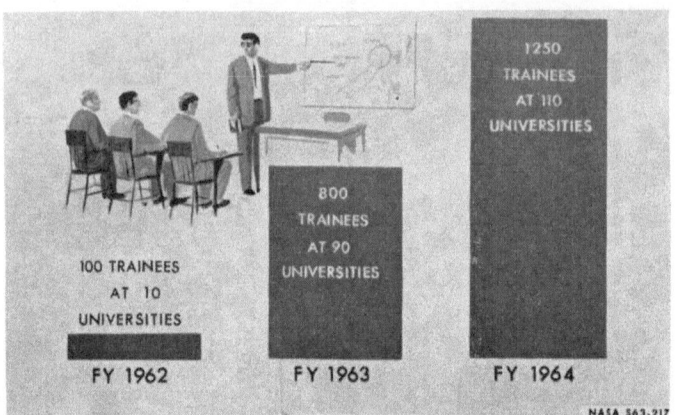

Fig. 5.   Graduate scientists and engineers entering training.

at a cost of nearly 15 million dollars. The 25 million dollars which we have requested this year will increase the number of universities to around 110 and the number of new predoctoral students to approximately 1250. With this growth rate, we would therefore expect to have over 2100 students in training after the third year. The selection of recipients for traineeships is the responsibility of the institutions receiving the training grants. The main element of the training program provides three-year predoctoral training opportunities to selected graduate students at qualified universities offering Ph.D. degrees in space-related areas.

The first six grants for new research facilities were made in 1962 at a total cost of about 6.5 million dollars. This year an additional 11 million dollars is being used to provide laboratories at nine more institutions. The 18 million dollars requested for FY-1964 should provide facilities at approximately 15 more institutions so that by the end of 1964 some 30 universities should have grants for new facilities (see Figure 6). A major criterion for selecting the location of these facilities is the relative importance to the national space program of the particular field or combination of fields of research for which the university will use the laboratories.

The research supported by the Sustaining University Program is only a very small part of the total supported by NASA at the universities. In the past year, for instance, if we include all the research supported at universities by all the offices at Headquarters and by the NASA centers together with the funds provided for flight experiments, NASA spent about one hundred million dollars at universities, whereas only 5.6 million dollar was supported through the Sustaining University Program. Why do we need this mechanism to support university research? There are two reasons.

The other research programs support only those groups with a demonstrated interest and competence in the space program. The research funds in the Sustaining University Program are used to encourage new and untried groups to develop a research capability that will be of interest and value to the research programs. The second reason for such a program is to provide funds for the consolidation of related research

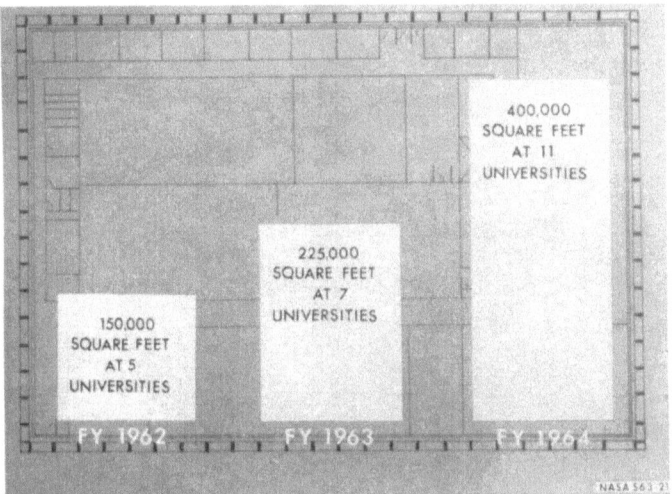

Fig. 6.   University Research facilities.

projects into a unified interdisciplinary activity. In summary, the Sustaining University Program was established to help train new people, to provide new laboratory facilities and to bring new research groups into the space program.

Bioscience Program

A Bioscience Program has rather recently been initiated under the direction of Dr. Orr E. Reynolds. Because biologists have had a difficult time visualizing important problems in their field of science which were both related to the space program and also capable of solution at the present state of technological development, the biological aspects of NASA's programs have been slow in getting started. Additional problems were presented by the difficulty in securing satisfactory experimental apparatus, even when appropriate areas for investigation in biology were defined.

NASA's current Bioscience Programs include one in environmental biology which has as an objective the extension of knowledge of the effect of the changed environmental conditions in space on earth organisms. Which way will a plant grow in the absence of gravity? What will be the effect of prolonged weightlessness on organisms, particularly man? There is also an Exobiology Program whose objective is the search for life on other planets and elsewhere in the solar system. One of the most important discoveries which could come from the space program would be the discovery of life away from the earth which had developed in an environment different from that of the earth.

Figure 7 is a picture of an experiment which has been developed to search for life on the moon and other planets. It is called "Gulliver". A "sticky string" is shot

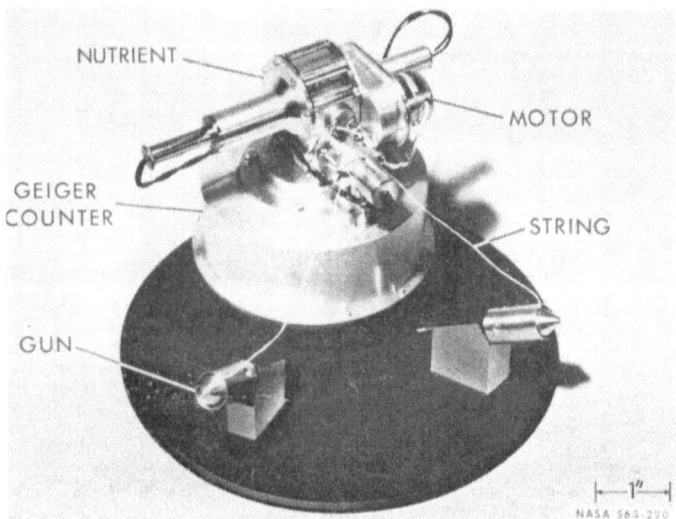

Fig. 7. A planetary life detector.

out by the guns and then withdrawn into a nutrient which contains radioactive carbon. Radioactive CO_2 is given off by the metabolism of the organisms collected, and its rate of evolution is measured by a geiger counter. This apparatus has worked well in tests on the earth. Development of suitable devices such as this to detect life on Mars, for instance, is not our major problem. The difficult problem is that of placing an experiment such as this on a planet in such a way that it does not contaminate the planet with organisms from earth. This requires a degree of sterility of equipment which far exceeds that of a modern operating room. The necessary sterilization procedures and the space equipment which can stand these procedures must both be developed. The equipment must not only withstand the sterilization procedure, but must also be able to work six months later when it reaches Mars.

Twenty-six million dollars have been requested from Congress this year to support a flight program in Biosciences. The Biosatellite shown in Figure 8 will be a self-contained laboratory for performing a variety of experiments. Three firms have study contracts for the Biosatellite. It has been announced that NASA will negotiate a contract with General Electric (Philadelphia). Some 150 experiments have been proposed for flight on Biosatellite.

In addition to the Biosatellite Program the balloon-borne Stratoscope II is being used to measure the infrared spectra of the planets in a search for life-related molecules.

In summary, there are many worthwhile and important bioscience experiments appropriate to NASA's mission. Some Interesting results have been obtained, but none as yet offer significant conclusions. A flight program has been initiated and is in its beginning phase, with expectations of significant results in two or three years. NASA biologists and engineers together are facing the demand for technology ap-

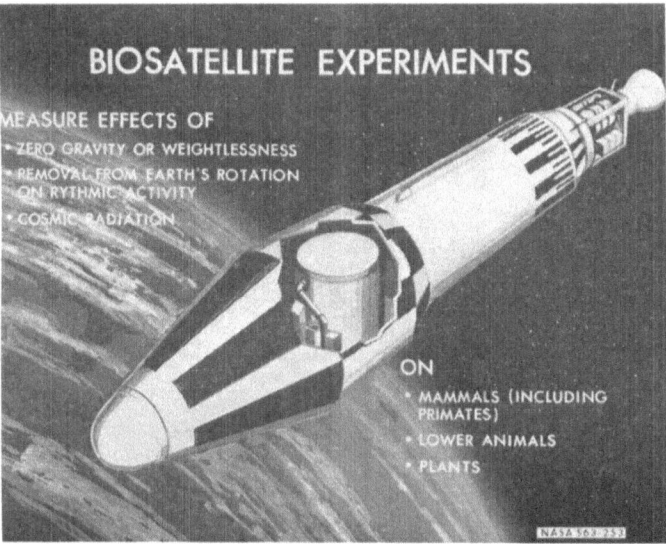

Fig. 8.   Biosatellite experiments.

propriate to the solution of biological problems so significant in space exploration. The task will require the best that both engineers and biologists can offer in talent and in cooperation.

Geophysics and Astronomy

The fundamental mission of the Geophysics and Astronomy Program is to learn more about the earth and its space environment; to investigate solar terrestrial relationships; and study the nature and history of the solar system and the universe. A secondary purpose is to provide needed information in support of both civilian and military requirements in space. Balloons, sounding rockets and satellites are used to carry instruments into space to accomplish these objectives.

Figure 9 shows the organisation of the Geophysics and Astronomy Program Office. It is divided into two groups, a group of engineers responsible for the on-going flight projects and a group of scientists responsible for establishing the objectives of the program and planning for future flight projects. For administrative reasons the scientists are grouped into four major scientific disciplines: Astronomy, Solar Physics, Physics, and Chemistry.

Astronomy may well be the most important program we have. We think that the study of the stars in the ultraviolet and in the infrared will provide fundamental data which will help to understand stellar processes and the nature and history of the universe. You will see that we are also putting the major share of our resources in the Astronomy Program.

Solar Physics and Solar-Terrestrial Relations is likewise a most important program. You have, I believe, already heard the very significant and interesting results which have been obtained from the Orbiting Solar Observatory, OSO. The ability to observe the solar radiation and to monitor continuously the flux from the sun over the entire spectrum from X-rays to the infrared region will certainly provide new fundamental

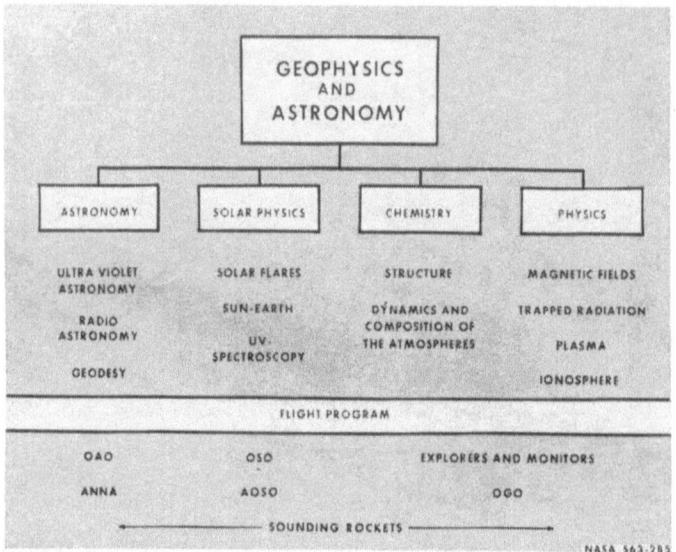

Fig. 9.

data on the processes which take place in the sun. Knowledge of this flux and of its variation, combined with results from simultaneous measurements of the properties of the magnetosphere, ionosphere and atmosphere of the earth will provide new insight into the control which the sun exerts over our terrestrial environment.

While I have indicated that I think that most significant results may come from solar physics and astronomy, the major share of our results to-date have come from the studies of the magnetosphere, the trapped radiation, solar and galactic cosmic rays, and the ionosphere.

Supporting each of these disciplines is a Flight Program. The Orbiting Astronomical Observatory (OAO) carries the experiments for the Astronomy Program. ANNA, the Geodetic Satellite, is also a part of the Flight Program of astronomy. The Orbiting Solar Observatory (OSO) is the major flight project in solar physics. We have asked Congress for funds to initiate an Advanced Orbiting Solar Observatory (AOSO) which I will discuss later. There are a number of explorers and monitor satellites which support both chemistry and physics. Explorers XII, XIV, XV and XVII were all a part of this program. The Orbiting Geophysical Observatory (OGO) will carry experiments for both chemistry and physics as well as solar physics and astronomy.

We have a broad program of sounding rockets which supports all disciplines. Sounding rockets are used for two major reasons. There is an altitude in the atmosphere between about 50 and 200 kilometers which is accessible only to rockets; balloons cannot go above about 40 kilometers and satellites will not operate continuously below about 180 or 200 kilometers. Sounding rockets are also used to test out experiments before they are flown on more expensive spacecraft. We follow a general policy of asking an experimenter to fly his experiment on a sounding rocket before it is flown in a spacecraft. Some experimenters have objected, but I know of no case in which, after the fact, an experimenter has complained that it was a waste of time. In all cases, experimenters have acquired some very valuable information about the characteristics of their apparatus, and in many instances have had to go back and redesign their experiment on the basis of data obtained from sounding rockets.

Table II shows the budget of the Geophysics and Astronomy Program and how it is spent. You will note that the two big observatory programs, OAO and OGO, take almost half of the budget. Where are we in the Geophysics and Astronomy Program? What are we doing now? Table III shows the scientific satellites which

TABLE II

GEOPHYSICS AND ASTRONOMY
IN MILLIONS OF DOLLARS

	FY 1962	FY 1963	FY 1964
OSO	4.3	10.1	15.8
OAO	35.9	35.1	43.5
OGO	23.1	32.2	46.7
Explorers, monitors, etc.	14.7	20.8	27.6
Sounding rockets	12.1	17.2	19.6
Supporting research	11.3	19.1	21.2
Launch vehicles	18.4	39.7	58.2
Total	119.8	174.2	232.6

TABLE III

SUCCESSFUL SCIENTIFIC SATELLITES LAUNCHED BY NASA SINCE 1962

Name	Purpose	Launch date
OSO I	Solar physics	Mar. 7, 1962
Ariel	Ionospheric physics	Apr. 26, 1962
Alouette	Ionospheric physics	Sept. 29, 1962
Explorer XIV	Energetic particles and fields	Oct. 2, 1962
Explorer XV	Artificial belt	Oct. 27, 1962
Explorer XVII	Atmospheric physics	Apr. 2, 1963

have been launched by NASA since the beginning of 1962. You will note that in 1962 we had six successful satellites out of six attempted launches. Five of these were launched with a Delta rocket. The sixth was launched with a Thor-Agena, the Alouette Satellite. All of the satellites achieved their objectives. At the present time, we seem to have solved the problem of making small spin-stabilized satellites work. We have just begun to use the large observatories. The Orbiting Solar Observatory was the first of these, and it was a success.

Figure 10 shows the scientific satellite summary. It shows that we are launching more successful satellites each year. In about two more years we will be heavily

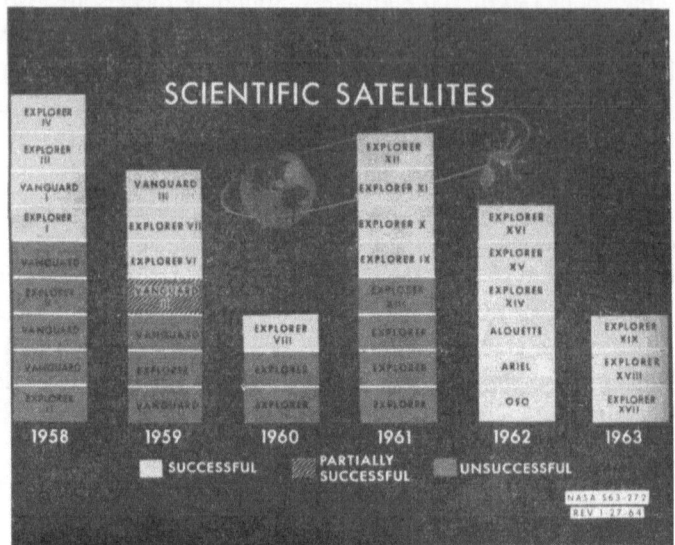

Fig. 10.   Scientific satellites. Summary of launches.

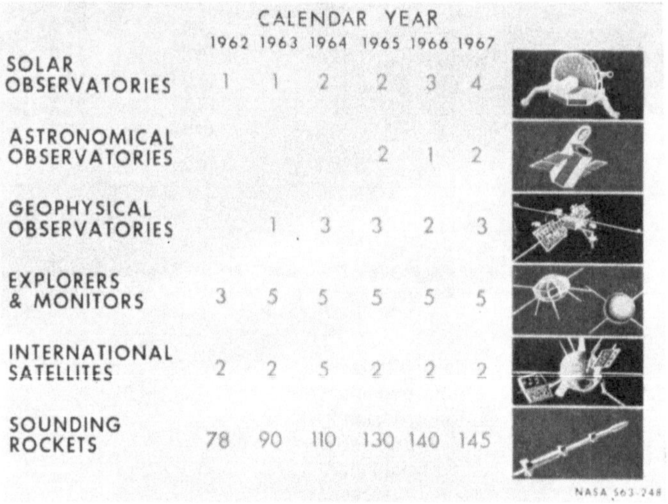

		CALENDAR YEAR					
	1962	1963	1964	1965	1966	1967	
SOLAR OBSERVATORIES	1	1	2	2	3	4	
ASTRONOMICAL OBSERVATORIES				2	1	2	
GEOPHYSICAL OBSERVATORIES		1	3	3	2	3	
EXPLORERS & MONITORS	3	5	5	5	5	5	
INTERNATIONAL SATELLITES	2	2	5	2	2	2	
SOUNDING ROCKETS	78	90	110	130	140	145	

NASA S63-248

Fig. 11.   Geophysics and astronomy programs.

involved in our large complex observatory program. We have already selected experiments for all the spacecraft to be flown through 1965. At the present time any experimenter looking for space on a spacecraft has to wait until 1966 before he will have a chance to fly an experiment. Figure 11 is a summary of the spacecraft which we will

be flying in the next five years. There will be about two Solar Observatories per year.*
Beginning in 1965, we will fly about one Astronomical Observatory per year. About
two Geophysical Observatories will be launched per year, beginning in early 1964.
One of these Geophysical Observatories will be placed in a highly eccentric orbit
and the other in a low altitude polar orbit. We will fly about five of the small spin-
stabilized explorers per year. We have had a very vigorous and successful International Program. Two international satellites have been flown, Ariel and Alouette.
Both were highly successful. A second United Kingdom satellite is scheduled for later in
1963 and we are negotiating with the Canadian Government for a continuing program
of ionospheric studies in which Canada will supply the spacecraft and we will supply
the launch vehicle. Experiments will come from both Canada and the United States.

Now I would like to describe briefly the various spacecraft which we use or are
developing in this program. Figure 12 shows a picture of the Orbiting Solar Ob-

NASA S63-241

Fig. 12.   The first solar observatory, weighing 450 pounds and launched on March 7, 1962. Its orbit
is a nearly perfect circle. Its primary missions are directed on the spectrum and intensity of solar
ultraviolet and on solar gamma rays.

servatory, OSO-1. This satellite has two parts, a sail, which is pointed continuously
at the sun as long as the sun is visible from the satellite and a rotating section to
provide stability. The major experiments are carried in this rectangular box mounted
on the sail. These experiments are directed at the sun to an accuracy of about one

* The second Orbiting Solar Observatory was launched February 3, 1965. Altogether there will be
eight such observatories in the series in order to cover a complete solar cycle.

minute of arc. The first OSO is still working in June 1963. It is still pointing at the sun and is still giving some data. We will launch the second Solar Observatory later this year. The third and fourth will be launched in 1964. Although we are well satisfied with OSO, it is limited in its pointing accuracy, in the weight of the experiments which it can direct at the sun, and in the volume of space which it has available for experiments. In order to have a spacecraft available to make detailed studies of solar phenomena at the next solar maximum we are beginning work on an Advanced

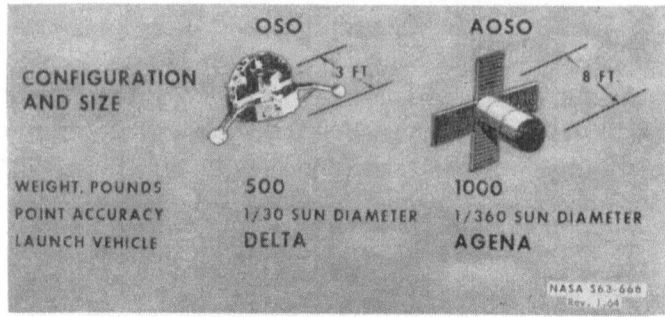

Fig. 13.   Orbiting solar observatory and advanced orbiting solar observatory.

Fig. 14.   Orbiting astronomical observatory. The OAO is containing (a) Four 12-inch Smithsonian telescopes to map the entire sky in ultraviolet; (b) Four 8-inch and one 16-inch Wisconsin telescopes to study bright stars and nebulae; (c) A 3-foot Goddard telescope for studying 5000 stars and nebulae; (d) A 32-inch Princeton telescope studying interstellar matter. The ultimate pointing accuracy is of 1/36 000 of one degree. Its weight is 3600 pounds and it will orbiting at a distance of 500 km.

Solar Observatory shown in Figure 13. Three industrial firms have completed preliminary studies of an AOSO. If Congress approves funds for this project we will be starting design studies this year. An AOSO will be launched into a polar orbit using either an Atlas-Agena or a Thor-Agena. It will be capable of pointing about 1000 pounds of instruments at any position on the sun to an accuracy of five seconds of arc. It will be able to accommodate optical instruments up to eight feet long. It will have a much increased telemetry capability. With a start now, AOSO will be ready for its first launch in 1957, and should be fully operational by next solar maximum.

Orbiting Astronomical Observatory

The Orbiting Astronomical Observatory (OAO) shown in Figure 14 is easily the most complex and most expensive spacecraft which we have under development. It

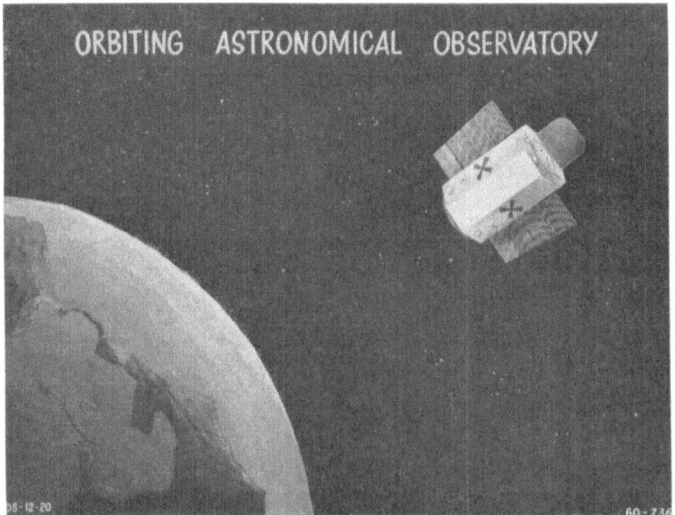

Fig. 15.   Artist's conception of the Orbiting Astronomical Observatory in orbit.

will weigh 3600 pounds and is designed to point experiments to an accuracy of a tenth of a second of arc. At present three spacecraft and three separate sets of experiments are being built. The first OAO, which will be launched in 1965*, will carry two sets of experiments. One experiment supplied by the Smithsonian Institution Astrophysical Observatory consists of four 12-inch telescopes to map the sky in the ultraviolet. A second experiment, from the University of Wisconsin, consists of four 8-inch and one 16-inch telescope to study bright stars and nebulae. The second OAO is planned to carry a three-foot telescope from the Goddard Space Flight Center and will be used to study some 5000 stars and nebulae. The third OAO will carry a 32-inch Princeton University telescope to study interstellar matter. Figure 15 is an

---

* Now planned for launching in early 1966.

artist's conception of the OAO in orbit. I like to look at his picture. It reminds me that some day an OAO will actually go into orbit and will give us some new and very exciting data. Occasionally we need this help to give us courage to continue. Right now all we see is that OAO gives us some very high costs and some very difficult technical problems. The first three OAO's will cost about 200 million dollars to build and fly. There are about 200 people at the Goddard Space Flight Center involved in directing the development of this spacecraft, and there are about a thousand people at the prime contractor, Grumman, doing the actual design, construction and testing of the spacecraft. At the present time, a complete guidance and control system together with all the other spacecraft systems have been assembled and placed on an airbearing table. This entire system has worked, and the assembly was able to locate stars and to point in the proper direction to the proper accuracy. We still have a long way to go in this program before we have a telescope working in space, but so far we have been able to solve hie problems that we have met, and the spacecraft and its experiments are on schedule.

Orbiting Geophysical Observatory (OGO)

Figure 16 is a picture of OGO, the Orbiting Geophysical Observatory. OGO is a large, standardized spacecraft suitable for a wide variety of missions and capable of supporting up to 50 separate experiments. There were two major requirements for

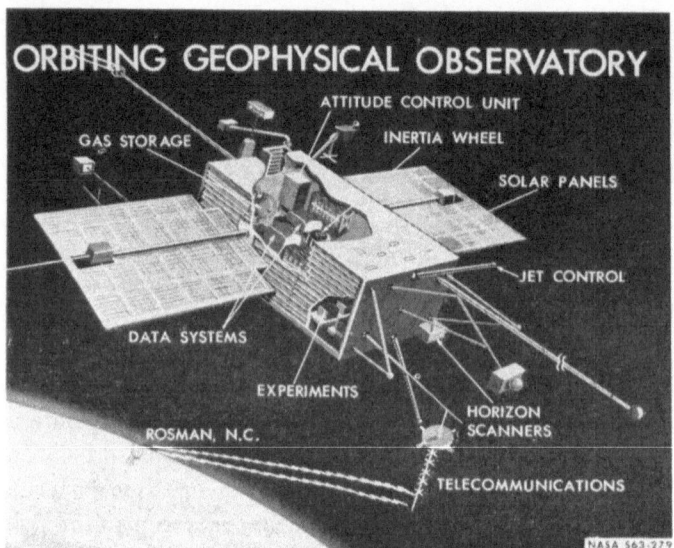

Fig. 16.

such a spacecraft. We needed a spacecraft which could carry a large variety of scientific experiments to make correlated measurements of the magnetic field, trapped radiation, plasmas, the ionosphere and atmosphere of the earth. In addition to accommodating a large number of different kinds of experiments a vehicle was wanted which could be used by a large number of individual experimenters from universities and NASA laboratories. Finally, many measurements must be made at regular intervals over a complete solar cycle. Therefore, we plan to launch an OGO on a regular schedule of one about every six months.

Experiments are carried on the top and bottom surfaces of the main body, which is oriented so that this surface is always normal to the local vertical. There is also space on the solar panels for experiments which require pointing at the sun. The odd looking appendages in the figure are experiments which require a large separation from the main body. On the present OGO's, magnetometers and VLF detectors are carried on such booms. The OGO's will be launched into two kinds of orbits. One orbit will be highly eccentric with an apogee of about 100000 km and a perigee of about 300 km and inclined about 31° to the equator. The other orbit will be a nearly circular polar orbit with an apogee of about 900 km. We call the first of these two missions, EGO, for Eccentric Orbiting Geophysical Observatory and the second POGO for Polar Orbiting Geophysical Observatory.

Each observatory will weigh about 1000 pounds and will carry a scientific payload of 150 pounds.

The first launch, an EGO, will be in the second quarter of 1964. The first POGO will be launched in late 1964.*

Experiments have been chosen for the first two EGO's and the first two POGO's. Table IV shows their breakdown by scientific discipline as of October 1962.

EGO carries primarily particles and fields experiments, whereas POGO has a larger proportion of planetary atmosphere experiments.

TABLE IV

ORBITING GEOPHYSICAL OBSERVATORIES EXPERIMENT SUMMARY

Scientific discipline	Number		Weight, Lbs	
	EGO I	POGO I	EGO I	POGO I
Planetary atmospheres	3	7	14.3	74.8
Ionospheres and radio physics	6	4	39.0	17.0
Particles and fields	9	7	89.4	42.0
Astronomy	2	1	13.3	7.6
Solar physics	0	0	0	0**
Total	20	19	156.0*	141.4**

* Includes an 8.1 Lb. backup experiment.
** 8.6 Lbs. held in reserve.

Explorers and Monitor Program

The small explorer class satellites, weighing about 100 to 400 pounds and launched with Delta and Scout vehicles, have contributed most of the data from space so far. We will continue to use these at the rate of five to seven per year.

* The first EGO was launched September 4, 1964. The first POGO is planned for launching in the fall of 1965.

Figure 17 shows the small satellites we plan to launch in 1963. One of these, the Interplanetary Monitoring Platform, carries instruments to measure the interplanetary magnetic field, the plasma, and solar and galactic cosmic rays. The first of these will be placed in a highly eccentric orbit the latter part of 1963*. We plan to repeat this at regular intervals to insure that we continuously monitor the conditions in interplanetary space. The second United Kingdom satellite will be launched this year. There are two ionospheric physics satellites scheduled for launch in 1963. One of these, S-66, is truly an international satellite. NASA is placing a satellite in orbit

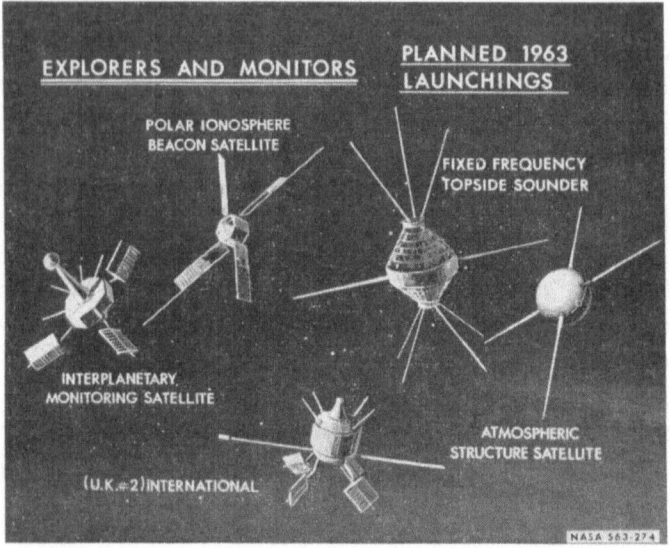

Fig. 17.

which will broadcoast at 20, 40, 41, and 360 Mc/sec. Some 66 experimenters around the world will measure the signal strength and phase changes of these signals and use this information to measure the electron content of the ionosphere. The other ionosphere satellite is the Fixed Frequency Topside Sounder.

We are presently reviewing the results from Explorers XV and XVII, the artificial radiation belt satellite and the atmosphere structure satellite, to determine if we should repeat either of these. In each case, there is either a back-up spacecraft available or sufficient parts to assemble one. We are also reviewing the results from the first geodetic satellite, ANNA, to determine when and how NASA should continue with the program.

We would like to start a small radio astronomy satellite. At present we do not have one. We have, however, begun a radio astronomy program with sounding rockets and OGO.

It is clear that there are ample uses for these small satellites.

* The first IMP (Explorer XVIII) was launched November 26, 1963. It was followed by the second (Explorer XXI) on October 4, 1964 and the third (Explorer XXVIII) on May 29, 1965.

Sounding Rocket Program

We fire about 80 sounding rockets per year. Figure 18 is a picture of those we commonly use. There are about fifteen launch sites for sounding rockets. We recently had an international series of shots in which sodium vapor was injected into the

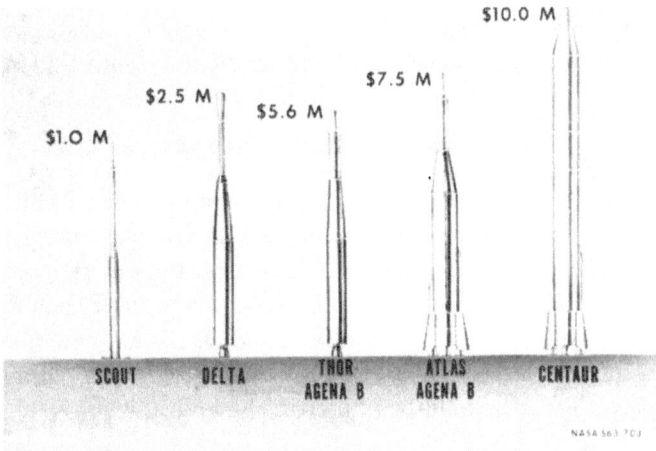

Fig. 18.   Typical sounding rockets.

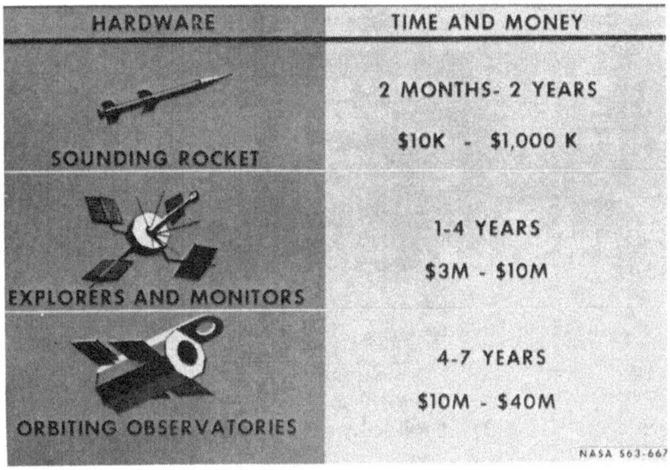

HARDWARE	TIME AND MONEY
SOUNDING ROCKET	2 MONTHS- 2 YEARS $10K - $1,000 K
EXPLORERS AND MONITORS	1-4 YEARS $3M - $10M
ORBITING OBSERVATORIES	4-7 YEARS $10M - $40M

Fig. 19.   Map of the time and money required for various kinds of projects in geophysics and astronomy.

atmosphere to measure the winds aloft. Twenty six (26) rockets were fired in a two-week period from a half dozen sites.

We encourage new groups interested in starting a program of space research to start with a sounding rocket program. Sounding rockets are cheap. An experiment can be prepared in a relatively short length of time. At the same time the experimenter learns all the harsh facts about shock, vibration, miniaturization, telemetry and field operations. All of our successful experimenters have used either balloons or rockets in their research programs prior to a satellite or spacecraft flight.

Figure 19 is a map of the time and money required for various kinds of projects. A sounding rocket project may take from two months to two years to develop and launch, and costs from 10 thousand to 2 million dollars. An explorer satellite takes from one to four years to launch, and costs from 3–10 million dollars. An observatory takes four to seven years to design and build, and costs from 10–40 million dollars per shot.

Lunar and Planetary Program

The basic objective of the Lunar and Planetary Program, under the direction of Mr. Oran Nicks, is to explore, in situ, the moon, the planets and interplanetary space. We have requested 331 million dollar from Congress for this program.

The flight projects in this program can be divided into three main groupings: (1) Lunar Exploration; (2) Planetary Exploration; and (3) Exploration of interplanetary space and the sun.

Very roughly a little less than two thirds of the budget is for lunar exploration;

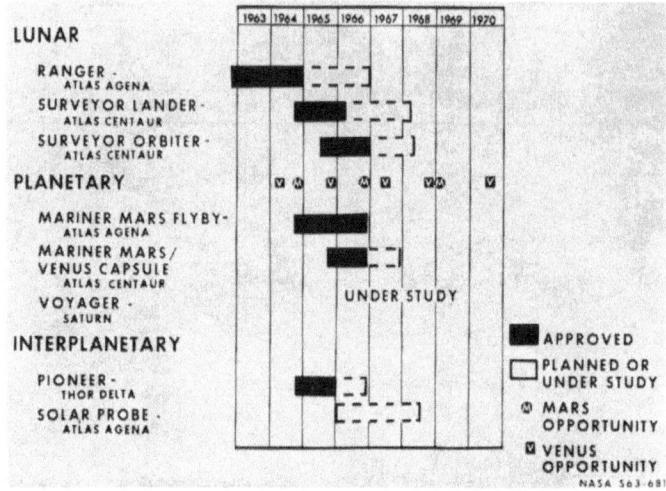

Fig. 20.   Lunar and planetary flight schedule.

there is about one third for planetary exploration, and about 5% for interplanetary exploration. However, as you know from the Mariner II results, a great deal of interplanetary work is done while a planetary spacecraft is on its way to the planet.

Figure 20 is a flight schedule for the Lunar and Planetary Program. The small *V* and *M* show the times when Venus and Mars launch opportunity occurs.

One of the major problems which this program has had has been the availability of a suitable launch vehicle. An Atlas-Agena can carry a 750 pound Ranger to the moon or a 570 pound Mariner on a Mars flyby. However, it cannot soft land a

2100 pound Surveyor on the moon or land a capsule on Mars. These missions require the capability of a Centaur. The first Surveyor is scheduled to be flown on Centaur in 1964. In the Lunar Exploration there are three separate flight projects, Ranger, Surveyor and Orbiter. I will discuss the objectives of each of these and show you what the spacecraft looks like. I will do the same for the two planetary projects, the Mars flyby and Mars/Venus capsule or lander. I will also describe Pioneer, the Interplanetary project which is underway and which is the partner to IMP in the Geophysics and Astronomy Program.

RANGER

Figure 21 is a picture of Ranger. It weighs 750 pounds and is launched with an Atlas-Agena. So far five Rangers have been fired. The first three were unsuccessful because of failure in the launch vehicle. The last two which were fired in April and

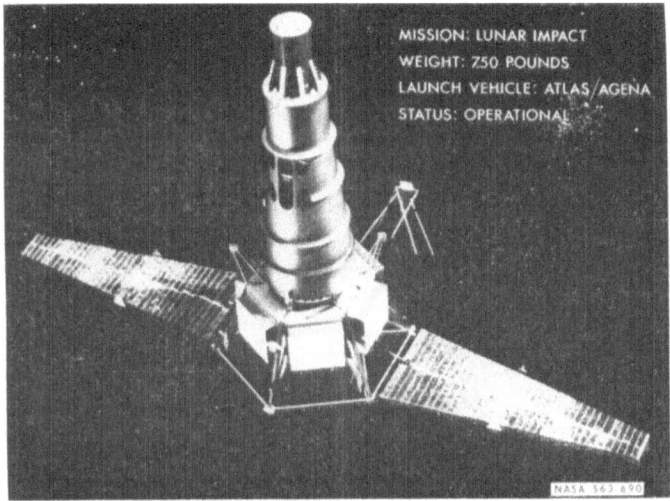

Fig. 21.   Ranger.

October 1962, were cases in which the launch vehicle performed satisfactorily but the spacecraft failed after being ejected into the proper trajectory. As a result of these troubles with the spacecraft the launch of future Rangers was delayed pending an intensive review of the program. A number of improvements have been made to increase the reliability of the spacecraft. The next flight of a Ranger will take place late in 1963. We are confident that the Ranger spacecraft series will succeed and yield valuable data to science*, and also to the Apollo Program. The Ranger spacecraft was designed for hard landings of scientific payloads and for taking close-up pictures of the lunar surface. Figure 22 shows the kind of pictures which Ranger will transmit back. When the first picture is taken it will include a square of lunar surface 125 miles

* Rangers VII, VIII and IX launched in 1964 and 1965 took more than 17 700 photographs of excellent quality of the lunar surface, introducing a new era in lunar exploration.

on a side and have about a 60 mile resolution. The last picture will be of a square of 60 feet on a side and will have about a three foot resolution. In between these pictures Ranger will take some 3250 pictures.

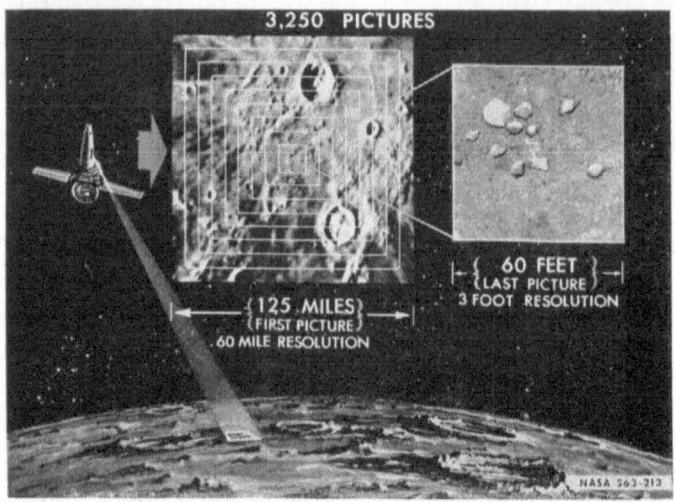

Fig. 22.   Ranger TV photos.

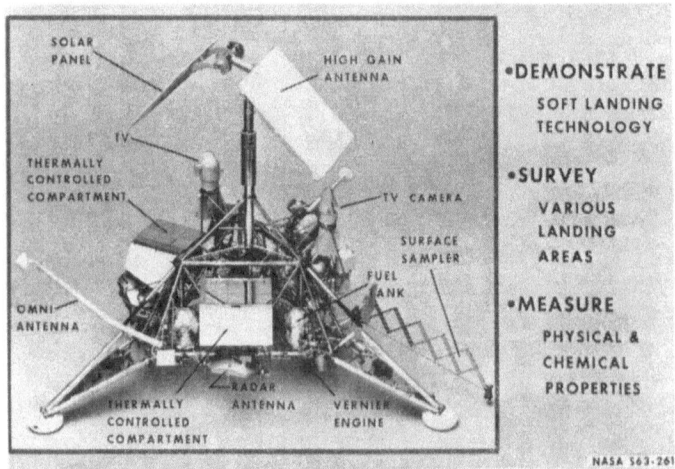

Fig. 23.   Surveyor spacecraft.

SURVEYOR

Figure 23 is a picture of the "hard mock-up" of the Surveyor spacecraft. It weighs about 2100 pounds and will be soft-landed on the moon. Surveyor will also obtain some TV pictures during its final decent to the moon; however, its greatest contribution from the standpoint of pictures will come after Surveyor is safely on the lunar surface. Its television system will provide eye level viewing of the local lunar landscape in the vicinity of the spacecraft and visual examination of the texture and form of the material comprising the lunar surface. Surveyor will be used to survey the various landing sites for Apollo. It will also be used to measure the physical and chemical properties of the lunar surface in the immediate vicinity of the spacecraft. In addition, Surveyor carries devices to sample the lunar surface and to analyse it, and instruments to measure the local mechanical properties of the lunar surface.

Surveyor instruments

Figures 23 and 24 show some of the instruments which may be carried on Surveyor. In the upper right hand corner of Figure 23 is a Lunar Surface Sampler operated by remote control from the earth which will be used to pick up samples of the lunar surface for analysis. The bevameter will be used to measure the bearing strength of the lunar surface. The X-ray Diffractometer will use X-rays scattered by materials collected with the Lunar Surface Sampler to determine their composition. A seismometer will be used to measure vibrations of the lunar surface.

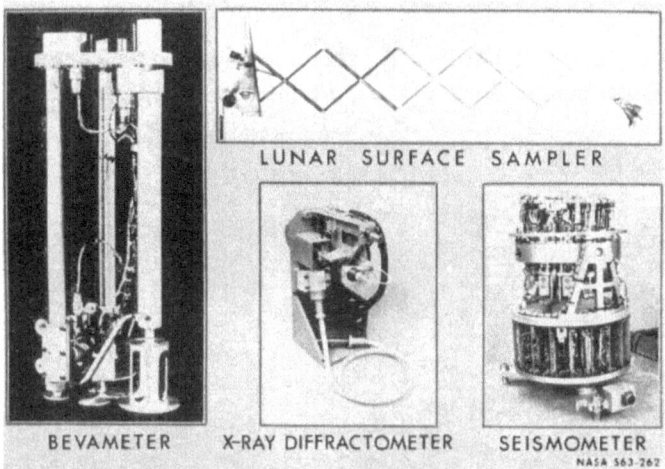

Fig. 24.   Instruments of the Surveyor.

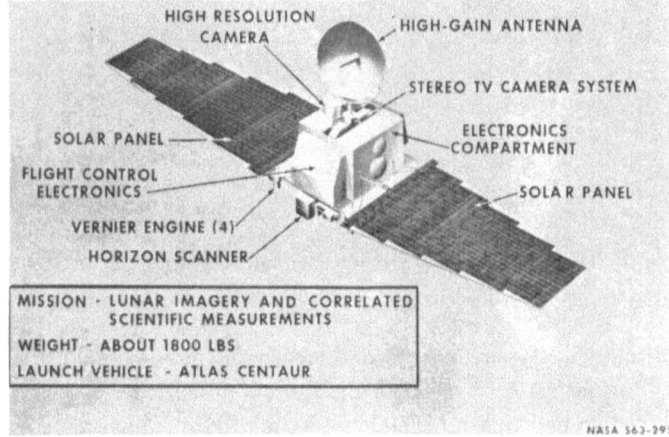

Fig. 25.   Picture of the Surveyor Orbiter.

Surveyor orbiter

Figure 25 is a picture of one version of the Surveyor Orbiter. This version would be launched with the Atlas-Centaur and would weigh about 1800 pounds. We are still studying the Surveyor Orbiter, and the spacecraft ultimately may be launched with an Atlas-Agena. The Lunar Orbiter will be used to provide photo-reconnaissance coverage which will help us to extrapolate to other lunear areas the knowledge gained about local spots from landings of Rangers or Surveyors. This information will help the Apollo project in selection of manned lunar landing sites. In addition, the Orbiter will enable us to determine the gravity field of the moon, information which will be needed to plan Apollo orbits.

MARINER SPACECRAFT

Mariner II

Figure 26 is a picture of Mariner II. The latter weighed 447 pounds and carried 41 pounds of instruments. Some of the more significant and interesting results which

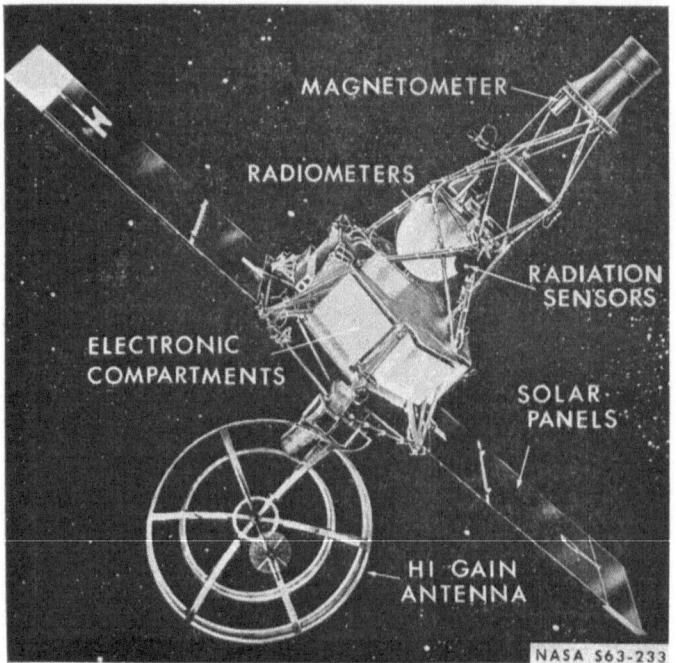

Fig. 26.   The Mariner II, launched August 27, 1962, flew by Venus on December 14, 1962.

you will be discussing this week were obtained with the magnetometer and particle detectors on Mariner II. As a result of the success of the measurements which were made we decided to cancel a repeat of the Venus Flyby which had been scheduled for 1964.

Mariner C

Figure 27 is a picture of Mariner C, a spacecraft which is scheduled to fly by Mars in 1964.* It will carry a complement of instruments similar to that carried on Mariner II to measure the solar plasma in the interplanetary magnetic field and the energetic particles. One item of interest is the small solar sails located at the ends of the solar

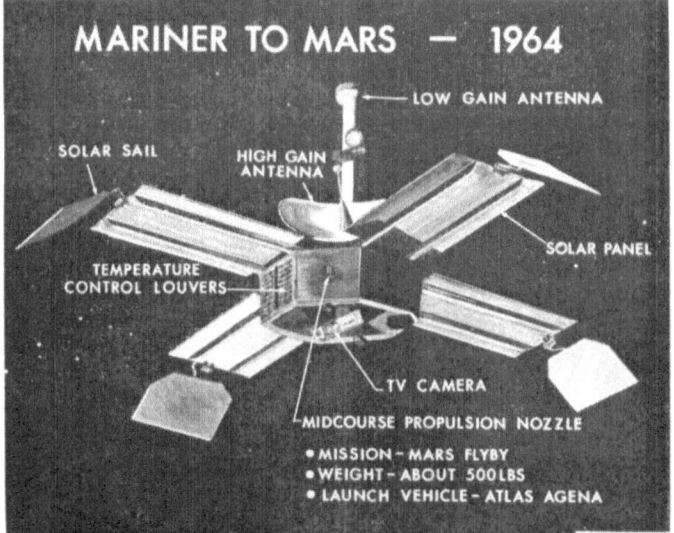

Fig. 27.

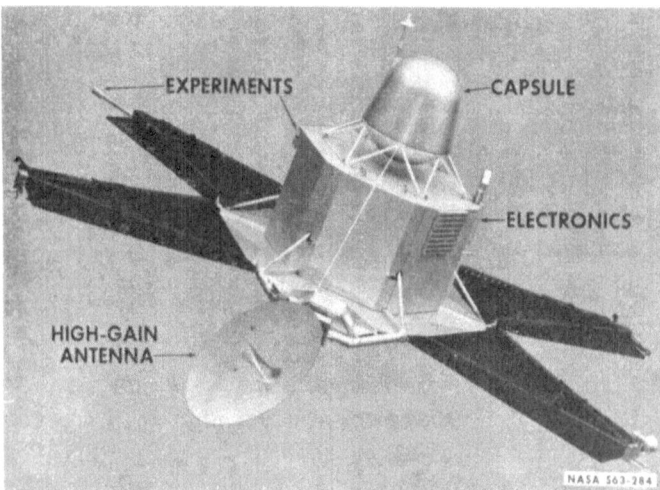

Fig. 28. Mariner B. Flyby and land capsule.

* Mariner IV launched November 28, 1964; its major mission to Mars ended triumphantly on July 14, 1965 by acquiring close photos of this planet.

panels on Mariner C. These small panels can be moved and will use solar radiation to help control the orientation of the spacecraft.

Mariner B

Figure 28 shows our concept of Mariner B, the spacecraft in which there is a bus or main portion which is designed to fly by the planet, and as it passes to eject a capsule which will land on the planet. This Mariner will be launched with an Atlas-Centaur and the first such flight will probably be to Mars in 1966.

Pioneer

Figure 29 is an artist's conception of Pioneer for the International Year of the Quiet Sun, or PIQSY, as it is called. It will weigh 120 pounds and will be launched with a

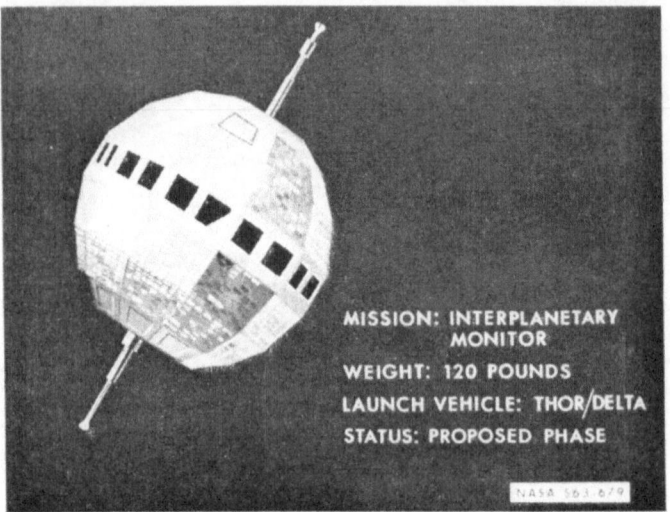

Fig. 29.   Artist's conception for the Pioneer for the International Year of the Quiet Sun.

Thor-Delta vehicle. It will be spin-stabilized, with its spin axis perpendicular to the plane of the ecliptic. The boom sticking out of the top will carry a magnetometer. Pioneer is an approved project. The payload has not been finalized, but will include a magnetometer, plasma probe and energetic particle detectors.

Pioneer orbits

Figure 30 shows the orbits which Pioneer will fly, depending upon whether it is fired backward or forward in the earth's orbit. If it is fired backward it will move inward to 75 million miles from the sun. If it is fired in the direction the earth is going, it will move outward to about 101 million miles from the sun. It will be instrumented with a magnetometer, plasma probes and galactic cosmic ray detectors.

From Pioneers and IMP's, we will begin to be able to plot the interplanetary

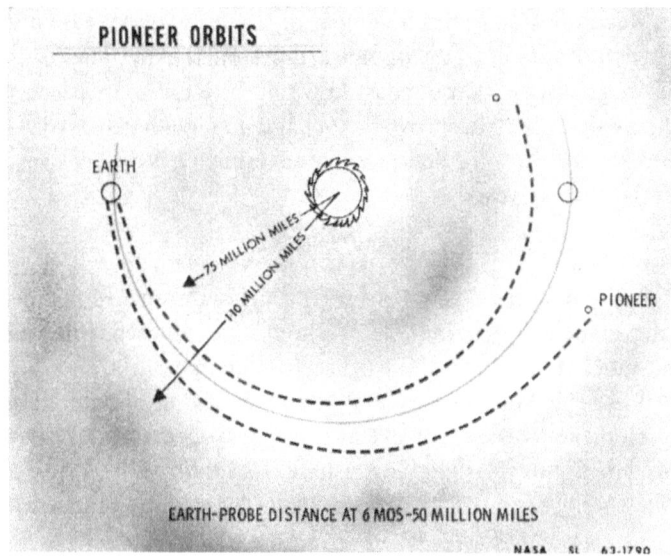

Fig. 30.

field and to observe the size of the plasma streams which are ejected from the sun, and perhaps able to measure incubation times of solar disturbances.

Missions Under Study

Table V lists some of the missions which are under study in the Lunar and Planetary Program. These are mission which we would like to do if we had the money and the necessary vehicles to do them. Voyager would be a vehicle capable of orbiting Mars or Venus and landing. It would weigh somewhere between 3000 and 8000 pounds and would require the launch capability of a Saturn.

TABLE V

MISSIONS UNDER STUDY

Voyager
Mission: Mars and Venus, orbit and land
Weight: 3000–8000 Lbs.
Launch vehicle: Saturn

Solar probe
Mission: Probe of the sun to about 1/3 AU
Weight: About 300 Lbs.
Launch vehicle: Atlas Agena

Other
Outer planets
Out of the plane of the ecliptic
Comets
Asteroids
Escape from the Solar System

The Ames Research Center has a number of studies underway on a vehicle which would go inward to about ⅓ of an astronomical unit from the sun. They estimate that this vehicle would weigh about 300 pounds and would be launched with an Atlas-Agena. Neither of these missions are approved as yet and no flight hardware is being built for them. It would be 1969 or 1970 before we could launch a Voyager even if we started work on it in the coming year.

Problems

There are a number of specific problems with which we are wrestling at present. One of these is the planetary program. How should we proceed with this program? Should we do more research on the earth before going to Mars or Venus? Should we continue to use the Mariner "C" 600 pound flybys, or should we use those resources to develop the Voyager for use as a heavy spacecraft, and orbit the planets and land capsules. Figure 31 illustrates a basic fact that we have to face. This is a plot of relative energy required to carry a given payload to Mars or Venus at the

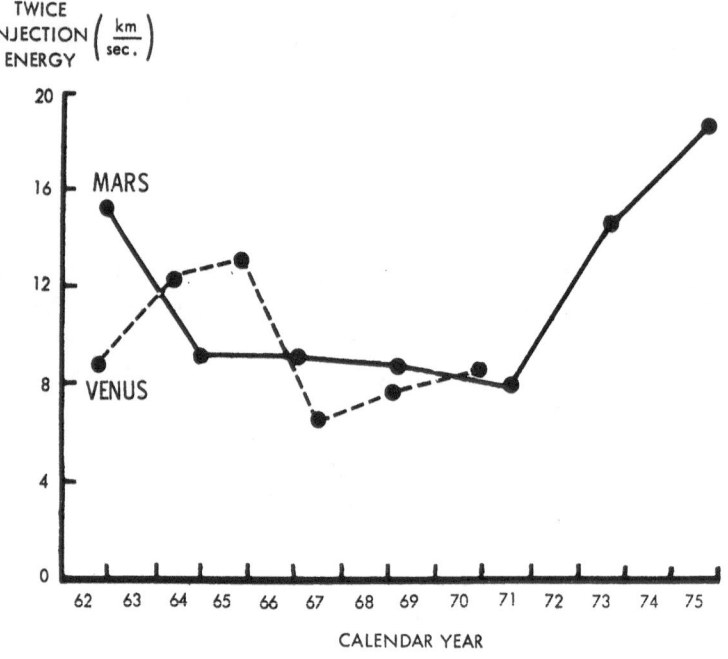

Fig. 31.   Energy requirements for Mars and Venus.

different planetary opportunities in the next decade. You will note that there is a sharp increase in the energy requirement after 1971. Or in other words, the payload, which a given vehicle can carry, decreases sharply after 1971 and remains low for the next several years. If we start development on a Voyager spacecraft and a third stage for Saturn in FY 1965 they would both be ready in 1969. If we delay they won't be ready until 1971 or later.

You will have noted that all interplanetary flights presently scheduled remain between the orbits of Mars and Venus. We are not making large radial excursions as yet. We have a probe towards the sun under study, but we have not made a decision to start the design of the spacecraft yet. For a long time there was a continuing debate as to whether it would be of more scientific value to venture in toward the sun or to go out away from the sun. That question was generally resolved in favor of going in toward the sun first on the basis that conditions ought to change faster and there should be higher gradients of all properties as you go closer to the sun. That is why we are conducting more extensive studies on a solar probe rather than a probe to Jupiter, or a probe to try to reach out to the edge of the solar magnetosphere or a probe out of the ecliptic plane. There are still some fundamental questions about the solar probe. How far in do you need to go in order to get significant results? Should you try to carry a heavy load of instruments into the 0.3 of an astronomical unit or should you first send a small, simply-instrumented beacon around behind the sun and use the attenuation of its signal to measure the variation of the electron density around the sun? Or can you study the solar corona adequately by studying the light scattered by the electrons and use the scattered radiation to deduce the magnetic field configuration and the flow of plasma? I am sure that by the end of the week each of you will have an opinion on the best way to study solar physics and how important this question is.

We have another problem in connection with OAO. There is no question about the value of the astronomy, but the experiments are expensive. We have developed separate instruments for the first three OAO's. Should we develop yet another instrument for the fourth OAO or should we arrange to use again one of the instruments that we have already developed?

The foregoing are not all of the questions and problems that we have, but they are typical.

Summary

In summary I want to emphasize:

1. The NASA Space Science Program is a logical extension of astronomical research which has been going on since Galileo.

2. There is an opportunity for fundamental discoveries – discovery of life – new insights into stellar processes and stellar evolution.

3. There is a strong university participation in all phases, program formulation, experiment selection, supporting research, data analysis and interpretation, and theoretical studies.

4. NASA has established a sustaining university program to provide new scientists, new facilities, new research groups and interdisciplinary research.

REMARKS ON SCIENCE AND TECHNOLOGY

LELAND J. HAWORTH

Director, National Science Foundation, Washington, D.C.

Dean Marlowe, Professor Chang, and Guests of the University. I would like to discuss tonight some of the interesting trends that are taking place in science and technology, and some of its general impacts on society. First, let me discuss some of the aspects of cooperation in research. Until fairly recent times the extent of scientific knowledge was so limited that it could all be encompassed in a single mind such as that of Newton. With the advent of research and the resulting growth of knowledge, fields of specialization began to appear and barriers arose between the various fields so that only a rare individual pretended to have even general knowledge of scientific subjects very different from his own. Not only did this result in intellectual narrowness, it also inhibited exchanges of knowledge and understanding that could result in mutual help across the scientific disciplines.

Happily, in my opinion this trend has been reversed. Through a variety of circumstances a dramatic change has taken place in the past few years in this and other aspects of the conduct of scientific and technical research and development. Let me give a few examples.

Fifty years, or even a shorter time ago, basic research – the exploration of nature's laws – and engineering – the practical application of those laws – were pursued quite independently. Physicists, chemists and other scientists, usually in the universities, carried on research in pursuit of knowledge for its own sake. Their never-ending quest was to observe how nature behaves under given situations and, from this information, to deduce the laws underlying that behavior. Engineers, mostly in industry, sought to develop new devices and processes of practical utility. The lapse of time between the making of some scientific discovery and its practical application was likely to be very long; for example, it was many decades between the fundamental discoveries of Oersted, Faraday and Ampère, and the large-scale practical use of electric power.

This situation began to change during the second and third decades of this century, perhaps first in the fields of chemistry and chemical engineering. Extraordinary impetus was given to it, however, by the advent of World War II. The establishment of large laboratories endeavoring to develop on a "crash" basis such military tools as radar, proximity fuses, bomb-sights, and atomic weapons brought scientists and engineers together on a scale that had never before occurred. The resulting rubbing of elbows brought about a mutual appreciation and a degree of cooperation that has continued to grow to the great benefit of both research and application.

For example, before the war, physicists built their own accelerators. During the war they developed an appreciation of the usefulness of engineers in such endeavors.

Chang & Huang (eds.), Proc. Plasma Space Sci. Symp. All rights reserved.

Now all sizeable accelerators are developed and designed by large teams of physicists and engineers working in close cooperation. Correspondingly, the war-time experience demonstrated to the engineers, on a scale they had not theretofore appreciated fully, the vast gold mine of useful information accumulated by the research scientists, and, in addition, the usefulness of the scientists themselves in the field of applications. Industrial and other applied laboratories now employ large numbers of scientists and are careful to keep abreast of all results being obtained in basic research. As a result, the time span between basic discoveries in science and their applications in technology has been greatly shortened.

A second, overlapping trend toward cooperation in research has resulted from the increasing complexity of the problems and of the requisite equipment. In many fields the advances of the last few decades have resulted in the need for large and complicated devices such as nuclear reactors, large accelerators, radio-telescopes, special laboratories, and the tools used for space research, whose construction and operation involve vast expenditures in money and human effort, and whose effective utilization can be accomplished only by the cooperative efforts of large groups of scientists and engineers working in many fields. At a large research reactor, for example, physicists, chemists and metallurgists – biologists and medical scientists, chemical, mechanical and other engineers, work side by side in performing experiments, running the gamut from very fundamental studies to the testing of practical devices. Similarly the instruments of many kinds of scientists are carried together into space. This rubbing of elbows leads to a close relationship and mutual help that enhances all the programs in both tangible and intangible ways. For example, research in biology and medicine benefits greatly because of the many contacts by groups with special skills and overlapping interests among the physical scientists and engineers. Not only are instruments and techniques often interchangeable, but there are many contributions of ideas.

This coming together of the various disciplines has given added impetus to still another trend – new disciplines that bridge across the old ones. Among the most important present fields are those of biophysics and biochemistry, which, as their names imply, involve the application of physical understanding and techniques to the study of living systems. Similar to these are the fields of geophysics and astrophysics, relating respectively to our own planet and to the universe.

In still a different context is the cooperation now quite common between the scientists in different institutions and even in different countries.

A special type of inter-institutional cooperation concerns the regional or national laboratory. Many of the largest tools are beyond the scope of most universities and private institutions. They must be placed in very large centers such as the national laboratories of the Atomic Energy Commission. It is, however, essential both to the effective advancement of science and to the health and vigor of the traditional institutions that they continue to participate in the advanced fields utilizing such devices.

Consequently, it is a definite policy of the AEC and of all its national laboratories to make their facilities available on a cooperative basis to scientists of other institutions. Consequently these laboratories have added a new dimension to the concept of a national laboratory, namely cooperation in research on a nation-wide basis. The universities and other institutions from whence the visitors come are benefited greatly by the opportunity given to their staffs to work in fields not otherwise available. The national laboratories benefit by the contributions of the visitors. Both these groups, science and the Nation at large, benefit from the direct results and from the cross-fertilization and exchange of new ideas resulting from the meetings of the scientists from many institutions.

This idea has spread to other fields. National observatories for optical and radio-astronomy are now in operation. Starts are being made in such fields as meteorology and geophysics.

The idea has even been extended on an international scale. The laboratory known as CERN (the initials for *Conseil Européenne pour la Recherche Nucleaire*) at Geneva, Switzerland, was established in 1952 by 12 Western European nations who pooled their efforts to construct and operate very high energy accelerators.

There are many other international scientific cooperative efforts. The United States cooperation in international science covers a broad spectrum of activities, ranging through participation in international organizations and their operations, the direct support of research abroad, training, exchange of technical publications, exchange of scientists, and many other relationships. Government agencies, private foundations and organizations, scientific groups and industry all participate.

By far the largest international effort has been the International Geophysical Year with which this group is quite familiar. For a period of 18 months, scientists from all nations, both East and West, were given an opportunity to participate in certain generally defined areas of science including meteorology, geophysics, oceanography and astrophysics. Although in general, they did not work together at the same locations, the planning and the results of all their observations were closely coordinated. By such planned procedures and simultaneous observation of given phenomena at many points throughout the world, it has been possible to reach a greater understanding of many large-scale phenomena which cannot be successfully observed from any single point. The practical results, together with the added spirit of cooperation, were valuable indeed. The forthcoming year of the quiet sun should also have great value.

Thus a good beginning has been made in international cooperation on unclassified scientific matters. While contacts are necessarily limited, the important thing is that a start has been made. A still more definite type of mutual cooperation has also been suggested – direct cooperation between East and West in large and expensive projects in a manner similar to that of CERN. Indeed, informal discussions have been held between American and Soviet scientists concerning the possibilities of jointly building and operating a very large accelerator.

It is to be hoped that in some such ways as these the universality and wide appeal of science may help to point the way toward mutual understanding. It is perhaps not too much to hope that in the future the barriers that seperate our worlds can partially be broken by greater exchanges of technical information and longer working visits.

Science provides a powerful device for building understanding among the peoples of the world. We must take advantage of it.

Another important impact that large-scale research can have is its role in education, and, hence, in our cultural and intellectual life. About 10 percent of all scientific and technical support goes into basic research, the objective of which is to develop a fundamental understanding of nature and her laws. Such research has important intellectual and cultural values and is at the same time the source from which all technological progress ultimately springs.

The heart and soul of basic research rests in the universities. In recent years there is growing recognition that research and graduate education go hand in hand and complement each other. This proposition is well discussed in a report by a panel of the President's Science Advisory Committee, *Scientific Progress, the Universities and the Federal Government* that I urge you all to read. To quote from that report: "Science and the making of scientists go best together".

Farsighted men within the Federal Government recognize the duality of benefits derived from the support of universities. We can have our cake and eat it, too. We can obtain research results while increasing the potential for further advance by creating more scientists.

Historically, the Government first started intellectual intercourse with science in the field of agriculture, at the land-grant institutions. During World War II and particularly in the field of nuclear physics, American science showed the desirability of its practical results. In the years after World War II, first the defense agencies and then those in the field of health entered into research relations with the Universities. Still later came the National Science Foundation. Our common interests have led us into formal bonds that will not be dissolved, for as long as our American Government and scientific community endure.

As in the course of any healthy marriage, we are still growing toward a better understanding of our own and our partner's needs. We have learned that any real problem is a mutual problem. Either partner can damage their union; both must give it care and attention, and work together if it is to be a really fruitful union. Both the university and the Government must give firm support to excellence in basic research.

To create more good scientists we must make scientific careers attractive to our talented young people. We must give them the best possible opportunity to experience meaningful scientific inquiry in the college laboratory. In science, and indeed in all academic fields, we must work for decent salaries, time and facilities for good research, and continuous modernization of curricula. We must enable more men and women to complete all the advanced training the Ph.D. degree symbolizes. Too many of our best students simply cannot find the money for sustained full-time study and drop out, or are forced to take part-time jobs that delay their progress and flatten their spirits. I deeply hope that *we* in the Government can, in time, expand our graduate fellowship programs enough to really approach a solution to this problem.

The dramatic expansion in research in this country is outrunning our capacity to conduct it. In many fields and in many universities, the immediate bottleneck is in buildings and equipment. As you know the National Science Foundation, the Advanced Research Projects Agency, and the Atomic Energy Commission and other agencies are trying their hardest to help meet these needs. But the States, the Foundations and the universities themselves must bear a heavy burden. The National Laboratories or similar institutions can be of real help here by continuing to extend the opportunities they give to university faculties and students.

An important consideration is that more universities must become first-class institutions for research and graduate education in the sciences. We must have more centers of excellence in this respect. Toward this end the National Science Foundation is seeking special funds to assist universities to improve their quality.

From all of this it is quite clear that the nation's scientific program involves innumerable questions of public policy. It is vital to our national defense. It has other important international implications. It determines in significant measure our mode of living and the direction of our industrial economy. It expends large quantities of public funds.

It is thus important that it be scrutinized by an informed public.

A great deal more effort and attention should, indeed *must*, be given by all to give the public a true picture of science, in the most meaningful form. The aspirations, the potentialities and the limitations of science must be understood by all people so that they can meet the responsibilities for progressive citizenship called for in the world today. They must be made known to government so that decision-makers can act rationally within the framework of reality.

Additionally and importantly, we must instill in the individual citizen the knowledge that an understanding of the fundamental truths of nature yields intellectual and cultural rewards that are among the greatest of the lasting human values.

In this connection I had an interesting experience about two weeks ago. I spoke at dinner on a scientific subject to an annual "Alumni Institute" at the University of Indiana of which I am a graduate. In my talk I alluded to the intellectual and cultural values of basic scientific research, pointing out that the objective of the fundamental scientist is an understanding of nature without regard to possible practical application, that the reward is one of the mind and of the spirit rather than of material things.

Later in the evening an artist on the faculty gave a very interesting illustrated talk on what interests and inspires the artist. Quoting from my talk he concluded that the artist and the basic scientist are not so far apart as is generally supposed, that their objectives and their rewards are much the same. I thoroughly agree. It seems to me the only basic difference is that the scientist seeks to learn about and understand the intrinsic beauty to be found in nature and her laws, which are immutable, whereas the artist is free to create beauty by combining nature with his own imagination.

I hope that the artist's view will spread. We cannot afford a two-culture society, even in people's minds.

I have spoken of two things – of how cooperation in science hopefully can provide a catalyst to assist in bringing the world into more sympathetic relationships and understanding, and *second* of how science is an important integral part of our intellectual and cultural life.

It seems to me that these two facts might well be welded into a cohesive whole to serve as a beacon to a troubled world.

I believe we all agree that the major problems of the world result primarily from the crass and practical – from the understandable yearnings of all men to live in comfort and security – and not to want for food and fuel – for shelter and the niceties of life.

But there are other yearnings. As man develops and solves his practical problems, he realizes more and more that the greatest human triumphs and the greatest human joys are of the mind and spirit, that they can best be realized through cultural and intellectual means. These yearnings are shared by all men on this earth, without regard to nationality, or creed or color. All men deserve to have them satisfied.

Perhaps science can form a bridge uniting these deep and basic yearnings. For, uniquely among all activities of men, it is a source of deep and lasting cultural and intellectual satisfaction, and on the other hand the source of all our technological and hence material progress.